INTELLIGENT INFRASTRUCTURE

Neural Networks, Wavelets, and Chaos Theory for Intelligent Transportation Systems and Smart Structures

Hojjat Adeli
Xiaomo Jiang

CRC Press
Taylor & Francis Group
Boca Raton London New York

CRC Press is an imprint of the
Taylor & Francis Group, an **informa** business

T0200354

CRC Press
Taylor & Francis Group
6000 Broken Sound Parkway NW, Suite 300
Boca Raton, FL 33487-2742

First issued in paperback 2019

© 2009 by Taylor & Francis Group, LLC
CRC Press is an imprint of Taylor & Francis Group, an Informa business

ISBN-13: 978-1-4200-8536-5 (hbk)
ISBN-13: 978-0-367-38671-9 (pbk)

Library of Congress Cataloging-in-Publication Data

Adeli, Hojjat, 1950-
    Intelligent infrastructure : neural networks, wavelets, and chaos theory for intelligent transportation systems and smart structures / Hojjat Adeli, Xiaomo Jiang.
        p. cm.
    Includes bibliographical references and index.
    ISBN 978-1-4200-8536-5 (alk. paper)
    1. Civil engineering--Data processing--United States. 2. Smart structures--United States. 3. Intelligent Vehicle Highway Systems--United States. 4. Expert systems--United States. I. Jiang, Xiaomo. II. Title.

TA345.A329 2008
624.0285--dc22   2008036600

Visit the Taylor & Francis Web site at
http://www.taylorandfrancis.com

and the CRC Press Web site at
http://www.crcpress.com

INTELLIGENT INFRASTRUCTURE

Neural Networks, Wavelets, and Chaos Theory for Intelligent Transportation Systems and Smart Structures

Dedicated to

Nahid, Anahita, Amir, Mona, and Cyrus Dean
Adeli

Xueyu and Claire
Jiang

TABLE OF CONTENTS

PREFACE

According to a report published by the Federal Highway Administration in 2007, an estimated $1.6 trillion will be needed in the United States for rehabilitation, renewal, replacement, and maintenance of the existing infrastructure systems within the next 20 years. This book presents the foundations of intelligent infrastructure, a new vision and way of designing and managing the civil infrastructure of the nation. It presents novel technologies, methodologies, and detailed computational algorithms for creation of smart structures and intelligent freeways.

The two fields of structural engineering and transportation engineering are considered separate fields within the broader civil engineering discipline. This book crosses the two disciplines and provides a unique treatise for attacking and solving some of the most complex and intractable problems encountered in the emerging fields of smart structures and intelligent transportation systems (ITS). It is shown some of these problems can be solved effectively through ingenious integration of several different computing paradigms including chaos theory (based on nonlinear dynamics theory), wavelets (a signal processing method), and three complementary soft computing methods, that is, fuzzy logic, neural networks, and genetic algorithm.

The premise of ITS is intelligent and maximum utilization of the existing freeway systems rather than adding to them thereby minimizing the destruction of the nature and resulting in environmentally conscious and sustainable designs and solutions.

The authors envisage a new generation of civil smart structures where intelligent sensors and actuators are integrated for health monitoring and vibration control of structures under extreme external loading. The health of such a smart structure is continuously monitored in real time by sensors and its behavior is actively controlled in an effective and efficient way through properly designed actuators. In addition to the innovative hardware technologies such as smart materials, sensors and actuators, advanced computing technologies such as data mining and automatic feature extraction techniques play an important role in creating an effective smart structure technology, which is a focus of this book. Multidisciplinary methodologies are presented for health monitoring and nonlinear active control of tall building structures subject to dynamic loading such as those due to winds or earthquakes.

ACKNOWLEDGMENT

The work presented in this book was partially sponsored by the National Science Foundation, the Ohio Department of Transportation, and the Federal Highway Administration through research grants to Adeli. Parts of the work were published by the authors in several research journals: *Journal of Transportation Engineering* and *Journal of Structural Engineering* (both published by American Society of Civil Engineers), *International Journal for Numerical Methods in Engineering* and *Computer-Aided Civil and Infrastructure Engineering* (published by Wiley-Blackwell), *Integrated Computer-Aided Engineering* (published by IOS Press), and *International Journal of Neural Systems* (published by World Scientific), as noted in the list of references. Chapters 5 and 6 are based on the journal articles authored by Adeli and his former Research Associate, S. Ghosh-Dastidar, and are reproduced by permission of the publishers of the journals.

ABOUT THE AUTHORS

Hojjat Adeli received his Ph.D. from Stanford University in 1976 after graduating from the University of Tehran in 1973. He is currently Professor of Civil and Environmental Engineering and Geodetic Science and the holder of the Abba G. Lichtenstein Professorship at The Ohio State University. He has authored over 420 research and scientific publications in various fields of computer science, engineering, applied mathematics, and medicine including twelve books. His wide-ranging research has been published in 72 different journals. He has also edited thirteen books. He is the Founder and Editor-in-Chief of the international research journals *Computer-Aided Civil and Infrastructure Engineering*, in publication since 1986, and *Integrated Computer-Aided Engineering*, in publication since 1993. He is the quadruple winner of The Ohio State University College of Engineering Lumley Outstanding Research Award. In 1998 he received the *Distinguished Scholar Award*, The Ohio State University's highest research award "*in recognition of extraordinary accomplishment in research and scholarship.*" In 2005, he was elected Honorary Member (now Distinguished Member), American Society of Civil Engineers: "*for wide-ranging, exceptional, and pioneering contributions to computing in civil engineering disciplines and extraordinary leadership in advancing the use of computing and information technologies in many engineering disciplines throughout the world.*" In 2006, he received the ASCE Construction Management Award for "*development of ingenious computational and mathematical models in the areas of construction scheduling, resource scheduling, and cost estimation.*" In 2007, he received The Ohio State University College of Engineering Peter L. and Clara M. Scott Award for Excellence in Engineering Education "*for sustained, exceptional, and multi-faceted contributions to numerous fields including computer-aided engineering, knowledge engineering, computational intelligence, large-scale design optimization, and smart structures with worldwide impact,*" as well as the Charles E. MacQuigg Outstanding Teaching Award. He has presented Keynote Lectures at 64 conferences held in 38 different countries. He has been on the organizing or scientific committees of 295 conferences held in 57 countries. He holds a U.S. patent in the area of large-scale optimization.

Xiaomo Jiang received his Ph.D. from The Ohio State University in 2005, his M. Eng. from the National University of Singapore in 2000, and his M.S. and B.Sc. from the Northern Jiaotong University, Beijing, China in 1998 and 1995, respectively. He is currently a Lead Engineer in Condition Based Maintenance at General Electric Company. Prior to joining GE in 2008, he was a Research Engineer in Research and Advanced Engineering at Ford Motor Company, leading the project of model validation for dynamic systems. From 2005 to 2007 he was a Senior Research Associate in the Department of Civil and Environmental Engineering at Vanderbilt University. He conducts research in diverse areas of structural system identification, health monitoring, damage diagnosis and prognosis, risk and reliability analysis, model validation, and intelligent infrastructure systems. His research has been applied to a variety of aerospace, civil, and mechanical systems. He has authored two book chapters and more than 40 research articles in international journals and conference proceedings. He is listed in *Who's Who in America, Who's Who of Emerging Leaders*, and *Who's Who in Science and Engineering*. He is a registered professional engineer in the state of Ohio.

Chapter 1

INTRODUCTION

This book presents a new vision and way of designing and managing the civil infrastructure of the nation. Someday perhaps in the not-too-distant future our structures will be smart and our freeways will be intelligent. This book aims towards that goal and presents the foundations of intelligent infrastructure. It presents novel technologies, methodologies, and detailed computational algorithms for creation of smart structures and intelligent freeways.

The two fields of structural engineering and transportation engineering are considered separate fields within civil engineering. This book provides a unique treatise for attacking and solving some of the most complex and intractable problems encountered in the emerging fields of intelligent transportation systems (ITS) and smart structures. A thesis of this book is that some of these problems can be solved only through ingenious integration of several different computing paradigms including chaos theory (based on nonlinear dynamics theory), wavelets (a signal processing method), and three complementary soft computing methods, that is, fuzzy logic, neural networks, and genetic algorithm. The book advances a holistic approach to the creation of the intelligent infrastructure by covering both smart structures and ITS. It crosses the two disciplines and breaks their boundaries.

Chapters 2, 3, and 4 are introductory chapters covering three fields that have created a lot of excitement in the research community during the past two decades. Chapter 2 provides an introduction to the field of artificial neural networks. Commonly used neural network approaches such as backpropagation (BP), radial basis function (RBF), and Boltzmann neural networks are presented followed by a brief discussion of the dynamic neural network employed and advanced in several chapters in the book. Chapter 3 introduces the readers to wavelets for signal processing. The fundamental topic of multiresolution analysis is presented followed by the more recent and powerful concept of wavelet packets. The key ideas of the chaos theory such as state space, time delay, and embedding dimension along with three illustrative examples are presented in Chapter 4.

Adeli and coworkers have been advocating, creating, and advancing an innovative multi-paradigm and multi-disciplinary approach for the solution of complicated pattern recognition problems in various fields such as structural engineering (see, e.g., Adeli and Park 1998; Soegiarso and Adeli 1998; Saleh and Adeli 1998a&b; Adeli and Kumar 1999; Adeli and Saleh 1999; Adeli and Soegiarso 1999; Adeli 2000 & 2001; Adeli and Kim 2000 & 2004; Zhou and Adeli 2003; Kim and Adeli 2004 & 2005; Jiang and Adeli 2005; Adeli and Jiang 2006; Adeli and Sarma 2006; Park *et al*. 2007; Adeli and Panakkat 2007; Jiang and Adeli 2007; Jiang *et al*. 2007), ITS (e.g., Samant and Adeli 2000 & 2001; Adeli and Samant 2000; Adeli and Karim 2000 & 2005; Karim and Adeli 2002a, b, c&d; Karim and Adeli 2003; Ghosh-Dastidar and Adeli 2003 & 2006; Adeli and Jiang 2003; Dharia and Adeli 2003; Jiang and Adeli 2003, 2004a, b&c, 2005; Adeli and Ghosh-Dastidar 2004; Hooshdar and Adeli 2004), construction engineering (Adeli and Karim 1997 & 2001; Adeli and Wu 1998; Karim and Adeli 1999a, b&c; Senouci and Adeli 2001), and biomedical engineering, computational neuroscience, and neurology (Adeli *et al*. 2003; Adeli *et al*. 2005a&b; Ghosh-Dastidar and Adeli 2007; Ghosh-Dastidar *et al*. 2007; Ghosh-Dastidar *et al*. 2008). The book follows the same ideology.

Chapters 5 to 12 deal with topics in intelligent freeways or ITS. These chapters build upon and extend the previous work of Adeli and coworkers which culminated in the 2005 book titled *Wavelets in Intelligent Transportation Systems* (Adeli and Karim 2005). The premise of ITS is intelligent and maximum utilization of the existing freeway systems rather than adding to them thereby minimizing the destruction of the nature and resulting in environmentally conscious and sustainable designs and solutions.

Chapter 5 presents automatic detection of incidents in freeways based on speed, volume, and occupancy data from a single detector station using a combination of wavelet de-noising, statistical clustering analysis, and neural network pattern recognition. An improved feature extraction model is presented using fourth-order Coifman wavelets to reduce the effects of noise in the data from various sources including erroneous sensor data. Statistical analysis based on the Mahalanobis distance is employed to perform data clustering and parameter reduction in order to reduce the size of the input space for the subsequent step of classification by the Levenberg-Marquardt BP neural network.

Freeway work zones result in congestion and traffic delays leading to increased driver frustration, increased traffic accidents, and increased road user delay cost. Chapter 6 presents a neural network-wavelet micro-simulation model for tracking the travel time of each individual vehicle for traffic delay and queue length estimation at freeway construction work zones. The extracted congestion characteristics obtained from a mesoscopic-wavelet model are used in a Levenberg-Marquardt BP neural network for classifying the traffic flow as free flow, transitional flow, or congested flow with stationary queue. The model incorporates the dynamics of a single vehicle in changing traffic flow conditions and yields values for queue lengths and user delay which can be used to study their impacts on construction planning and management.

Chapter 7 presents a macroscopic model for freeway work zone traffic delay estimation and total work zone cost optimization. The model uses hourly traffic flow and takes into account a large number of factors such as number of lane closures, length of the work zone segment, anticipated hourly traffic flow of the freeway approaching the work zone, starting time of the work zone (time of the day in hours), darkness, seasonal variation in travel demand, and duration of the work zone in hours. A total work zone cost function is defined as the sum of the user delay, accident, and maintenance costs. The work zone traffic delay and cost optimization model is applicable for both short-term (less than one day) and long-term (more than day) work zones. The model yields the global optimum values for the work zone segment length and the starting time of the work zone. A Boltzmann-simulated annealing neural network model is developed to solve the resulting mixed real variable-integer work zone cost optimization problem. A work zone traffic engineer can use the model to find the answer to important what-if questions, such as one-lane closure versus two-lane closure or selection of the starting time of the day systematically and quickly.

The freeway work zone capacity cannot be described by any mathematical function because it is a complicated function of a large number of interacting variables. Chapter 8 presents an adaptive neuro-fuzzy logic model for estimating the freeway work zone capacity. Eighteen different factors impacting the work zone capacity are included in the model. A BP neural network is employed to estimate the parameters associated with the bell-shaped Gaussian membership functions used in the fuzzy inference mechanism. The proposed model can be implemented into an intelligent decision support system a) to estimate the work zone capacity in a rational way, b) to perform scenario analysis, and c) to study the impact of various factors influencing the work zone capacity.

In Chapter 9, two additional models are presented for estimating the freeway construction work zone capacity. The subtractive clustering approach is judiciously integrated with the RBF and BP neural networks to create the clustering-RBF and clustering-BF neural network models. The results of a parametric study of the factors impacting the work zone capacity can assist work zone engineers and highway agencies to create effective traffic management plans (TMPs) for work zones quantitatively and objectively.

Chapter 10 presents an object-oriented model for freeway work zone capacity and queue delay estimation. The model is implemented into an interactive software system, called *IntelliZone* (Intelligent decision support system for work zone traffic management). The ideal goal of an effective TMP is to minimize travelers' delays and construction and operation costs while enhancing the safety of the travelers and highway workers. *IntelliZone* can be used as a powerful tool to achieve that goal.

Accurate and timely forecasting of traffic flow is of paramount importance in ITS in order to manage the traffic congestion in the freeway network effectively. A detailed understanding of the properties of traffic flow is essential for creating a reliable forecasting model. Chapter 11 investigates wavelet analysis of traffic flows. The goal is to identify important characteristics of traffic flow using wavelets and statistical autocorrelation function (ACF) analysis. A hybrid wavelet packets-ACF method is presented for analysis of traffic flow time series and determining its self-similar, singular, and fractal properties. The method provides a powerful tool in removing the noise and identifying the singularity in the traffic flow.

Chapter 12 presents a dynamic time-delay wavelet neural network (WNN) model for forecasting traffic flows in freeways through adroit integration of a pattern recognition paradigm, neurocomputing, a signal processing methodology, wavelets, and a statistical tool, autocorrelation function. The model incorporates both the time of the day and the day of the week of the prediction time. As such, it can be used for long-term traffic flow forecasting in addition to short-term forecasting. Short-term traffic flow forecasting would be of great interest in on-line ITS applications and long-term forecasting would be of great interest in planning applications.

Chapters 13 to 17 deal with smart structures. These chapters lay the foundation for development of a new generation of smart structures. The authors envisage a new generation of civil smart structures where intelligent sensors and actuators are integrated for health monitoring and vibration control of structures under extreme external loadings. The health of such a smart structure is continuously monitored in real time by sensors and its behavior is actively controlled in an effective and efficient way through properly designed actuators. In addition to the innovative hardware technologies such as smart materials, sensors and actuators, advanced computing technologies such as data mining and automatic feature extraction techniques play an important role in creating an effective smart structure technology, which is a focus of this book. Multidisciplinary methodologies are presented for health monitoring and nonlinear active control of tall building structures subject to dynamic loadings such as those due to winds or earthquakes.

Structural health monitoring requires detecting changes in the global or local conditions of structures under extreme dynamic loadings such as severe earthquakes, strong winds, and impact. Its goals are to (1) provide an early warning of structural failure in order to reduce the loss of life and property, (2) improve understanding of structural behavior under extreme loadings, and (3) facilitate proactive, rational decisions with respect to maintenance and repair. Accurate health monitoring of large structures is a particularly challenging problem due to the complicated nonlinear behavior of the structural system and the incomplete and noisy nature of the sensed data.

Accurate system identification needed for health monitoring of structures requires capturing the dynamic characteristics of time series response data. Chapter 13 presents a multi-paradigm dynamic time-delay fuzzy WNN model for nonparametric identification of structures using the nonlinear autoregressive moving average with exogenous inputs (NARMAX) approach. The model is based on the integration of four different computing concepts: dynamic neural network, wavelet, fuzzy logic, and chaos theory. The goal is to improve the accuracy and adaptability of nonparametric system identification of structures under dynamic loadings. In order to preserve the dynamics of time series, the reconstructed state space concept from the chaos theory is employed to construct the input vector. Wavelets are employed in two different contexts. First, they are used for denoising the data. Second, they are used in combination with two soft computing techniques, neural networks and fuzzy logic, to create a new pattern recognition model to capture the characteristics of the time series sensor data accurately and efficiently. The model a) balances the global and local influences of the training data due to the complementary properties of two soft computing techniques, b) incorporates the imprecision existing in the sensor data effectively due to the use of a fuzzy clustering technique, c) provides more accurate system identifications, and d) results in fast training convergence thus significantly reducing the computational requirements.

The training of a dynamic neural network is substantially more complicated and time-consuming than the training of conventional neural networks because in the former both input and output are not single-valued but are in the form of time-series. An effective training algorithm is essential for identification accuracy and real-time implementation of the dynamic fuzzy WNN model for health monitoring or control of large-scale structures. Chapter 14 presents a hybrid adaptive learning algorithm, called the Levenberg-Marquardt-Least-Squares (LM-LS) algorithm, for training and adjusting the parameters of the dynamic fuzzy WNN model. This chapter also presents applications of the dynamic time-delay fuzzy WNN model to three-dimensional high-rising building structures taking into account their geometric nonlinearity. The results of this chapter can be used for real-time health monitoring, damage detection, and control of large structures such as high-rising building structures.

Chapter 15 presents a nonparametric system identification-based model for damage detection of high-rising building structures subjected to dynamic loading such as strong ground motions using the dynamic fuzzy WNN model with an adaptive LM-LS learning algorithm described in the previous two chapters. The method does not require complete measurements of the dynamic responses of the whole structure. This is achieved by dividing a large structure into a series of sub-structures around selected floors where measurements are made. A new damage evaluation method is presented based on a power density spectrum method, called *pseudospectrum*. The multiple signal classification (MUSIC) method is employed to compute the pseudospectrum from the structural response time series. The structural damage is identified explicitly through comparison of the pseudospectra of the predicted output of the trained model and the measured output of the structure. It is shown that the proposed power density spectrum approach provides an effective method for damage detection in high-rising building structures.

Following the book *Control, Optimization, and Smart Structures - High-Performance Bridges and Buildings of the Future* (Adeli and Saleh, 1999) published nearly a decade ago, Chapter 16 presents active nonlinear control of large three-dimensional (3D) building structures based on the WNN model described in earlier chapters. Both material and geometrical nonlinearities are considered in modeling the structural response under strong earthquake loadings. Furthermore, the structural modeling takes into account two coupling actions: the coupling action between the actuators and the structure and the coupling between the lateral and torsional motions of 3D irregular structures. A dynamic fuzzy WNN is developed as a fuzzy wavelet *neuroemulator* to predict structural responses in future time steps from the immediate past structural responses and actuator dynamics.

Chapter 17 presents an intelligent control algorithm for nonlinear control of structures using a floating point genetic algorithm (GA). In this chapter, fuzzy logic, wavelets and genetic algorithm are integrated for nonlinear active control of structures. It is shown that the algorithm is particularly effective for active control of large 3D high-rising building structures. The dynamic fuzzy wavelet neuroemulator presented in Chapter 16 is used to predict future nonlinear structural displacement responses. A genetic algorithm is developed to find the optimal control forces for any time step during the earthquake excitation. Validation results demonstrate that the proposed control methodology is effective in significantly reducing the response of large three-dimensional building structures subjected to seismic excitations, including structures with plan and elevation irregularities.

Chapter 2

ARTIFICIAL NEURAL NETWORKS

2.1 Introduction

An artificial neural network (ANN) is a functional abstraction of the biologic neural structures of the central nervous system (Aleksander and Morton 1993; Rudomin *et al.* 1993; Arbib 1995; Anderson 1995). It is essentially a simple mathematical model representing a nonlinear mapping function from a set of inputs **x** to outputs **y**, i.e., *f*: **x** → **y**. The most commonly used ANN model is the feedforward neural network. This network usually consists of input, hidden, and output layers. The information moves only in the forward direction, from the input nodes, through the hidden nodes and to the output nodes. There are no loops in the network.

As an example, Fig. 2.1 shows a widely used three-layer feedforward ANN model, with arrows depicting the dependencies between variables at each layer. Its single output $\hat{y}$ (i.e., model prediction) is the nonlinear weighted sum of the inputs **x**. In Fig. 2.1, the input vector **x** consists of p variables x_i ($i = 1, \ldots, p$). The parameter a_{ij} ($i = 1, \ldots, p; j = 1, \ldots, K$) represents the weight of the link connecting the input node i to node j in the hidden layer, in which K is the number of nodes in the hidden layer. The parameter w_j represents the weight of the link connecting the hidden node j to the node in the output layer, and the variable d represents the weight of the bias. The bias term (a constant value, typically one) allows the neural network to return a nonzero value at the origin. The parameters a_{ji}, w_j, and d need to be estimated using a learning algorithm which will be discussed later.

Let $\boldsymbol{\theta} = \{a_{ji}, w_j, d\}$ represent the group of parameters to be estimated. The model output $\hat{y}$ is conceptually expressed as

$$\hat{y} = f(\mathbf{x}, \boldsymbol{\theta}) \tag{2.1}$$

In training an ANN model, a measure of the error e is often used as a model-quality index. The error is defined as the difference between the actual and predicted values, expressed as

$$e = y - \hat{y} \qquad\qquad (2.2)$$

A well-trained model should yield small prediction errors, which is often used to assess the validity of the trained model.

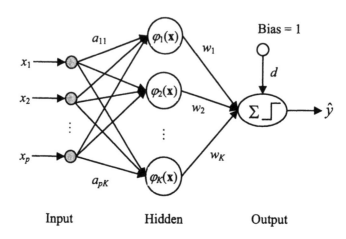

$\mathbf{x} = [x_1, ..., x_p] = $ Input vector containing p variables
$a_{ij} = $ Weight of the link connecting the input node i to node j in the hidden layer
$w_j = $ Weight of the link connecting the hidden node j to the node in the output layer
$d = $ Weight of the bias
$\hat{y} = $ Model output

Figure 2.1 An illustrative example for feedforward three-layer ANN model

The ANN has been widely applied for pattern recognition and classification. It operates as a black-box, model-free, and adaptive tool to capture and learn significant structures in data. The ANN can be trained to learn to perform a particular task. In addition to pattern recognition, its computing abilities have been proven in the fields of prediction, estimation, and optimization (see, e.g., Adeli and Hung 1995; Golden 1996; Mehrotra *et al.* 1997; Adeli and Park 1998; Haykin 1998). The ANN is suitable particularly for problems too complex to be modeled and solved by classical mathematics and traditional procedures, such as engineering design and image recognition (Adeli and Hung 1995). Example ANN models include backpropagation (BP) neural network (Hagan *et al.* 1996; Mehrotra *et al.* 1997), radial basis function (RBF) neural network (Karim and Adeli 2003a), Boltzmann neural network (Jiang and Adeli 2003a), and dynamic neural network (Hagan *et al.* 1996; Mehrotra *et al.* 1997; Pham and Karaboga 2000; Adeli and Jiang 2006; Jiang and Adeli 2007). These ANN models will be introduced briefly in the following sections.

2.2   Backpropagation Neural Network (BPNN)

One of the reasons for popularity of the neural network is the development of the simple error backpropagation (BP) training algorithm (Rumelhart *et al.* 1986), which is based on a gradient descent optimization technique. The BP algorithm is now described in many textbooks (e.g., Adeli and Hung 1995; Mehrotra *et al.* 1997; Topping and Bahreininejad 1997; Haykin 1998). A review of the BP algorithm with suggestions on how to develop practical neural network applications is presented by Hegazy *et al.* (1994). The great majority of civil engineering applications of neural networks is based on the use of the BP algorithm primarily because of its simplicity. Training of a neural network with a supervised learning algorithm such as BP means finding the weights of the links connecting the nodes using a set of training examples. An error function in the form of the sum of the squares of the errors between the actual outputs from the training set and the computed outputs is minimized iteratively. The learning or training rule specifies how the weights are modified in each iteration.

The momentum BP learning algorithm (Rumelhart *et al.* 1986; Adeli and Hung 1995) is widely used for training multilayer neural networks (Fig. 2.1) for classification problems. This algorithm, however, has a slow rate of learning. The number of iterations for learning an example is often in the order of thousands and sometimes more than one hundred thousands (Carpenter and Barthelemy 1994). Moreover, the convergence rate is highly dependent on the choice of the values of learning and momentum ratios encountered in this algorithm. The proper values of the two parameters depend on the type of the problem (Adeli and Hung 1994; Yeh 1998). As such a number of other neural network learning models, such as the adaptive conjugate gradient neural network algorithm (Adeli and Hung 1994), have been proposed. In addition, the stochastic optimization algorithm (Masri *et al.* 1999), genetic algorithms (Hung and Adeli 1994), and fuzzy logic (Adeli and Hung 1993; Hurson *et al.* 1994; Furuta *et al.* 1996; Nikzad *et al.* 1996) have been applied to train the neural networks.

On the other hand, neural network models may lose their effectiveness when the patterns are very complicated or noisy. For example, traffic data collected from loop detectors installed in a freeway system and transmitted to a central station present such patterns. Neural networks have been used to detect incident patterns from nonincident patterns with limited success. The dimensionality of the training input data is high, and the embedded incident characteristics are not easily detectable. The wavelet transform has been used for feature extraction, denoising, and effective preprocessing of data before a neural network model is used for effective traffic incident detection (Samant and Adeli 2000; Adeli and Karim 2000), and crack identification (Liew and Wang 1998), and damage identification (Marwala 2000) of structures. Wavelets are discussed in Chapter 3.

2.3   Radial Basis Function Neural Network (RBFNN)

RBF neural networks (Yen 1994; Moody and Darken 1989; Poggio and Girosi 1990; Hassoun 1995; Amin *et al.* 1998; Haykin 1998; Jayawardena and Fernando 1998; Adeli and Karim 2000) are powerful techniques for interpolation in multidimensional space. An RBF is a function with a distance criterion with respect to a center. Figure 2.2 shows a sample RBFNN. It has a simple topology consisting of an input layer, a hidden layer of nodes with radial basis transfer functions, and an output node with a linear transfer function. A Gaussian (bell-shaped) function is usually chosen as the processing element in the hidden layer. An RBFNN creates clusters for similar patterns in training data. Each cluster has a center (represented by a hidden layer node). Thus, a radial unit is defined by the center and a radius (or deviation) of the training data. The RBFNN is widely applied in regression and classification problems. In regression problems, the output layer is a simple linear combination of hidden layer values representing the mean predicted output. The interpretation of this output layer value is similar to a regression model in statistics. In classification problems the output layer is typically a sigmoid function of a linear combination of hidden layer values, representing a posterior probability.

The RBF network in Fig. 2.2 is often complemented with a linear part to account for any existing linear relationship between the input data and the model output beyond

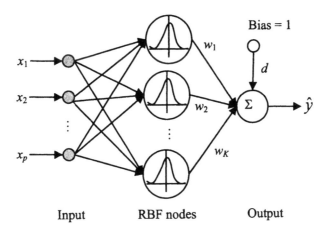

$$\mathbf{x} = [x_1, \ldots, x_p] = \text{Input vector containing } p \text{ variables}$$
w_j = Weight of the link connecting the RBF node j to the node
      in the output layer
d = Weight of the bias
$\hat{y}$ = Model output

Figure 2.2 An illustrative example for RBFNN model

and above the common nonlinear relationship. Mathematically, the RBF network with a linear part produces an output as follows:

$$y = \sum_{j=1}^{K}\left(w_j e^{-\lambda_j^2(\mathbf{x}-c_j)^2} \right) + \sum_{i=1}^{P} b_i x_i + d \qquad (2.3)$$

where K is the number of neurons (RBF nodes) in the hidden layer, each containing a basis function. The parameters of the RBF network consist of the positions of the basis functions c_j, the width of the centers of the basis functions λ_j, the weights in output sum w_j, the parameters of the linear part b_i, and the weight of the bias d.

In comparison to a multilayer feedforward neural network such as a BPNN, an RBFNN has three main advantages. First, it does not suffer from entrapment in a local minimum in the same way as a BP network. This is because the shape parameters c_j and

λ_j of the Gaussian function are determined from the training data by using one of the clustering techniques to be discussed later. The only parameter (w_j in Fig. 2.2) to be adjusted in the learning process is the linear mapping from the RBF layer to the output layer. Linearity in RBFNN ensures that the error surface is quadratic and therefore has a single easily found minimum. In regression problems this can be found in one matrix operation. In classification problems the fixed non-linearity introduced by the sigmoid output function can be efficiently handled using the iterative least squares algorithm.

Second, in an RBFNN model, similarity of any new pattern to the training patterns is measured by its proximity to the centers of the clusters. As such, the RBFNN is a regularization or generalization network (Moody and Darken 1989; Poggio and Girosi 1990). It is most suitable for estimation problems where limited data are available and overfitting needs to be avoided. The danger of overfitting is reduced by the local nature of the transfer functions that allow only a fraction of the nodes to participate in the mapping of a given pattern. When data are limited, the effect of noise becomes significant and some patterns may not be sufficiently represented in the training. Generalization in the vicinity of cluster centers is maintained by the graded nature of the transfer functions. The generalization properties of RBFNNs are discussed in detail by Poggio and Girosi (1990) and Adeli and Wu (1998). In contrast, the BP neural network is susceptible to the overfitting problem when training data are limited and noisy.

The third advantage of the RBFNN over the BPNN is its quick convergence in training. Information in an RBFNN is locally distributed. As such, only a few weights have to be modified in each iteration during the training process. Because of these advantages, the RBFNN is found to be most suitable for learning a complex problem for which only limited data are available.

In RBFNN, the centers should be assigned to reflect the natural clustering of the training data. Sub-sampling and K-means algorithms have been commonly used to determine the centers. In the sub-sampling algorithm (Haykin 1998), randomly-chosen training points are assigned to the radial unit to represent the distribution of the data in a statistical sense. This algorithm is suitable for determining the centers of the RBFNN with a relatively large number of radial units.

The K-Means algorithm (Bishop 1995) is one of the simplest unsupervised learning algorithms for solving the clustering problem. It tries to select an optimal set of points that are placed at the centroids of clusters of training data. Given K clusters or radial units, the positions of the centers are adjusted so that (1) each training point belongs to a cluster center, and is nearer to this center than to any other center, and (2) each cluster center is the centroid of the training points in the cluster.

Recently, a fuzzy c-means clustering (FCM) algorithm (Bezdek 1981; Cannon *et al.* 1986) was developed by Adeli and Karim (2000) to improve the performance of RBFNN for pattern recognition problems. This algorithm performs a fuzzy partitioning of the data set into classes. This is in contrast to crisp assignment of data vectors to distinct classes employed in classical statistical clustering techniques such as the K-Means algorithm. The prefix c in the fuzzy c partitions refers to the number of classes in each partition.

2.4   Boltzmann Neural Network

The aforementioned feedforward networks propagate data only from input to output, while a recurrent network also propagates data from subsequent processing stages (e.g., output layer) to earlier stages (e.g., hidden layer). A Hopfield neural network (Hagan *et al.* 1996; Mehrotra *et al.* 1997; Pham and Karaboga 2000) is a kind of recurrent network where all connections have bi-directional data flow. It is known to converge to a local optimum, thus suffering from the same hill-climbing problem as the BPNN. To overcome this shortcoming, Ackley *et al.* (1985) introduced the so-called Boltzmann machine by introducing *noise* in the network trajectory to avoid the problem of entrapment in a local optimum. The concept of noise in the Boltzmann machine is analogous to the concept of temperature in the simulated annealing algorithm. The magnitude of the noise is reduced steadily based on a probability distribution till the network converges to the global optimum. A second distinction between the Hopfield network and the Boltzmann machine is that the former has no hidden layer but the latter does.

The Boltzmann neural network is often applied in association with the simulated annealing algorithm to avoid the so-called hill-climbing problem in model learning (Jiang and Adeli 2003a). The neural network training is a nonlinear programming problem with many hills and valleys, where the solution can get stuck in a local optimum, say one of the valleys in the minimization problem. A number of approaches have been proposed in the recent literature to overcome this problem and find the true global optimum solution such as genetic algorithms (Adeli and Cheng 1993; Adeli and Hung 1995) and simulated annealing (Kirkpatrick *et al.* 1983). Simulated annealing is inspired by the metallurgical process of annealing where a metal is heated to near melting point and then cooled slowly and intermittently until an *equilibrium* is achieved for an *optimum* material microstructure with desirable structural properties such as ductility. The material microstructure may be changed easily and rapidly at high temperatures with high kinetic energy. But, sudden cooling of the material can result in undesirable brittleness. In contrast, a gradual and carefully controlled cooling operation can result in a material with optimum microstructure. This process may be explained as removing *local pockets of stress energy* to allow the metal to escape from local *elevated energy minima* and reach a global energy minimum (Aleksander and Morton 1991).

Metaphorically, simulated annealing for solution of nonlinear programming problems with multiple local optima can be considered as maximizing *strength* and minimizing *brittleness* by minimizing an *energy* functional (Mehrotra *et al.* 1997). As such, this approach requires the definition of an energy function and a temperature parameter to be lowered gradually during the optimization iterations. The selection of a candidate solution in successive iterations and the corresponding modification of the temperature parameter are guided by a probabilistic distribution. This process helps the solution to jump from one *valley* to the next *valley* in search of the true global optimum. Thus, a typical simulated annealing algorithm is implemented in two nested loops, an outer loop where the temperature parameter is reduced and an inner loop where direction of iterations is determined. In every iteration, solutions in the vicinity of the current solution are explored. Solutions that decrease the energy functional are maintained for additional moves. Further, solutions that increase the energy function with an acceptable selection probability, expressed as a function of the temperature parameter and the change in the energy from the previous iteration, are also maintained for additional move in search of the global optimum. The probability of acceptance is chosen to be larger with a larger value of the temperature parameter, similar to the metallurgical annealing process where the material microstructure is modified more easily at higher temperatures.

2.5   Dynamic Neural Network

There is a special type of function approximation problem where the input data are time dependent such as a time series. This means that the function is dynamic and has timewise "memory", which is therefore referred to as a dynamic system. For such systems, past information can be used to predict future behavior. Two engineering examples of dynamic system problems are: (1) predicting the freeway traffic flow and (2) approximating the dynamic response of a building structure as a function of the external loading or ground motion. In both of these examples the output signal at some time instant depends on what has happened earlier. The first example is a time-series problem modeled as a system involving no inputs. In the second example there is a time series input, i.e., the applied external loading. Examples of these kinds will be presented in Chapter 12, *Traffic Flow Forecasting* and in Chapter 13, *Structural System Identification*.

A dynamic neural network is defined as the network which models a dynamic system by employing memory in its inputs; specifically, storing a number of past input and output data. Such neural network structures are also referred to as *tapped-delay-line neural networks, neural autoregressive with exogenous inputs* (NARX), or *neural autoregressive* (NAR) models. A dynamic neural network consists of a combination of feedforward or RBF neural networks and a specified time-delay input vector to accommodate memory for past information. The time-delay input vector or regressor vector is constructed to connect the neural network with a dynamic system. Note that the lagged output values are often fed back to the input layer; the resulting network is also called recurrent neural network. Therefore, the input vector usually contains lagged input and output values of the system specified by three indices: n_a, n_b, and n_k, to be defined in the next paragraph.

Let $u(t)$ and $y(t)$ be the input and output time series of a dynamic system, respectively. For a single-input single-output (SISO) model the input vector to a dynamic neural network is expressed as

$$\mathbf{x}(t) = [y(t-1), \cdots, y(t-n_a), u(t-n_k), \cdots, u(t-n_k-n_b+1)]' \qquad (2.4)$$

where indices n_a and n_b represent the numbers of lagged output and input values, respectively. The index n_a is often referred to as the order of the model. Index n_k is the input delay relative to the output. The prime indicates the transpose of a vector. In a multiple-input multiple-output (MIMO) case each individual lagged signal is a vector of appropriate length.

Using Eq. (2.1), the dynamic neural network is conceptually expressed as

$$y(t) = f[\mathbf{x}(t), \boldsymbol{\theta}] \qquad\qquad (2.5)$$

Equation (2.5) defines a mapping from the regressor space to the output space. A dynamic neural network model based on a one-hidden-layer feedforward network is shown in Fig. 2.3.

Unlike the BP or RBF neural network model, the dynamic neural network preserves the time sequence of the input vectors and memorizes the past of data. Therefore, it is suitable for modeling problems involving time-dependent data.

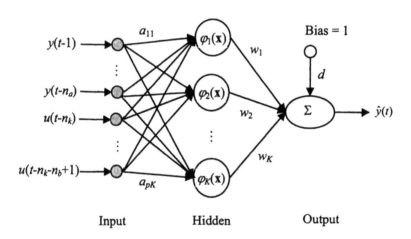

$\mathbf{x} = [y(t-1),\ldots, y(t-n_a), u(t-n_k),\ldots, u(t-n_k-n_b+1)]$ = Input vector
a_{ij} = Weight of the link connecting the input node i to node j in the hidden layer
w_j = Weight of the link connecting the hidden node j to the node in the output layer
d = Weight of the bias
$\hat{y}(t)$ = Model output

Figure 2.3 An example of a dynamic feedforward ANN model

Chapter 3

WAVELETS

3.1   Introduction

Wavelets consist of a family of mathematical functions used to represent a signal in both time and frequency domains. A *wavelet transform* (WT) decomposes a temporal signal into a set of time-domain basis functions with various frequency resolutions. The WT is computationally similar to the windowed short-term Fourier Transform (WFT). However, unlike the sine and cosine functions used in the WFT, the wavelet functions used in the WT are localized in spaces. Thus, the wavelet transform can decompose a signal into two sub-signals: *approximations* and *details*. The *approximations* represent the high-scale, low-frequency components of the signal, while the *details* represent the low-scale, high-frequency components. Therefore, compared with the Fourier transform, the wavelet transform provides a more effective representation of discontinuities in signals and transient functions.

If the time series with N measured data points, $f(t)$, is considered to be a square integrable function (i.e., the integral of its square is finite), both *continuous wavelet transform* (CWT) and *discrete wavelet transform* (DWT) can be used to analyze the time series data. The CWT of $f(t)$ is defined as (Chui 1992)

$$W_f(a,b) = \int_{-\infty}^{\infty} f(t)\psi_{a,b}(t)dt \tag{3.1}$$

and the two-dimensional wavelet expansion functions $\psi_{a,b}(t)$ are obtained from the basic function (also known as *mother* or *generating* wavelet) $\psi(t)$ by simple scaling and translation

$$\psi_{a,b}(t) = \frac{1}{\sqrt{|a|}}\psi(\frac{t-b}{a}), \ a, b \in \Re, \ \psi \in L^2(\Re) \tag{3.2}$$

where $\in$ denotes membership, t is the time variable, the parameters $a \neq 0$ and b denote the frequency (or scale) and the time (or space) location, respectively, and $\Re$ is the set of real numbers. The notation $L^2(\Re)$ represents the square summable time series space of the data, where the superscript 2 denotes the square of the modulus of the function.

To avoid intensive computations for every possible scale a and dilation b, the dyadic values are often used for both scaling and dilation in discrete wavelet transform as follows:

$$a_j = 2^j, \ b_{j,k} = k2^j, \ j,k \in Z \tag{3.3}$$

where k and j denote the time and frequency indices, respectively, and Z is the set of all integers.

Substituting Eq. (3.3) into Eq. (3.2), the following wavelet expansion function is obtained:

$$\psi_{j,k}(t) = 2^{-j/2}\psi(2^{-j}t - k), \ j,k \in Z, \ \psi \in L^2(\Re) \tag{3.4}$$

and Eq. (3.1) is rewritten as

$$W_f(j,k) = 2^{-j/2} \int_{-\infty}^{\infty} f(t)\psi(2^{-j}t - k)dt, \ j,k \in Z, \ \psi \in L^2(\Re) \tag{3.5}$$

which is the DWT of the time series $f(t)$.

The DWT represented by Eq. (3.5) aims to preserve the dominant features of the CWT represented by Eq. (3.1) in a succinct manner. Conceptually, the DWT can be considered as a judicious sub-sampling of CWT coefficients with just dyadic scales (i.e., 2^{j-1}, $j = 1, 2, \ldots L$, where L is the maximum number of the decomposition level). Compared with the DWT, the CWT may represent the physical system more accurately as it makes very subtle information visible, but it requires more intensive computations for integrating over every possible scale, a, and dilation, b.

The time series can be reconstructed by inverse of the CWT of Eq. (3.1) in the double-integral form or the DWT of Eq. (3.5) in the double-summation form. In terms of DWT, the time series $f(t)$ is reconstructed by

$$f(t) = \sum_j \sum_k a_{j,k}\psi_{j,k}(t) \tag{3.6}$$

where the coefficients of the series $\{a_{j,k}\}$ are calculated as follows:

$$a_{j,k} = \langle f(t), \psi_{j,k}(t) \rangle \qquad (3.7)$$

in which $\langle f(t), \psi_{j,k}(t) \rangle$ represents the inner product of two functions $f(t)$ and $\psi_{j,k}(t)$.

3.2   Multiresolution Analysis

The *multiresolution* analysis (MRA) provides a powerful tool for time-frequency analysis of signals. In this analysis, the time series is decomposed in terms of different levels of details using scaling and wavelet functions (Mallat 1989 & 1999; Holschneider 1995; Burrus *et al.* 1998). A nesting relationship is defined for various scaling function subspaces, $\mathbf{V}_i$ ($i \in Z$) as:

$$\cdots \mathbf{V}_{-2} \subset \mathbf{V}_{-1} \subset \mathbf{V}_0 \subset \mathbf{V}_1 \subset \mathbf{V}_2 \subset \cdots \subset L^2(\Re) \qquad (3.8)$$

where the notation $\mathbf{V}_i \subset \mathbf{V}_{i+1}$ means that $\mathbf{V}_i$ is a subspace of $\mathbf{V}_{i+1}$. The subspace $\mathbf{V}_i$ is defined by spanning $L^2(\Re)$ by the scaling functions, $\varphi_k(t) = \varphi(k-t)$, $k \in Z$ as:

$$\mathbf{V}_i = \overline{\underset{k}{Span}\{\varphi_k(t)\}} \qquad (3.9)$$

in which the over-bar indicates closure of the subspace (i.e., boundaries are included in the subspace).

The wavelet subspace $\mathbf{W}_i$ is defined as the orthogonal complement of $\mathbf{V}_i$ in $\mathbf{V}_{i+1}$ such that

$$\mathbf{V}_1 = \mathbf{V}_0 \oplus \mathbf{W}_0, \qquad \mathbf{V}_2 = \mathbf{V}_0 \oplus \mathbf{W}_0 \oplus \mathbf{W}_1 \qquad (3.10)$$

and the entire time series space of the square summable integrals is represented as the summation of all subspaces:

$$L^2(\Re) = \mathbf{V}_{j0} \oplus \mathbf{W}_{j0} \oplus \mathbf{W}_{j0+1} \oplus \cdots \qquad (3.11)$$

where $\oplus$ represents a direct space sum and the integer subscript j_0 is any general starting scaling parameter index from negative infinity to positive infinity including zero.

For the DWT, the subspace, V_i, satisfies a natural scaling condition (Daubechies 1988 & 1992):

$$f(t) \in V_i \iff f(2t) \in V_{i+1} \tag{3.12}$$

where $\iff$ denotes the bi-directional inference. Equation (3.12) implies that the time series in a subspace is scaled by a dyadic factor in the next subspace.

The multiresolution analysis requires that a space containing the higher resolution signals also contain the lower resolution signals (Burrus *et al.* 1998). Thus, a new scaling function, $\varphi(t)$, can be obtained by a weighted sum of scaling functions $\psi(2t)$ shifted by n (Burrus *et al.* 1998; Goswami and Chan 1999) as

$$\varphi(t) = \sum_n h_0(n)\sqrt{2}\psi(2t - n) \tag{3.13}$$

where $n \in Z$ is the order of the wavelet function (or number of wavelet coefficients) and $h_0(n)$ is a sequence of n real or complex numbers called the scaling function coefficients (or scaling filters).

The wavelet function, $\psi(t)$, can be represented by a weighted sum of scaling functions $\psi(2t)$ shifted by n, similar to Eq. (3.13) as

$$\psi(t) = \sum_n h_1(n)\sqrt{2}\psi(2t - n) \tag{3.14}$$

where $h_1(n)$ represents the sequence of wavelet function coefficients (or wavelet filters). For orthogonal wavelets, the scaling coefficients $h_0(n)$ and wavelet coefficients $h_1(n)$ for a filter of length N are related by the following equation (Burrus *et al.* 1998):

$$h_1(n) = (-1)^n h_0(N - 1 - n) \tag{3.15}$$

Figure 3.1 shows the scaling, $\varphi(t)$, and wavelet, $\psi(t)$, functions for the Daubechies wavelet of order 4 (Daubechies 1992). It is a compactly supported wavelet function with an orthonormal basis.

The values of $h_0(n)$ and $h_1(n)$ are obtained for any given type of wavelet function by satisfying two basic wavelet properties: the integral of $\psi(t)$ is zero and the integral of the squared $\psi(t)$ is unity (Daubechies 1988 & 1992). Equations (3.13) and (3.14) provide the prototype or mother wavelet $\psi(t)$ needed in Eq. (3.6). Thus, the time series, $f(t)$, can be represented by the inverse wavelet transform and the coefficients obtained from the multiresolution analysis as follows:

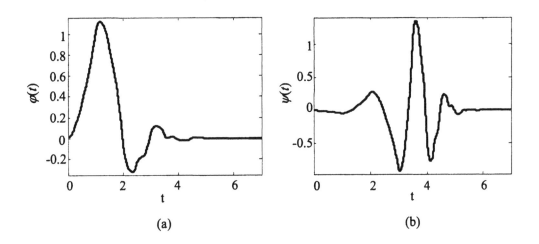

Figure 3.1 Daubechies wavelet of order 4: (a) Scaling function, and (b) wavelet function

$$f(t) = \sum_{k=-\infty}^{\infty} s_{j_0,k}\varphi_{j_0,k}(t) + \sum_{j=j_0}^{\infty}\sum_{k=-\infty}^{\infty} w_{j,k}(t)\psi_{j,k}(t) \qquad (3.16)$$

where the first term is a coarse resolution at scale j_0 (or approximations) and the second term provides the frequency and time breakdown of the signal (or details). The j_0 level scaling coefficients $s_{j_0,k}$ are obtained from Eqs. (3.7) and (3.13) as follows:

$$s_{j_0,k} = \langle f(t),\varphi_{j_0,k}(t)\rangle = \sum_{m} h_0(m-2t)s_{m,j_0+1} \qquad (3.17)$$

where $m = 2t + n$. Similarly, the j level wavelet coefficients $w_{j,k}$ are obtained from Eqs. (3.7) and (3.14) as follows:

$$w_{j,k} = \langle f(t), \psi_{j,k}(t) \rangle = \sum_m h_1(m - 2t)a_{m,j+1} \qquad (3.18)$$

The discrete values of the original signal $f(t)$ are used for initial values of the scaling coefficients $s_{j_0,k}$. Equations (3.17) and (3.18) provide a recursive way to compute the discrete wavelet transform of the time series data. They are used to perform digital filtering which involves the convolution of the original time series with the filter coefficients. This process transforms the original signal into two high and low resolution signals which doubles the number of the original data points. After digital filtering, downsampling is performed in order to reduce the number of data points by half and obtain the same number of data points as in the original signal.

Figure 3.2 shows the decomposition of the scaling function space V_j in a three-level DWT multiresolution analysis. The orthogonal nesting relationship of subspaces is shown in Fig. 3.2a where no subspace overlaps with another (nonoverlaping functions are always orthogonal). Figure 3.2b shows the decomposition tree for the three-level DWT wavelet multiresolution analysis where a space vector is decomposed into two subspaces V_j (an approximation) with filter coefficients h and W_j (a detail) with filter coefficient h_1. The approximation subspace V_j itself is then split into a second-level approximation V_{j-1} and detail W_{j-1}, and the process is repeated. The down-pointing arrows in the left side of Fig. 3.2b denote down-sampling by a factor of two (Eq. 3.3).

3.3 Wavelet Packets

For the conventional DWT and wavelet multiresolution analysis described in the previous sections, the resulting time-frequency resolution has narrow bandwidths in the low frequencies and wide bandwidths in the high frequencies. It is not sufficient for analyzing the time series displaying fractals. Coifman and Wickerhauser (1992) proposed the *wavelet packet* analysis to allow for a finer and adjustable resolution of frequencies at high frequencies (details). In the conventional wavelet transform, only the scaling functions or approximations are decomposed into subspaces, as shown in Fig. 3.2(b). In wavelet packet analysis, both scaling functions representing the approximations and the wavelet functions representing the details are decomposed into subspaces. In that case, Eqs. (3.10) and (3.11) are rewritten as

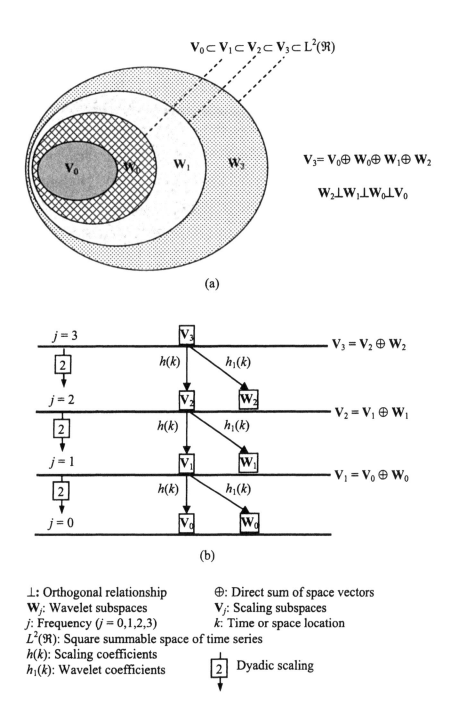

$$V_0 \subset V_1 \subset V_2 \subset V_3 \subset L^2(\Re)$$

$$V_3 = V_0 \oplus W_0 \oplus W_1 \oplus W_2$$

$$W_2 \perp W_1 \perp W_0 \perp V_0$$

(a)

$V_3 = V_2 \oplus W_2$

$V_2 = V_1 \oplus W_1$

$V_1 = V_0 \oplus W_0$

(b)

$\perp$: Orthogonal relationship $\oplus$: Direct sum of space vectors
W_j: Wavelet subspaces V_j: Scaling subspaces
j: Frequency ($j = 0,1,2,3$) k: Time or space location
$L^2(\Re)$: Square summable space of time series
$h(k)$: Scaling coefficients
$h_1(k)$: Wavelet coefficients $\boxed{2}$ Dyadic scaling

Figure 3.2 Three-level DWT decomposition of the scaling function space: (a) Nested vector spaces, and (b) decomposition tree

$$\mathbf{V}_1 = \mathbf{V}_0 \oplus \mathbf{W}_0, \qquad \mathbf{V}_2 = \mathbf{V}_0 \oplus \mathbf{W}_0 \oplus \mathbf{W}_{10} \oplus \mathbf{W}_{11} \tag{3.19}$$

and

$$L^2(R) = \mathbf{V}_{j0} \oplus \mathbf{W}_{j0} \oplus \mathbf{W}_{j10} \oplus \mathbf{W}_{j11} \oplus \cdots \tag{3.20}$$

The discrete wavelet packet transform (DWPT) provides greater flexibility for detecting the oscillatory or periodic behavior and the fractal properties of time series than the conventional DWT. Figure 3.3 shows the three-level DWPT decomposition of the scaling function space $\mathbf{V}_j$. The orthogonal nesting relationship of subspaces is shown in Fig. 3.3a. In contrast to the conventional wavelet multiresolution analysis shown in Fig. 3.2a, the wavelet subspaces in DWPT are also split in the successive decompositions. Figure 3.3b shows the decomposition tree for the three-level DWPT, where a space vector is decomposed into two subspaces $\mathbf{V}_j$ (approximation) and $\mathbf{W}_j$ (details). The scaling and wavelet subspaces are simultaneously split into second-level subspaces, and the process is repeated until the desirable decomposition level is reached.

In practical applications of DWPT decomposition, a given time series, $f(t)$, is simultaneously decomposed into a series of scaling coefficients, $s_j(k)$, and wavelet coefficients, $w_j(k)$. The time series can then be represented by the inverse wavelet transform and the DWPT coefficients, i.e., $s_j(k)$ and $w_j(k)$, as follows:

$$f(t) = \sum_{k \in Z} \sum_{j \in Z} \left[s_j(k)\varphi_{j,k}(t) + w_j(k)\psi_{j,k}(t) \right] \tag{3.21}$$

where the double summation indicates that the scaling and wavelet subspaces are simultaneously split into second-level subspaces to provide both frequency and time breakdown of the signal.

Figure 3.4 shows the three-level DWPT decomposition of the time series data $f(t)$, where the approximation is represented by A (subspace $\mathbf{V}_j$) and the detail is represented by D (subspace $\mathbf{W}_j$). Figure 3.4a shows the decomposition tree of the scaling and wavelet function spaces and coefficients, and Fig. 3.4b shows the relationship of wavelet subspaces. The coefficients corresponding to A and D are simultaneously split into the next level. For example, the data are resolved into eight series corresponding to the eight decomposition subspaces in the third level, and identified as AAA$_3$, which represents the

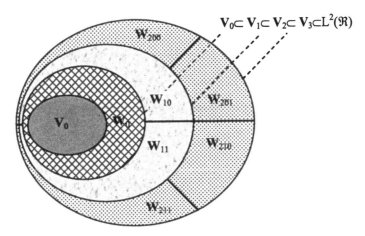

$$\mathbf{V}_0 \perp \mathbf{W}_0 \perp \mathbf{W}_{10} \perp \mathbf{W}_{11} \perp \mathbf{W}_{200} \perp \mathbf{W}_{201} \perp \mathbf{W}_{210} \perp \mathbf{W}_{211}$$

$$\mathbf{V}_3 = \mathbf{V}_0 \oplus \mathbf{W}_0 \oplus \mathbf{W}_{10} \oplus \mathbf{W}_{11} \oplus \mathbf{W}_{200} \oplus \mathbf{W}_{201} \oplus \mathbf{W}_{210} \oplus \mathbf{W}_{211}$$

(a)

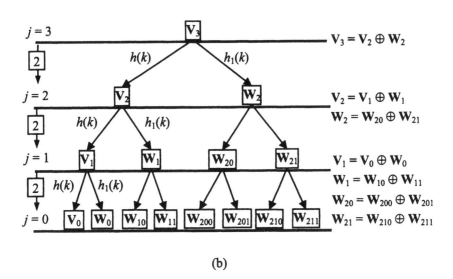

(b)

⊥: Orthogonal relationship   ⊕: Direct sum of space vectors
$\mathbf{W}_j$: Wavelet subspaces $\mathbf{V}_j$: Scaling subspaces
j: Frequency ($j = 0,1,2,3$) k: Time or space location
$L^2(\Re)$: Square summable space of the time series
$h(k)$: Scaling coefficients
$h_1(k)$: Wavelet coefficients   ⎍2 Dyadic scaling

Figure 3.3 Three-level DWPT decomposition of the scaling function space: (a) Nested vector spaces, and (b) decomposition tree

28

Intelligent Infrastructure

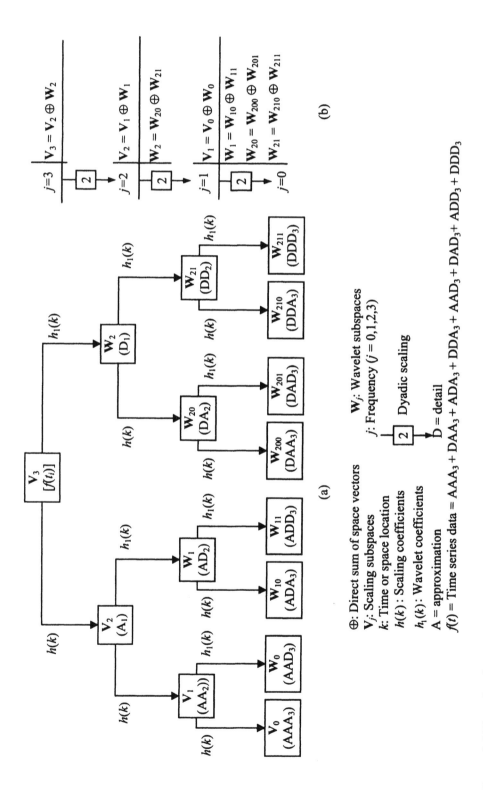

Figure 3.4 Three-level DWPT decomposition: (a) Decomposition tree of the scaling and wavelet function spaces and coefficients, and (b) relationship of wavelet subspaces

third-level approximation coefficients (A$_3$) resulting from the second-level approximation (AA$_2$) to DDD$_3$, which represents the third-level details coefficients (D$_3$) resulting from the second level details (DD$_2$). Thus, based on Eq. (3.21), the time series data, $f(t)$, can be represented mathematically by the third-level DWPT decomposition coefficients as follows:

$$f(t) = AAA_3 + AAD_3 + ADA_3 + ADD_3 + DAA_3 + DAD_3 + DDA_3 + DDD_3 \qquad (3.22)$$

Equation (3.22) mathematically relates the original time series data to the wavelet decomposed coefficients.

Chapter 4

CHAOS THEORY

4.1   Introduction

Chaos theory is the study of unstable aperiodic behavior in deterministic nonlinear dynamic systems (Kellert 1993). It stems partially from the discovery of Edward Lorenz, a meteorologist at MIT. In 1960, when Lorenz simulated weather patterns on a computer, he found that a small change in the initial condition produced a remarkable deviation in the simulation results (Lorenz 1963). Figure 4.1 illustrates Lorenz's model in the three-dimensional state space (equations representing this model will be described later in this section). Because the attractor (a set of data describing the evolution of a dynamic system over a sufficiently long time) of Lorenz's system looks like a butterfly, Lorenz's model has been well known as the "butterfly effect" and is often used to illustrate the complexity and unpredictability of nonlinear dynamics. The butterfly effect technically reflects the essence of chaos, i.e., its sensitive dependence on initial conditions. In meteorology it reflects how small changes in the initial condition can cause large changes in the atmospheric motion. For example, a butterfly flapping its wings somewhere in Shanghai may result in a tornado in Florida. This is also known as the butterfly effect. Therefore, long-term weather forecasting becomes impossible.

In addition to the Lorenz chaotic system shown in Fig. 4.1, the Rössler system (Rössler 1976) (described mathematically in Section 4.4.2) and the Hénon map (Hénon 1976) (described mathematically in Section 4.4.3) are the other two most studied examples of dynamic systems that exhibit chaotic behavior. Figures 4.2 and 4.3 show the attractors for the Rössler system and the Hénon map, respectively, both produced by a set of non-linear equations. These continuous-time dynamic systems exhibit chaotic dynamics.

In addition to chaotic dynamic systems shown in Figs. 4.1–4.3, chaotic behaviors are also observed in bifurcation and geometric fractals (i.e., fragmented geometric shape). Figure 4.4 shows the Feigenbaum fractal produced by a recursive logistic equation (May 1976). If we magnify the diagram, we can observe that the diagram consists of many smaller versions of itself. This is the well-known bifurcation fractal. The bifurcation fractal diagram shows the evolution and periodic orbits of the system defined by a logistic function with a bifurcation parameter. It has been shown that the bifurcation fractal shown in Fig. 4.4 can be used as a simple mathematical model to represent the species population with no predators but limited food supply (Gleick 1987).

Figure 4.4 illustrates that fractals demonstrate a self-similar property at all levels of magnification. Many natural objects are observed to be approximate fractals, such as clouds, mountain ranges, coastlines, and snowflakes. Figure 4.5 shows the Sierpiński triangle, a well-known geometric fractal (Sierpiński 1915). This is one of the basic examples of self-similar sets. The self-similar set is referred to be a mathematically generated pattern that can be reproduced at any magnification or reduction. The Sierpinski triangle is created from a triangle in a plane. The triangle is then divided into four smaller triangles. The central one of the four triangles is removed.

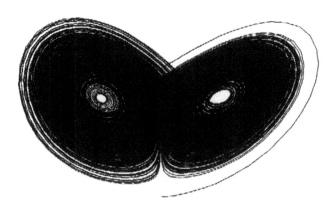

Figure 4.1 Attractor of the Lorenz system

Figure 4.2 Attractor of the Rössler system

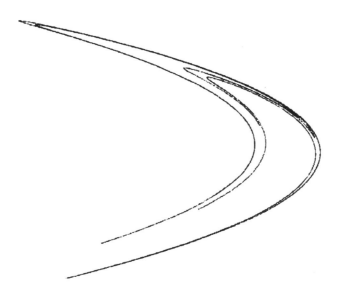

Figure 4.3 Attractor of Hénon's map

Intelligent Infrastructure

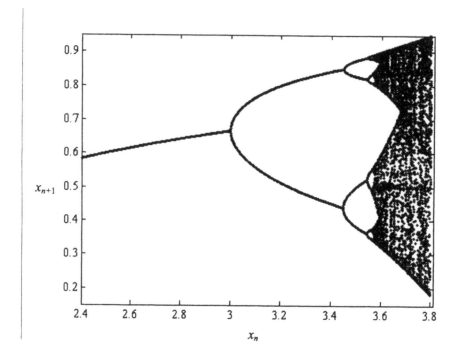

Figure 4.4 Bifurcation produced by a logistic equation

Figure 4.5 Sierpinski triangle

As another example of geometric fractals, Fig. 4.6 shows the Mandelbrot set, a very complex geometric figure (Gleick 1987). It is mathematically formed from the very simple formula $x_{n+1} = x_n^2 + c$ with complex c-values. Figure 4.6 indicates that as this figure is magnified, an increasingly intricate and beautiful pattern with an endless amount of detail is obtained.

Chaos theory has been widely applied to study turbulence and fractal geometry. It has been considered to be synonymous with complexity theory and used to describe and explain various natural and artificial time series. The characteristics of chaos theory can be summarized as follows: 1) Chaos is qualitative in the sense that it describes only the general character of a system's long-term behavior; 2) chaotic systems are unstable and incurred partly by an external disturbance; 3) chaotic systems are aperiodic in that the variables describing the state of a system do not demonstrate a regular repetition of values and have the appearance of noise; 4) chaotic systems are deterministic and often are expressed by a few, simple differential equations, and 5) a chaotic system is a nonlinear dynamic system with a strong sensitive dependence on initial conditions.

The dynamics of many physically complex systems in many engineering disciplines, however, cannot be represented by a set of differential equations. Recent research on chaotic time series shows that irregular and unpredictable or seemingly unpredictable time evolutions can be the result of completely deterministic dynamic systems rather than pure randomness (Abarbanel 1993). Chaos theory provides novel methodologies for analysis of chaotic time series and prediction of chaos encountered in the real world that previously had been regarded as stochastic signals and discarded as *noise*. Examples of chaotic time series may be seen in traffic flow (Disbro and Frame 1989; Dendrinos 1994), earthquake accelerograms (Turcotte 1992; Kanamori 1996), weather patterns (Benzi *et al.* 1999; Patil *et al.* 2001), population dynamics (Thunberg 2001; Billings *et al.* 2002), stock markets (Hsieh 1991), cardiological arrhythmias (Galka 2000), electroencephalograph (EEG) records (Galka 2000; Sarbadhikari and Chakrabarty 2001; Adeli *et al.* 2007), and epidemics (Bjørnstad *et al.* 2002).

When a time series is produced from a nonlinear dynamic system in the real world, the physical processes giving rise to the chaotic behavior are usually unknown and presumably multivariate. In order to characterize the chaotic behavior of a time series, a *state space* is reconstructed from the observed time series to simulate effectively

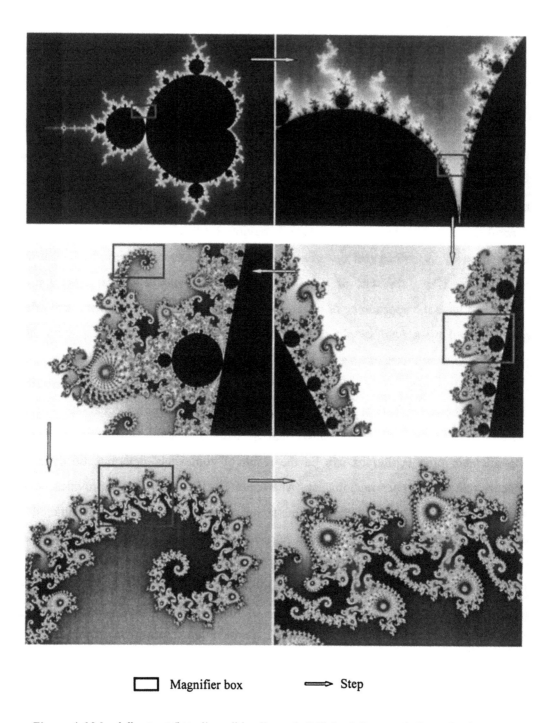

Magnifier box   ⟹ Step

Figure 4.6 Mandelbrot set (http://en.wikipedia.org/wiki/Mandelbrot_set). See color insert following Chapter 15.

the multivariate properties of the nonlinear system. For convenience of formulation, a time series is denoted by $\{y_n\}$, $n = 1, 2, ..., N$, where N is the number of data points. The n-th point in the state space is represented by the column vector, $\mathbf{y}_n$, as follows:

$$\mathbf{y}_n = [y_n, y_{n+\tau}, y_{n+2\tau}, ..., y_{n+(d_E-1)\tau}]' \tag{4.1}$$

where the prime indicates the transpose of a vector and the parameters d_E and τ are the embedding dimension and the lag time index used to reconstruct the state space, respectively. The data point subscript, n, satisfies the following condition:

$$n < N - \tau(d_E - 1), n \in Z \tag{4.2}$$

in which Z is the set of all integers. That means the maximum number of state space points, N_a, is equal to $N - \tau(d_E - 1)$.

Equation (4.1) is the delay time coordinates for representing chaotic motions proposed by Takens (1981). Takens proved that some dynamic invariants of the dynamic system producing the time series are preserved if the time series is transformed into a sufficiently large reconstructed state space (defined in terms of its dimension). This is known as Takens' theorem, which has been widely used for chaos analysis of time series (Fraser and Swinney 1986; Theiler 1990; Casdagli *et al.* 1991; Buzug and Pfister 1992; Kennel *et al.* 1992; Abarbanel 1993 & 1996). The delay time coordinate concept is employed to create a topological equivalent *attractor*, which is a set of points used to simulate the evolution trajectory in the original state space. The purpose of the attractor is to unfold the time series back to a multivariate state space representing the original physical system. An attractor is the geometric invariance in the state space representation of a time series. Chaotic or strange attractors often indicate chaos in the physical system represented by time series. A proper choice of the parameters d_E and τ plays an important role in accurately reconstructing the multivariate state space and calculating the invariants in the chaotic system, which will be discussed later.

Figure 4.7 illustrates the general schema for reconstructing the state space from a measured time series. Figure 4.7a shows a hypothetical physical dynamic system with two state variables and a two-dimensional *manifold* (a coordinate space or a non-self-intersecting geometric model used to describe a physical phenomenon), $\aleph = \Re^2$, where

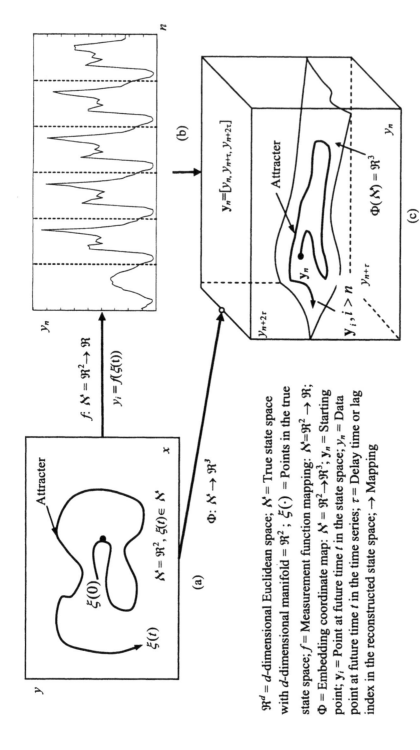

$\mathfrak{R}^d$ = d-dimensional Euclidean space; $\aleph$ = True state space with d-dimensional manifold = $\mathfrak{R}^2$; $\xi(\cdot)$ = Points in the true state space; f = Measurement function mapping: $\aleph = \mathfrak{R}^2 \to \mathfrak{R}$; Φ = Embedding coordinate map: $\aleph = \mathfrak{R}^2 \to \mathfrak{R}^3$; $\mathbf{y}_n$ = Starting point; $\mathbf{y}_i$ = Point at future time t in the state space; y_n = Data point at future time t in the time series; τ = Delay time or lag index in the reconstructed state space; $\to$ = Mapping

Figure 4.7 Schematic illustration of the reconstructed state space: (a) True state space; (b) original time series, and (c) reconstructed state space

$\mathfrak{R}^2$ represents a 2-dimensional Euclidean space. It represents a true state space where each state of the system corresponds to a point $\xi(t) \in \aleph$. The true state space is not known. Instead, a time series of actual measurements, $y_n = f(\xi(t))$, is obtained using a measurement function which maps a two-dimensional manifold to a one-dimensional manifold denoted as $f: \aleph = \mathfrak{R}^2 \to \mathfrak{R}$ (Fig. 4.7b). Usually applying a measurement function causes the loss of some information, which makes it impossible to reconstruct the true evolution of $\xi(t)$ based on the measurement series $\{y_n\}$. When the measurements are transformed into reconstructed vectors $y_n \in \mathfrak{R}^{d_E}$ through applying an embedding coordinate map $\Phi: \aleph \to \mathfrak{R}^{d_E}$ (shown as a three-dimensional surface in Fig. 4.7c), the reconstructed state space usually demonstrates some unknowable distortion instead of the original one (Fig. 4.7c).

The Lorenz system of equations describing the thermal driving of convection in the lower atmosphere (Lorenz 1963; Schuster 1995) is used in this chapter as an example to illustrate the chaotic phenomenon of a dynamic system. It is defined by three differential equations as follows:

$$\dot{x} = a(y - x) \qquad\qquad (4.3)$$

$$\dot{y} = x(b - z) - y \qquad\qquad (4.4)$$

$$\dot{z} = xy - cz \qquad\qquad (4.5)$$

where the variables x, y, and z are proportional to the intensity of the convection rolls, the horizontal temperature variation, and the vertical temperature variation, respectively. The overdot refers to time differentiation, and the parameters a, b, and c are constants representing the properties of the system. The fourth-order Runge-Kutta integrator is used to solve the Lorenz Eqs. (4.3) to (4.5). The initial conditions used in this work for solution of the differential equations are $x = 0$, $y = 1$, and $z = 1$ and the parameter values used are $a = 16$, $b = 45.92$, and $c = 4.0$, following Abarbanel (1993). The tolerances for relative and absolute values of the three variables are 0.001 and 10^{-6}, respectively. There are 27,667 data points created for the time interval [0 300]. Figure 4.8 shows the time series plot of the first 5,000 x data. The Lorenz attractor shown in Fig. 4.1 is built on the obtained 27,667 sets of three-dimensional data (x, y, z).

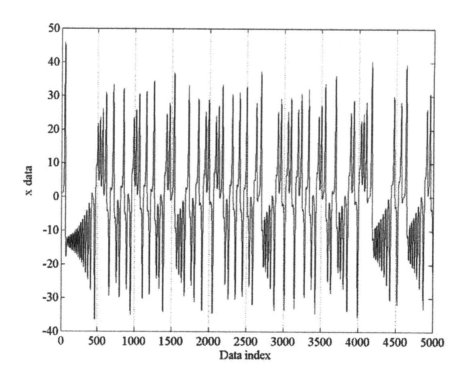

Figure 4.8 Time series plot of 5,000 x data from the Lorenz system with the initial conditions $x = 0$, $y = 1$, and $z = 1$ and the parameter values $a = 16$, $b = 45.92$, and $c = 4.0$

4.2   Delay Time

The delay time or lag time index, τ, is an important parameter that is required to be chosen carefully for reconstructing the state space. If τ is too small, the adjacent coordinates in the reconstructed state space, $y_{n+m\tau}$ and $y_{n+(m+1)\tau}$, will be too close to be distinguished from each other (m is a variable representing the embedding dimension, $m = 0, 1,..., d_E - 1$). If τ is too large, the adjacent coordinates become completely independent of each other statistically, and the projections of the evolution trajectories in the attractor move in two unrelated directions.

There are two widely used methods for selecting the magnitude of the time lag, namely, the linear autocorrelation function (ACF) and the average mutual information

(AMI) methods. The first method is based on the statistical concept, autocorrelation, which measures the degree of association between data in a time series separated by different time lags. The value of ACF for a time series $\{y_n\}$ with N data points at any lag time index τ is estimated as follows (Brockwell and Davis 2002):

$$\hat{\rho}(\tau) = \sum_{n=1}^{N-\tau} [y_{n+\tau} - \bar{y}][y_n - \bar{y}] \Big/ \sum_{n=1}^{N-\tau} [y_n - \bar{y}]^2 \tag{4.6}$$

where $\bar{y}$ is the average of the ordinates of time series. The ACF is evaluated for various values of the lag time, τ, and the results are plotted. Wherever the ACF curve intersects the lag time axis its value is zero, indicating that y_n and $y_{n+\tau}$ are linearly independent. The lag time corresponding to the first point of intersection is chosen as the *optimum* lag time. The ACF curve for the Lorenz system with $N = 27,667\ x$ data points is displayed in Fig. 4.9. This method yields a lag time of 17.

The second method is based on Shannon's idea of *mutual information* between two measurements (Gallager 1968), which is used to measure the generally nonlinear dependence of two variables. The mutual information for two measurements y_n and $y_{n+\tau}$ from the same time series, $\{y_n\}$, is expressed as

$$I_n(\tau) = \log_2 \left[\frac{P(y_n, y_{n+\tau})}{P(y_n)P(y_{n+\tau})} \right] \tag{4.7}$$

where the individual probability densities, $P(y_n)$ and $P(y_{n+\tau})$, are equal to the frequency with which the data points y_n and $y_{n+\tau}$ appear in the time series, respectively. The frequency can be obtained directly through tracing the data points in the entire time series. The joint probability density, $P(y_n, y_{n+\tau})$, is obtained by counting the number of times the values of the y_n and $y_{n+\tau}$ pair are observed in the series.

An average mutual information is computed for all data points in the following manner (Gallager 1968):

$$I(\tau) = \sum_{n=1}^{N-\tau} P(y_n, y_{n+\tau}) \log_2 \left[\frac{P(y_n, y_{n+\tau})}{P(y_n)P(y_{n+\tau})} \right] \tag{4.8}$$

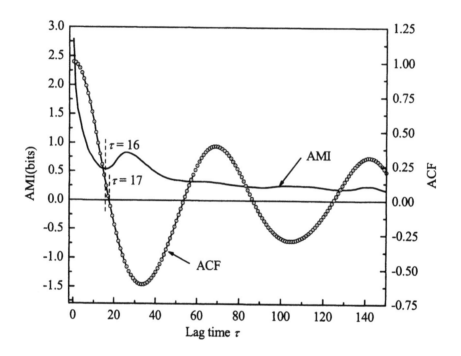

Figure 4.9 Lag time for the Lorenz system with 27,667 x data points using ACF and AMI methods (the right vertical axis represents ACF and the left vertical axis represents AMI)

When $P(y_n, y_{n+\tau}) = P(y_n) P(y_{n+\tau})$, and $I(\tau)$ approaches zero, the data points y_n and $y_{n+\tau}$ are completely nonlinearly independent. That means that the two data points are random with respect to each other. A zero value of $I(\tau)$ does not therefore provide any useful information for studying the given time series. A value of τ is desirable where two data points are somewhat independent, but not entirely nonlinearly independent. Fraser and Swinney (1986) suggest selecting the lag time corresponding to the first minimum of $I(\tau)$. The AMI curve for the Lorenz system with $N = 27,667$ x data points is displayed in Fig. 4.9. This method yields a lag time of 16. This value is close to the value of 17 obtained using the ACF method.

The AMI is a kind of nonlinear autocorrelation function in terms of τ and can measure the more general dependence of the two variables. Thus, in general, the AMI approach provides a more accurate criterion for the selection of lag time τ than the ACF method. However, the AMI approach may not provide any minima in some cases, for example, when the data are generated from the Hénon attractor (Abarbanel 1993). In such cases, the ACF approach can be used to find the optimum lag time.

4.3   Embedding Dimension

Over the past two decades, a number of approaches have been proposed for finding the optimal embedding dimension for the chaotic study of time series based on Takens' embedding theorem (Galka 2000). Among them are the fill-factor method (Buzug and Pfister 1992), the average integral local deformation (AILD) method (Buzug and Pfister 1992), and the false nearest neighbor (FNN) method (Kennel *et al.* 1992; Abarbanel 1993 & 1996). All three approaches are based on the fact that the geometrical invariants in an attractor usually remain unchanged when the estimated dimension is larger than the minimum value required for properly reconstructing the state space. All of them require a large number of data points for identifying the geometrical invariants in the chaotic time series. According to Theiler (1990), a data set with the minimum size of 10^{d_E} is required to estimate the minimum embedding dimension. These approaches are limited and can estimate the minimum embedding dimension accurately when 1) the physical system is relatively low dimensional (usually when $d_E \leq 5$) and 2) the dynamics of the system are known and represented by a set of differential equations explicitly. The three methods are described briefly and their drawbacks are demonstrated through studying the Lorenz system of differential equations.

4.3.1   Fill Factor Method

This approach is based on the assumption that the optimally reconstructed state space results in the maximum separation of the evolution trajectories of a physical system (Buzug and Pfister 1992). It attempts to simultaneously identify the optimum embedding dimension and lag time iteratively.

For any given embedding dimension, m, and the time lag, τ, a state space is reconstructed using Eq. (4.1). A sufficiently large number of data sets, N_r, are randomly selected from the reconstructed attractor as reference points, $\mathbf{y}_{r_i}$, $i = 0, 1, ..., m$. That is, each set of data consists of $m+1$ state space points, $\{\mathbf{y}_{r_0}, \mathbf{y}_{r_1}, \cdots, \mathbf{y}_{r_m}\}$. The index r_i indicates the reference points in the attractor. An $m \times 1$ displacement vector is defined for each set of data as follows:

$$\mathbf{d}_i(r_0, m, \tau) = [(y_{r_0} - y_{r_k})\ (y_{r_0+\tau} - y_{r_k+\tau})\ \cdots\ (y_{r_0+\tau(m-1)} - y_{r_k+\tau(m-1)})]', i{=}1,..., m \ (4.9)$$

Next, N_r parallelepipeds are constructed respectively from the N_r data sets. The edges of each parallelepiped are determined by the displacement vector, Eq. (4.9). The index r_i ($i = 0, 1, ..., m$) represents $m + 1$ corners of the parallelepiped in the m-dimension state space. The volume of the parallelepiped for the jth data set is equal to the absolute value of the determinant of its displacement matrix, $\mathbf{D}_j(r_0, m, \tau) = (\mathbf{d}_1\ \mathbf{d}_2\ \cdots\ \mathbf{d}_m)_j$, $j = 1, 2, ..., N_r$, expressed as

$$V_j(r_0, m, \tau) = |\det[\mathbf{D}_j(r_0, m, \tau)]|, \ j = 1, 2, ..., N_r \tag{4.10}$$

The volume represented by Eq. (4.10) is averaged over the N_r data set and normalized by the volume of the parallepiped formed with an edge length equal to the maximum difference between any two values of the time series:

$$\overline{V}(m, \tau) = \frac{\frac{1}{N_r}\sum_{j=1}^{N_r} V_j(r_0, m, \tau)}{(\max_i\{y_i\} - \min_i\{y_i\})^m}, \ i < N \tag{4.11}$$

where $\max_i y(t_i)$ and $\min_i y(t_i)$ are the maximum and minimum values of the time series, respectively. The so-called fill factor, $f(m, \tau)$, is equal to the logarithm of the normalized and averaged parallelepiped volume represented by Eq. (4.11) as follows:

$$f(m, \tau) = \log_{10} \overline{V}(m, \tau), \ i < N \tag{4.12}$$

Figure 4.10 shows an example two-dimensional state space with a two-dimensional attractor reconstructed from a given time series assumed to be in the form of a torus. The fill factor is computed for the parallelepiped created from the cross-hatched area defined by three reference points r_0, r_1, and r_2.

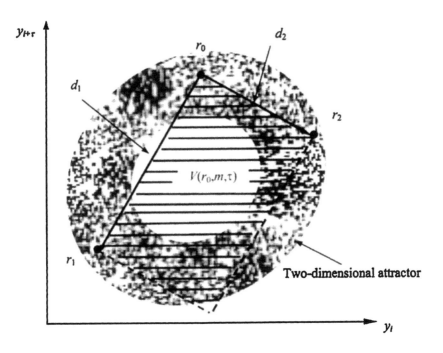

τ = Lag time; $d_j = j^{th}$ edge length of the parallelepiped (j = 1, ..., m); m = Embedding dimension (=2); $r_i = i^{th}$ reference point index (i = 0, 1,..., m); y_i = Data point (i = 1, 2, ..., N_a); N_a = Number of state space points

Figure 4.10 A two-dimensional reconstructed state space showing a two-dimensional attractor in the shape of a torus

For any given embedding dimension, m, the fill factor is computed for various values of the delay or lag time index, τ. The results are plotted as a series of curves. Figure 4.11 shows the fill factor curves versus the lag time for the Lorenz system reconstructed from N = 27,667 x data with m = 2 to 11. About two percent (2%) of the state space points (N_r = 550) are used as the reference data sets to calculate the fill factor. For low dimensions, the curves are generally smooth. As dimension, m, increases the curves display waves and kinks or *structure*.

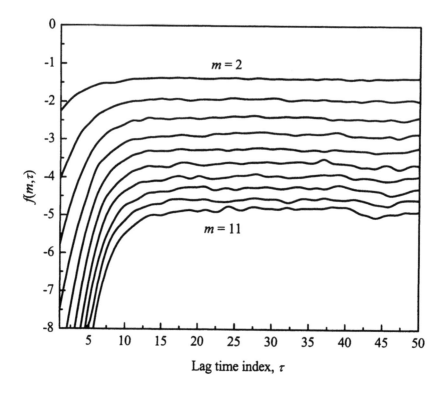

m = Embedding dimension τ = Lag time index $f(m,\tau)$ = Fill factor

Figure 4.11 Fill factor curves for estimating the optimum embedding dimension of the Lorenz system with $m = 2$ to 11 using x data

Buzug and Pfister (1992) suggest that the embedding dimension of the first curve showing visible structures is the optimum (minimum) embedding dimension, and the lag time corresponding to the first maximum of this curve is the optimum lag time for reconstructing the state space. Since differential equations are available for the Lorenz system its true attractor can be found. Its true attractor is three-dimensional because the Lorenz system has three variables. Additional numerical studies performed in this work indicate that increasing the size of the reference data sets does not help identify the optimum embedding dimension but increases the required computer processing time substantially. Consequently, the fill factor method is not a proper method for identifying the optimum embedding dimension and lag time for the solution of the Lorenz system of equations.

In conclusion, the fill factor approach used for finding the optimum embedding dimension in chaos analysis has several shortcomings: a) the arbitrary selection of the parameter N_r (number of reference data sets) results in various fill factor curves, b) it requires visual inspection and subjective judgment to identify the curve with multiple structures, c) it is computationally intensive, and d) in some cases the fill factor plots do not display any noticeable multiple structures at all.

4.3.2   Average Integral Local Deformation (AILD) Method

This approach is based on the assumption that the points on neighboring trajectories in the state space of a physical system remain neighboring points for small evolution times. The adjacent trajectories should stay close in a well-reconstructed state space without intersecting (Buzug and Pfister 1992; Galka 2000). The reconstructed state space with a sufficiently high dimension guarantees non-intersecting trajectories. In this approach, the evolution of the distance between a reference point in the reference trajectory and the center of mass of the spherical *neighboring cloud* (defined by a given radius) is used to evaluate the reconstructed state space (Fig. 4.12). A minimum deviation between the two evolutions of the adjacent trajectories indicates an optimal reconstruction.

For any given embedding dimension, m, and the time lag, τ, a state space is reconstructed using Eq. (4.1). A time step and N_r reference points at this time step, $\mathbf{y}_r$ (r = 1, 2, ..., N_r), are selected randomly. The value of N_r should be less than $N - \tau(m-1) - s$, where s is an input parameter representing the number of evolution time steps. For each reference point, a spherical neighboring cloud with N_n neighboring points, $\mathbf{y}_p$ (p = 1, 2, ..., N_n), is identified using a given radius, R_0, based on $\|\mathbf{y}_p - \mathbf{y}_r\| \le R_0$ (p = 1, 2, ..., N_n). The center of mass of the neighboring cloud for the reference point, $\mathbf{c}_r$, is obtained by averaging over all the points in the cloud (a circle in the two-dimensional case and a sphere in the three-dimensional case):

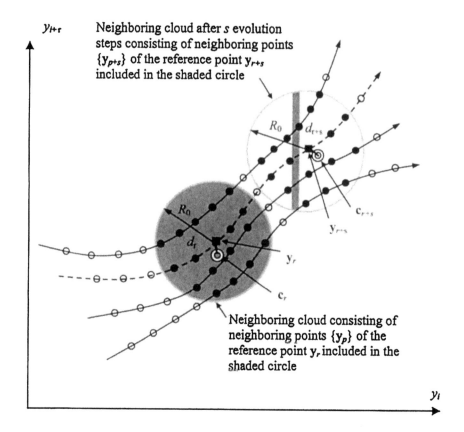

O State space point   ■ Center of neighboring cloud (a reference point)

● Points in a neighboring cloud   ◎ Center of mass of the neighboring cloud

⎯⎯▶ Evolution trajectory   ----▶ Reference trajectory

m = Embedding dimension (=2); τ = Lag time index; y_r, y_{r+s} = r^{th} reference point and its value after s evolution steps; c_r, c_{r+s} = Center of mass of the neighboring cloud and its value after s evolution steps; d_r, d_{r+s} = Distance between the reference point and the center of mass of the neighboring cloud; R_0 = Neighboring cloud radius

Figure 4.12 Two neighboring clouds in the two-dimensional reconstructed state space of the AILD method

$$\mathbf{c}_r = [c_r, c_{r+\tau}, c_{r+2\tau}, \ldots, c_{r+(m-1)\tau}]' = \frac{1}{N_n} \sum_{p=1}^{N_n} \mathbf{y}_p \qquad (4.13)$$

After s evolution steps of the reference point, the absolute growth of the distance between the reference point (y_r) and the center of mass of the neighboring cloud (c_r) is obtained by

$$\delta_{r+s}(m,\tau) = d_{r+s} - d_r = \left[\sum_{k=0}^{m-1}[c_{r+s+\tau k} - y_{r+s+\tau k}]^2\right]^{1/2} - \left[\sum_{k=0}^{m-1}[c_{r+\tau k} - y_{r+\tau k}]^2\right]^{1/2} \quad (4.14)$$

where d is the distance between the reference point and the center of mass of the neighboring cloud and $c_{r+s+\tau k}$ represents the coordinates of the center of mass of the neighboring cloud after s evolution steps computed by Eq. (4.13).

The integral local deformation, $\Delta_r(m,\tau)$, is obtained by the discrete integral of Eq. (4.14) over N_s evolution steps as follows ($N_s = 4$ is used in this chapter):

$$\Delta_r(m,\tau) = \sum_{s=1}^{N_s} \frac{\delta_{r+s-1}(m,\tau) + \delta_{r+s}(m,\tau)}{2}, \quad r = 1, 2, ..., N_r \quad (4.15)$$

The average integral local deformation, $\Delta(m,\tau)$, is computed through averaging the values obtained from Eq. (4.15) over the N_r reference points and then normalizing the result with the spatial extension of the attractor, $\max_i\{y_i\} - \min_i\{y_i\}$, expressed as

$$\Delta(m,\tau) = \frac{\sum_{r=1}^{N_r}\sum_{s=1}^{N_s}[\delta_{r+s-1}(m,\tau) + \delta_{r+s}(m,\tau)]}{2N_r(\max_i\{y_i\} - \min_i\{y_i\})} \quad (4.16)$$

For any given embedding dimension, m, the average integral local deformation is computed for various values of the delay or lag time index, τ. The results are plotted as a series of curves. Figure 4.13 shows the AILD curves versus the lag time for the Lorenz attractor reconstructed with $N = 27,667$ x data with $m = 2$ to 11. About two percent (2%) of state space points ($N_r = 550$) are used as the reference data sets to calculate the AILD curves. About four percent (4%) of the spatial extension of the attractor is used as the radius R_0 for creating neighboring clouds. Among the AILD curves with a local minimum, the one corresponding to the smallest embedding dimension yields the optimum embedding dimension and the lag time. In Fig. 4.13, the optimum embedding dimension is $d_E = 2$ and the corresponding delay time is $\tau = 4$. Consequently, it may be concluded that the AILD approach is more effective than the fill factor approach in identifying the minimum embedding dimension of the Lorenz system.

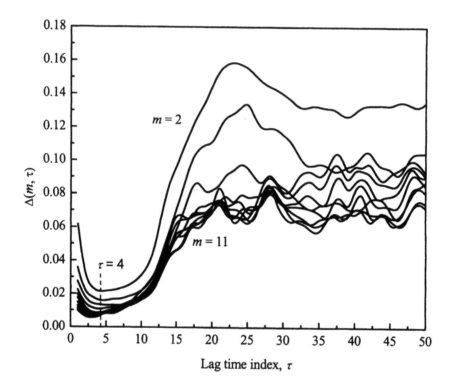

m = Embedding dimension τ = Lag time index
$\Delta(m,\tau)$ = average integral local deformation

Figure 4.13 AILD curves for estimating the optimum embedding dimension of the Lorenz system with m = 2 to 11 using x data

However, numerical studies performed in this work indicate that the AILD approach also suffers from two major shortcomings. First, the approach is computationally intensive. Second, two parameters, the number of reference points, N_r, and the radius of the neighboring cloud, R_0, are chosen arbitrarily without any theoretical basis.

4.3.3   False Nearest Neighbors (FNN) Method

This approach is based on the assumption that a small embedding dimension results in state space points that are far apart in the original state space to be considered neighboring points in the reconstructed state space (Kennel *et al.* 1992; Abarbanel 1993 & 1996). The previous two methods yield a time lag corresponding to the optimum embedding dimension. The false nearest neighbors method, however, yields the optimum embedding dimension only. The lag time index, τ, is computed using an existing method such as the ACF or AMI method. For any given embedding dimension variable, m, any reconstructed state space point, y_n, has $N_a - 1$ neighbors (N_a is the maximum number of state space points). The square of the Euclidean distance between this point and its rth neighboring point, y_n^r, is computed as

$$D_n^2(m,\tau) = \sum_{k=0}^{m-1}[y_{n+\tau k} - y_{n+\tau k}^r]^2 , \; n = 1, 2, ..., N_a \tag{4.17}$$

The dimension of the reconstructed state space is increased from m to $m+1$, the $(m+1)^{st}$ coordinate, $y_{n+\tau m}$, is added to the previous state space points based on Eq. (4.1), and the relative increase of the Euclidean distance for each point, δ_n, is computed as

$$\delta_n = \left[\frac{D_n^2(m+1,\tau) - D_n^2(m,\tau)}{D_n^2(m,\tau)}\right]^{1/2} = \frac{|y_{n+\tau m} - y_{n+\tau m}^r|}{D_n(m,\tau)}, n = 1, 2, ..., N_a \tag{4.18}$$

A false neighbor is identified when either of the following two conditions is satisfied:

$$\delta_n > \varepsilon \tag{4.19}$$

$$D_n(m+1) > \frac{\sigma}{\theta} \tag{4.20}$$

where $D_n(m+1)$ is obtained from Eq. (4.17) assuming τ is equal to one, σ is the standard deviation of the given time series, and the parameters, ε and θ, are some constant thresholds. Kennel *et al.* (1992) suggest a value of larger than 10 for ε but provide no suggestion for θ.

For any given embedding dimension, m, the proportion of the identified false nearest neighbors to all the neighbors is computed for the given time delay or lag time index, τ. The percentages of the false nearest neighbors are plotted as a function of the embedding dimension. A zero FNN percentage indicates the optimum (minimum) embedding dimension.

Figure 4.14 and Table 4.1 show the results of the FNN approach for determining the optimum embedding dimension of $N = 27,667$ x data of the Lorenz system using various values for the threshold parameters. The relation $\theta = 0.2\varepsilon$ is used in this chapter and the value of the parameter ε is varied from 30 to 700. Only those points with the percentage of FNN less than 3 are displayed in the plots of Fig. 4.14. The results summarized in Fig. 4.14 show that the optimum dimension obtained by the FNN approach varies from 2 to 8. The FNN approach is a simple approach for estimating the embedding dimension, but it has a major shortcoming: different values of the threshold parameters ε and θ result in completely different values for the optimum embedding dimension.

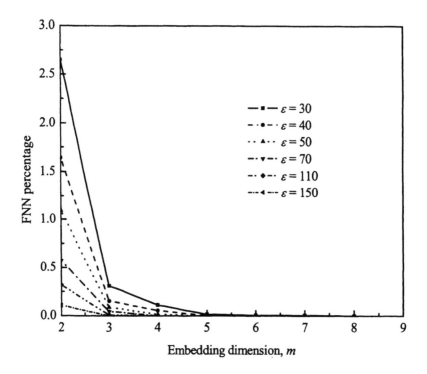

Figure 4.14 False nearest neighbor curves with various thresholds for determining the embedding dimension of the Lorenz system using x data

Table 4.1 Percentages of the false nearest neighbors in the FNN approach for the Lorenz system using various thresholds

Embedding dimension, m	$\theta=0.2\varepsilon$						
	$\varepsilon=30$	$\varepsilon=40$	$\varepsilon=50$	$\varepsilon=70$	$\varepsilon=110$	$\varepsilon=150$	$\varepsilon=700$
1	97.00	96.20	95.40	94.00	91.90	88.80	75.10
2	2.65	1.64	1.09	0.57	0.32	0.11	0
3	0.31	0.15	0.09	0.04	0.008	0	0
4	0.11	0.05	0.01	0	0	0	0
5	0.01	0	0	0	0	0	0
6	0.0078	0	0	0	0	0	0
7	0.0073	0	0	0	0	0	0
8	0	0	0	0	0	0	0

4.4   Illustrative Examples

The methods described in this chapter are further illustrated with three examples where the physical or chaos phenomenon is described by nonlinear equations and the exact embedding dimension is therefore known.

4.4.1   Example 1: Lorenz System

Figure 4.15 shows the Lorenz attractor reconstructed from three-dimensional state space $(x_n, x_{n+\tau}, x_{n+2\tau})$. These state space data are formed using Eq. (4.1) and the 27,667 x data created by the differential Eqs. (4.3) – (4.5). A lag time index of 17 is used to construct the three-dimensional state space vector based on Eq. (4.1). As described previously, the three methods, i.e., fill factor, average integral local deformation, and false nearest neighbors methods, are used to find the optimum embedding dimension ($d_E = 3$). Compared with the true attractor shown in Fig. 4.1, the reconstruction in Fig. 4.15 results in some distortion in the attractor. However, the reconstructed attractor provides sufficient information for characterizing the original system.

4.4.2   Example 2: Rössler System

Rössler's system is well known as a chaotic dynamic system defined by the following three differential equations (Rössler 1976):

$$\dot{x} = -(y+z) \tag{4.21}$$
$$\dot{y} = x + ay \tag{4.22}$$
$$\dot{z} = b + z(x-c) \tag{4.23}$$

It exhibits deterministic chaos for certain values of the parameters. The fourth-order Runge-Kutta integrator is used to solve the Rössler equations. The initial conditions for solution of the differential equations are $x = 0$, $y = 0$, and $z = 0$ and the parameter values used are $a = 0.38$, $b = 0.3$, and $c = 4.5$, following Tsonis (1992). The tolerances for relative and absolute values of the three variables are 0.001 and 10^{-6}, respectively. There are 33,933 data points created for the time interval [0 2,000].

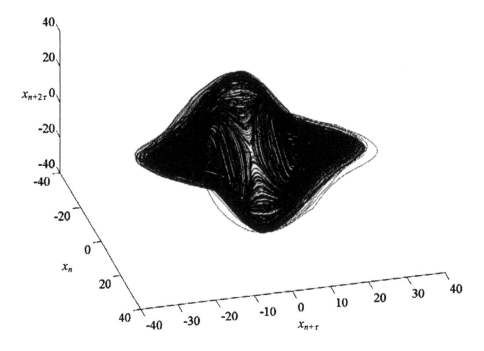

Figure 4.15 Reconstructed three-dimensional attractor of the Lorenz system with $a = 16$, $b = 45.92$, and $c = 4.0$: using the 27,667 x data in the state space $(x_n, x_{n+\tau}, x_{n+2\tau})$

The three methods described in Section 4.3 are used to determine the minimum embedding dimension of the Rössler's system using 33,933 x data. The fill factor and AILD methods can hardly identify the optimum embedding dimension. The results summarized in Table 4.2 show the optimum dimension obtained by the FNN approach, which varies from 2 to 6. Clearly, the optimum embedding dimension from the FNN approach depends on the selction of the threshold ε in Eq. (4.19).

Table 4.2 False nearest neighbor approach with various thresholds for determining the embedding dimension of the Rössler system using 33,933 x data points

m	$\theta = 0.2\varepsilon$				
	$\varepsilon = 40$	$\varepsilon = 60$	$\varepsilon = 100$	$\varepsilon = 150$	$\varepsilon = 500$
1	95.93	94.40	91.50	88.40	74.30
2	1.03	0.46	0.18	0.08	0
3	0.18	0.06	0.02	0	0
4	0.08	0.02	0	0	0
5	0.006	0.006	0	0	0
6	0	0	0	0	0

Figure 4.16 shows the Rössler attractor reconstructed from the three-dimensional state space $(x_n, x_{n+\tau}, x_{n+2\tau})$ formed by x data. A lag time index of $\tau = 20$ obtained from the ACF method is used to construct the three-dimensional state space vector. Its true attractor using the solutions of these differential equations is shown in Fig. 4.2. Again, compared with the true attractor shown in Fig. 4.2, the reconstruction in Fig. 4.16 results in some distortion in the attractor, but provides sufficient information for characterizing the original system.

4.4.3   Example 3: Hénon Map

The Hénon map is commonly used as an illustrative example in the literature on nonlinear chaotic time series analysis (Hénon 1976). The two-dimensional map is expressed as

$$x(n+1) = 1 + y(n) - 1.4x^2(n)$$
$$y(n+1) = 0.3x(n)$$

$$n = 1, 2, \ldots, 30{,}000 \qquad (4.24)$$

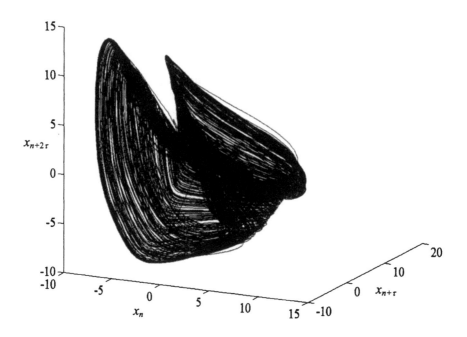

Figure 4.16 Reconstructed three-dimension attractor of the Rössler system with $a = 0.38$, $b = 0.3$, and $c = 4.5$ using 33,933 x data points in the state space $(x_n, x_{n+\tau}, x_{n+2\tau})$

Equation (4.24) is used to create 30,000 sets of data points (x, y). The same optimum embedding dimension of the Hénon map $(d_E = 2)$ is obtained from three different methods described in Section 4.3 using the 30,000 x data. Four percent of the spatial extension of the attractor is adopted as the radius for searching the neighboring points in the fill factor method, while the number of reference points is chosen to be $N_r = 600$ (2% of the number of data points) in both the fill factor and AILD methods. In the FNN method, a value of 10 is used as the threshold ε to identify the false nearest neighbors.

The 30,000 sets of x data points are used to plot the attractor of the Hénon map. The chaotic attractor reconstructed from the two-dimensional state space formed by x

data $(x_n, x_{n+\tau})$ is shown in Fig. 4.17. The AMI approach cannot provide any minimum in this example. The lag time of one $(\tau = 1)$ obtained from the ACF method is used to reconstruct the attractor shown in Fig. 4.17. Once again, compared with the true attractor shown in Fig. 4.3, the reconstruction in Fig. 4.17 results in slight distortion (rotation) in the attractor. However, the reconstructed attractor provides sufficient information for characterizing the original system.

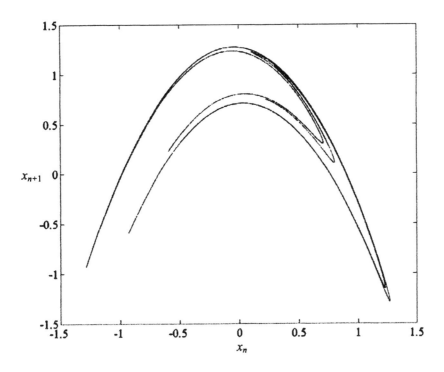

Figure 4.17 Reconstructed two-dimension attractor of Hénon's map using
30,000 x data

4.5   Final Comments

Selection of the optimum embedding dimension is the first step for chaos analysis of any physical phenomenon or system. In this chapter, three methods for determining the optimum embedding dimension were presented and studied using three different examples with available analytical equations where the exact value of the optimum embedding dimension is known. They are the fill-factor method, the average integral local deformation method, and the false nearest neighbors method. It is concluded that the FNN method is more accurate than the other two methods. The accuracy of the FNN method, however, depends greatly on the values selected for threshold parameters.

Chapter 5

FREEWAY INCIDENT DETECTION

5.1 Introduction

Freeway congestion can be divided into recurrent and non-recurrent congestion. Recurrent congestion is caused by traffic bottlenecks formed when the traffic volume on the freeway exceeds the freeway capacity. On the other hand, non-recurrent congestion may be caused by: (a) freeway incidents such as accidents, disabled vehicles, and spilled loads, (b) planned special events such as temporary maintenance and construction activities, and (c) unplanned events such as signal and detector malfunctions. Studies have shown that freeway incidents cause approximately 60% of all urban freeway delays in the United States (FHWA 1991).

The process of automated incident detection consists of data collection and classification. Data can be collected through a variety of methods such as satellite imaging, real-time cameras, highway patrol, and sensors such as point sensors, area sensors, and point-to-point sensors embedded in the freeway. The most commonly used data collection scheme is based on loop detectors which are a type of point sensor. They are used to collect traffic data averaged over 20- to 60-second intervals usually across all the lanes in one direction. Normally, two detectors are placed per mile of freeway. The compiled data are relayed to a central monitoring station where the traffic situation in an entire freeway system can be monitored. An incident detection algorithm can be used to detect an incident anywhere in the freeway system. Existing incident detection algorithms use different combinations of the three traffic parameters: traffic speed measured in miles per hour (mph) or kilometers per hour (kph), volume measured in number of vehicles per hour per lane (vphpl), and occupancy measured in number of vehicles per mile per lane (vpmpl) or per kilometer per lane (vpkpl) (usually as a percentage of the design

This chapter is based on the article: Ghosh-Dastidar, S. and Adeli, H., "Wavelet-Clustering-Neural Network Model for Freeway Incident Detection," *Computer-Aided Civil and Infrastructure Engineering* 2003 18(5), 325-338, and is reproduced by the permission of the publisher, Blackwell Publishing Company.

maximum value).

During the past three decades, a number of freeway traffic management systems have been developed to minimize the disruptive influence of incidents on normal traffic volume. The most widely used algorithms till the 80's were the California-type algorithms. These are variations of the basic California algorithm (Schaefer 1969) which is based on the logical assumption that traffic occupancy increases upstream and decreases downstream of a freeway incident significantly. Incident detection is based on three sequential tests on comparative occupancy changes using fixed predetermined thresholds at upstream and downstream detector stations.

A major shortcoming of the early California algorithms is their unreliability in differentiating between recurrent and non-recurrent congestion on a freeway, resulting in a high false alarm rate (false detection of incidents). Cook and Cleveland (1974) combine the energy distribution model of Courage and Levin (1968) and a California-type model into a composite algorithm with the goal of modeling the traffic dynamics more accurately. Payne and Tignor (1978) present another modification to the California algorithm by using a decision tree in an attempt to reduce the number of false alarms.

Another shortcoming of the California-type algorithms is missed or delayed detection. This occurs when the occurrence of an incident does not cause congestion in the freeway immediately because the freeway capacity of the remaining lanes is greater than the traffic demand. The incident is detected much later when the traffic demand exceeds the reduced freeway capacity. When the threshold is decreased to compensate for this, the number of false alarms increases. In order to solve this problem, the McMaster algorithm (Persaud et al. 1990) detects incidents based on the relationships between the three traffic parameters, speed, volume, and occupancy. This algorithm uses data from a single detector station instead of two based on the argument that volume changes due to an incident may not be evident at the upstream and downstream stations in the specified time window. Stephanedes et al. (1992) present a moving average method to compute the threshold instead of a single static threshold used in the California-type algorithm.

Recent computational paradigms such as fuzzy logic (Zadeh 1978) and artificial neural networks (Hua and Faghri 1994; Adeli and Hung 1995; Adeli and Park 1998) have also been used to recognize traffic incident patterns from incident-free patterns (Hsiao et al. 1994; Stephanedes and Liu 1995). In spite of improvements, these incident detection algorithms yield a large number of false alarms and missed detections. This is mainly due to random temporal fluctuations or traffic noise which make it difficult to differentiate between recurrent and non-recurrent congestion accurately. Similar to recurrent congestion, freeway incidents cause a reduction in average traffic speed and an increase in traffic occupancy. However, traffic volume patterns and the duration of incident-related effects are different. A more reliable detection algorithm can be created by more effective elimination of traffic noise and retention of relevant features signifying freeway incidents. The Minnesota Algorithm (Chassiakos and Stephanedes 1993) attempts to minimize false alarms and missed incidents by filtering out noise in traffic volume through averaging occupancy measurements over contiguous short-term intervals.

The computational complexity of neural network algorithms increases exponentially with an increase in the size of the network. Furthermore, with an increase in the size of the network, the size of the training set has to be increased exponentially in order to achieve the same level of accuracy (Wu and Adeli 2001). This *double exponential complexity* and the resulting *dimensionality curse* has been an impediment for practical application of neural networks to complicated real-life pattern recognition problems.

Adeli and coworkers have been working on creation of novel algorithms and computational models for reliable and efficient detection of incidents during the past few years. They introduced the concept of wavelets into the field of transportation engineering for the first time. Samant and Adeli (2000) present an effective traffic feature extraction model using discrete wavelet transform (DWT) and linear discriminant analysis (LDA). Adeli and Samant (2000) present a computational model for automatic traffic incident detection using discrete wavelet transform, linear discriminant analysis, and an adaptive conjugate gradient neural network model (Adeli and Hung 1994). Adeli and Karim (2000) present a multi-paradigm intelligent system approach to the solution of the problem employing advanced signal processing, pattern recognition, and classification techniques. The methodology consists of five steps: data preprocessing, data de-noising, clustering, classification, and decision making. It effectively integrates fuzzy, wavelet, and neural computing techniques to improve reliability and robustness. A wavelet-based de-noising technique is employed to eliminate undesirable fluctuations in observed data from traffic sensors. Fuzzy c-means clustering is used to extract significant information from the observed data and to reduce its dimensionality. They develop a radial basis function neural network (RBFNN) to classify the denoised and clustered observed data. Karim and Adeli (2002a) evaluate the performance of the fuzzy-wavelet RBFNN freeway incident detection model and compare it with the benchmark California algorithm #8 using both real and simulated data. The evaluation is based on three quantitative measures of detection rate, false alarm rate, and detection time, and the qualitative measure of algorithm portability. The proposed algorithm outperformed the California algorithm consistently under various scenarios.

In the previous work of Adeli and his co-workers, Daubechies wavelet was used for de-noising and feature extraction. In this chapter, an improved feature extraction model is presented using fourth order Coifman wavelets to reduce the effects of noise from various sources including erroneous sensor data. Statistical analysis based on the Mahalanobis distance is employed to perform data clustering and parameter reduction in order to reduce the size of the input space for the subsequent step of classification by the Levenberg-Marquardt backpropagation (BP) neural network. Figure 5.1 shows the outline of the new single-station incident detection algorithm schematically.

5.2   Feature Extraction using Coifman Wavelet

Wavelet transform described in Chapter 3 is applied individually to the three traffic data signals representing traffic speed, volume, and occupancy over the selected time duration. Since fast incident detection is essential, discrete wavelet transform is chosen over continuous wavelet transform to reduce the computational burden. The Coifman wavelet system is used because it provides a good approximation for high-resolution scaling coefficients by requiring that both the wavelet and the scaling function moments be zero (Daubechies 1992; Burrus et $al.$ 1998; Mathworks 2000a). The requirements for an Lth order Coifman wavelet are mathematically expressed as (Daubechies 1992)

$$m_0(0) = 1 \tag{5.1}$$

$$m_0(l) = 0 \qquad \text{for } l = 1, 2, ..., L\text{-}1 \tag{5.2}$$

$$m_1(l) = 0 \qquad \text{for } l = 0, 1, ..., L\text{-}1 \tag{5.3}$$

where $m_0(l)$ is the lth moment of the scaling function $\varphi(t)$, $m_1(l)$ is the lth moment of

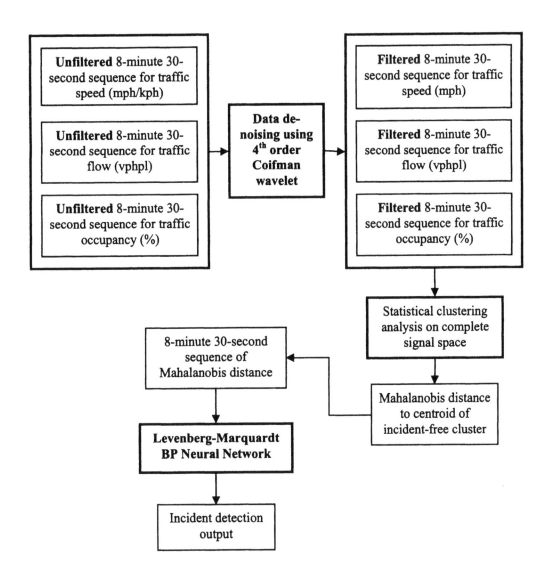

Figure 5.1   Overview of the incident detection model

the wavelet function $\psi(k)$, and L-1 is the minimum number of zero scaling and wavelet function moments in addition to the zero wavelet moment $m_1(0)$ needed for the orthogonality condition. These conditions are used to compute the coefficients for the filters $h_0(n)$ in Eq. (3.13) and $h_1(n)$ in Eq. (3.14). The filter coefficients for these wavelets are calculated using the quadrature mirror filter approach (Wickerhauser 1994), where the coefficients of one filter are the mirror image of the other as shown in Table 5.1.

Table 5.1 Scaling and wavelet filter coefficients for the fourth-order Coifman wavelet

n	$h_0(n)$	$h_1(n)$
-4	0.016387	0.000721
-3	-0.041465	0.001823
-2	-0.067373	-0.005611
-1	0.386110	-0.023680
0	0.812724	0.059434
1	0.417005	0.076489
2	-0.076489	-0.417005
3	-0.059434	-0.812724
4	0.023680	-0.386110
5	0.005611	0.067373
6	-0.001823	0.041465
7	-0.000721	-0.016387

Assuming the original traffic speed, volume, and occupancy sequences each contain N data points, the first-level wavelet transform after downsampling results in $N/2$ high resolution components and $N/2$ low resolution components. The $N/2$ high resolution components correspond to traffic fluctuations or traffic noise and are discarded. Single level decomposition may not always separate out all the desirable features. Therefore, the decomposition process is repeated, with successive approximations being decomposed in turn, so that a single original signal is decomposed into several lower resolution components. The $N/2$ low resolution components are subjected to second level wavelet transform to yield $N/4$ medium resolution components and $N/4$ lowest resolution components. Out of the $N/4$ medium resolution components, some need to be retained while others corresponding to traffic noise are discarded. Similarly, out of the $N/4$ low resolution components, some can be discarded as they may correspond to fluctuation due to the freeway geometry. Which components to be discarded is an algorithmic issue. In this chapter, it is decided based on two criteria: closeness of the regenerated signal to the original signal based on the mean squared error (MSE) value and the accuracy of the algorithm in terms of both detecting the incident and avoiding false alarms.

5.3   Parameter Reduction using Clustering Analysis

Although the wavelet transform denoises the signal by filtering out the effect of outliers and random temporal fluctuations, the size of the input space for the regenerated signals is the same as that of the original data signals for speed, volume, and occupancy. In order to reduce the size of the input space and overcome the *dimensionality curse* mentioned in Section 5.1, a statistical multivariate technique known as clustering analysis is employed. Clustering analysis is defined as classifying objects into groups based on inter-object similarities (Dillon and Goldstein 1984; Kachigan 1986; Fukunaga 1990). In contrast to the linear discriminant analysis used by Samant and Adeli (2000), it does not assume *a priori* well-defined groups to identify the difference-causing variables. The aim is to divide the given objects into predominantly two clusters representing incident-related and incident-free objects such that the variation within the cluster is small but the variation between the clusters is large.

The three traffic parameters for any given 30-second interval are represented as the x, y and z coordinates in a three-dimensional space with speed, volume, and occupancy as the X, Y and Z axes, respectively. Each data point in this space is considered an object for clustering analysis. All three traffic parameters are used in order to achieve an accurate representation of the input space. The variations of all three parameters approach a maximum simultaneously when an incident occurs. This characteristic is employed in the clustering algorithm in order to distinguish the incident-free group from the incident group effectively. The most essential step in clustering analysis is to define the similarity or proximity between each pair of objects, usually represented by a measure of distance, such as the Euclidean distance, the city block metric, or the Mahalanobis distance (Dillon and Goldstein 1984; Kachigan 1986; McLachlan 1992; Mathworks 2000b). The Mahalanobis distance is selected in the proposed incident detection model because it can represent the aforementioned correlation among the three traffic parameters accurately.

The position of the ith object in the three-dimensional traffic parameters space is defined by the ith row of the $K \times 3$ matrix of object coordinates, $\mathbf{X}_i$:

$$\mathbf{X}_i = [x_i \quad y_i \quad z_i], \; i = 1, 2, ..., K \qquad (5.4)$$

where K is the number of traffic parameter data points. The mean-corrected object coordinates or traffic parameter data matrix is computed as

$$\mathbf{X}_d = \mathbf{X} - \mathbf{I}\overline{\mathbf{X}} \tag{5.5}$$

where $\mathbf{I}$ represents a $K \times 1$ unit column vector and $\overline{\mathbf{X}}$ is the 1×3 row vector of the column means for K objects representing the centroid of the traffic parameters space. The sample 3×3 covariance matrix $\mathbf{C}$ is defined as

$$\mathbf{C} = \mathbf{X}_d'\mathbf{X}_d / (K - 1) \tag{5.6}$$

where the prime denotes the transpose of a matrix. The Mahalanobis distance d_{ij} between any two objects i and j is defined as (Dillon and Goldstein 1984)

$$d_{ij} = [\mathbf{X}_i - \mathbf{X}_j]\mathbf{C}^{-1}[\mathbf{X}_i - \mathbf{X}_j]' \tag{5.7}$$

Clustering the objects is based on the nearest-neighbor concept. The two objects closest to each other in terms of the Mahalanobis distance are paired together in one binary cluster. If any one of the remaining unclustered objects has a shorter distance to this cluster (defined by the distance to the nearest object in the cluster) than to the other unclustered objects, it forms a binary cluster with the same cluster. Otherwise, it is paired with the closest unclustered object into a separate binary cluster. A cluster can contain other clusters. If two clusters are closer to each other than to any other unclustered object, then they are grouped together in a bigger cluster. This process continues till all the objects are clustered in one big cluster that contains all the smaller clusters. Figure 5.2 shows the clustering method. The binary clustering based on the nearest neighbor concept is described in Fig. 5.2a. There are 10 original objects in the figure shown by black circles numbered 1 to 10. The closest objects are objects 9 and 10 which form the first cluster denoted as object 11 in Fig. 5.2a. In general, the binary clusters formed at each stage are considered to be objects for the next clustering stage and are numbered consecutively as $K + 1$, $K + 2$, ..., $2K - 1$ (K is the number of original objects). Then, objects 1 and 2 in Fig. 5.2a are clustered as object 12. Next, objects 6 and 11 are clustered as object 13, and so on. The total number of binary clusters created is $K - 1$. This binary clustering process can be represented in a tree hierarchy, dubbed a dendrogram

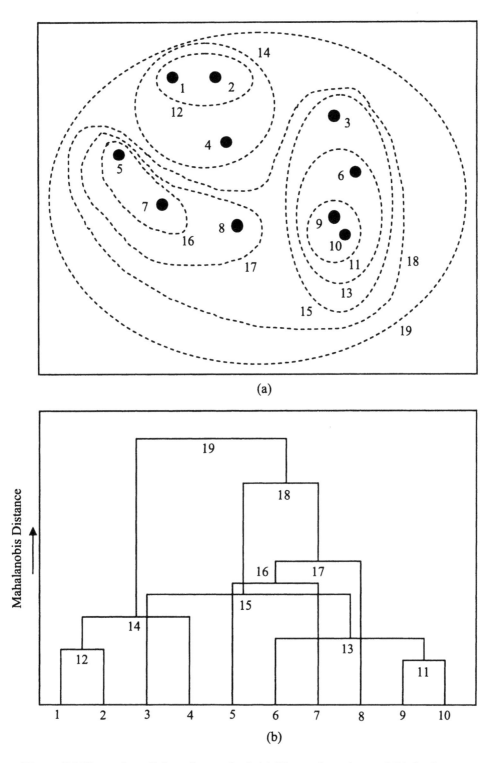

Figure 5.2 Illustration of clustering method: (a) Binary clustering, and (b) dendrogram for the binary clustering shown in (a)

(Fig. 5.2b) which summarizes the clustering stages. The numbers on the horizontal axis of the dendrogram refer to the original objects. The vertical axis represents the Mahalanobis distance. The numbers inside the dendrogram refer to the objects created through the binary clustering.

The results of the binary clustering process using the Mahalanobis distance are summarized in a $(K-1) \times 3$ linkage matrix $\mathbf{P}$ where each row represents a cluster. The first two elements in each row represent the two object numbers in the cluster and the third element represents the Mahalanobis distance between the two objects. A cluster is said to be inconsistent if the Mahalanobis distance between the two objects in the cluster differs significantly from the average Mahalanobis distance of the objects in clusters lower in the hierarchy (Fig. 5.2b). The number of clustering stages, identified by levels in the dendrogram (Fig. 5.2b), used to compute the average Mahalanobis distance is called the depth of the comparison and is chosen experimentally depending on the number of desired cluster types. Since in this chapter the focus is on two clusters – incident-free and incident-related, a two-level comparison is sufficient.

Mathematically, the dissimilarity among clusters is represented by an inconsistency coefficient Y_p for each cluster, p, defined in the following form:

$$Y_p = (P_p - S_p)/\sigma_p \tag{5.8}$$

where, for a two-level comparison, S_p is the average Mahalanobis distance of objects in cluster p and in the clusters immediately below cluster p in the hierarchy and is defined as

$$S_p = (P_{p,1} + P_{p,2} + P_{p,3})/3 \tag{5.9}$$

and σ_p is the standard deviation of the Mahalanobis distance of the objects in the clusters included in the computation and is defined as

$$\sigma_p = [(P_{p,1} - S_p) + (P_{p,2} - S_p) + (P_{p,3} - S_p)]^{1/2}/\sqrt{2} \tag{5.10}$$

A high inconsistency coefficient for any cluster implies that the distance between the objects contained in it is significantly large. Using an experimentally selected threshold level for the inconsistency coefficient, the data set is divided predominantly into two clusters. In addition to the two major clusters of interest there exist groups of outliers which are discarded. Since most of the traffic data are incident-free the Mahalanobis distance of each object from the centroid of the incident-free cluster is input to the Levenberg-Marquardt BP neural network for incident detection.

5.4   Incident Detection using Levenberg-Marquardt BP Neural Network

Features that show significant differences from the incident-free and incident-related situations are discarded from the original data using fourth-order Coifman wavelets to reduce the effect of outliers and map the original attributes into more effective features. To improve the performance and accuracy of the neural network classifier the dimension of the input space is reduced by a factor of three using clustering analysis described in the previous section (the Mahalanobis distance is used in place of three different traffic parameters). The resulting pattern recognition problem is solved using a multi-layer feedforward network trained by the Levenberg-Marquardt BP algorithm (Hagan *et al.* 1996).

The architecture of the neural network for incident detection is shown in Fig. 5.3. It consists of an input layer with one node representing the Mahalanobis distance of the data object from the centroid of the incident-free cluster. A network with a single hidden layer failed to produce satisfactory results because of the complexity of the pattern recognition problem. Consequently, the number of hidden layers was increased to two. The number of nodes in each hidden layer is chosen to be four after concluding that fewer nodes would not yield satisfactory convergence based on numerical experimentation. The output layer has a single node which returns a value of 1 if an incident is detected and 0 otherwise.

The input to the neural network is the Mahalanobis distance $d(n)$ and the output is $o(n)$ for the nth training instance. The layers are numbered from right to left starting with $l = 0$ for the output layer and ending with $l = 3$ for the input layer. The total input to the jth node in a hidden or output layer l is mathematically expressed as (Bose and Liang 1996)

$$i_j^l = \sum_i y_i^{l+1} w_{ij}^l , l = 0, 1, \text{ or } 2 \tag{5.11}$$

where y_i^{l+1} is the output of the ith node in layer $l + 1$ and w_{ij}^l is the weight of the link connecting the ith node in layer $l+1$ to the jth node in layer l. The output of the jth node in layer l is computed using the sigmoid function as follows:

$$y_j^l = 1/(1 + e^{-S})$$ (5.12)

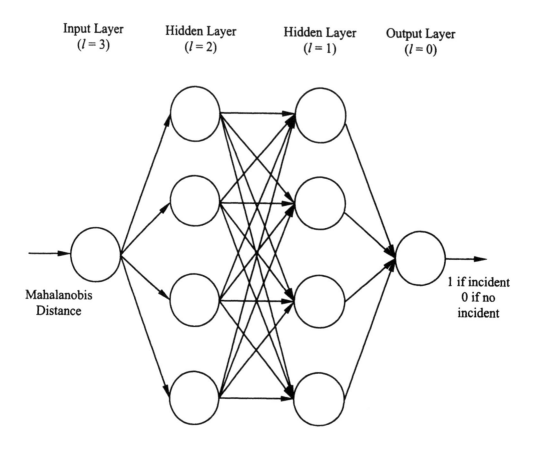

Figure 5.3 Architecture of the neural network for incident detection

where $S = i_j^l$ is the total input obtained from Eq. (5.11). Since there is only one node in the output layer, the total error function for M training instances is defined as

$$E = \sum_{n=1}^{M} [y(n) - o(n)]^2 = \sum_{n=1}^{M} [r(n)]^2$$ (5.13)

where $r(n)$ is the difference of the computed output, $y(n)$, and the desired or measured output, $o(n)$, for the nth training instance. In a vector form, the total error function is expressed as

$$E = \mathbf{r}'\mathbf{r} \qquad (5.14)$$

where $\mathbf{r}$ is an $M \times 1$ vector. This error is propagated backward through the network and the weights of the links are updated until the network converges within a given tolerance. This is basically an unconstrained minimization problem. In the simple BP algorithm, the weights of the links are updated in each iteration after all the training examples are applied using the generalized delta rule based on the computation of the gradient of the error function:

$$\mathbf{w}_n = \mathbf{w}_0 - \eta \mathbf{G} \qquad (5.15)$$

where $\mathbf{w}_n$ is the $m \times 1$ updated weight vector, $\mathbf{w}_0$ is the $m \times 1$ weight vector from the previous iteration, m is the number of weights, η is the learning rate, and $\mathbf{G}$ is an $m \times 1$ vector representing the gradients of the error function with respect to the weights:

$$\mathbf{G} = \nabla E = \partial E / \partial w = 2\mathbf{J}'\mathbf{r} \qquad (5.16)$$

In Eq. (5.16) $\mathbf{J}$ is the $M \times m$ Jacobian matrix and the prime represents the transpose of the matrix. The simple BP algorithm is shown to have a very slow rate of convergence (Adeli and Hung 1994). To improve the convergence of neural networks with quadratic error functions, quasi-Newton algorithms have been used to replace the fixed learning rate, η, used in the simple BP algorithm with an adaptive learning rate which is a function of the weights (Hagan *et al.* 1996). In this case, Eq. (5.15) is modified to

$$\mathbf{w}_n = \mathbf{w}_0 - \mathbf{H}^{-1}\mathbf{G} \qquad (5.17)$$

where $\mathbf{H}^{-1}$ is the inverse of the $m \times m$ Hessian matrix of the error function. The Hessian matrix is mathematically defined as

$$\mathbf{H} = \nabla^2 E = \partial^2 E / \partial \mathbf{w}^2 \qquad (5.18)$$

Computation of the Hessian matrix and its inverse introduces complexity and significant additional computational burden due to second-order derivatives of the error function. The Levenberg-Marquardt BP algorithm provides a simplified quasi-Newton approach for training of the feedforward neural networks by using the following approximation for the Hessian matrix whenever the error function is expressed as a sum of squares in the form of Eq. (5.14) (Hagan *et al.* 1996):

$$\mathbf{H} \approx \mathbf{J'J} \qquad (5.19)$$

Substituting Eqs. (5.16) and (5.19) into Eq. (5.17), the updating rule is modified to

$$\mathbf{w_n} = \mathbf{w_0} - (\mathbf{J'J} + \mu_n \mathbf{I})^{-1} \mathbf{J'r} \qquad (5.20)$$

where the term $\mu_n \mathbf{I}$ is added to insure that the Hessian matrix, $\mathbf{H}$, is invertible. To speed up convergence, the parameter μ_n is multiplied by 10 in the next iteration if the total error increases and divided by 10 if the total error decreases.

5.5   Comparing the Efficacy of Wavelets and Filtering Schemes

To determine the most appropriate type of wavelet for de-noising the traffic data, a comparative study of various wavelets and filtering schemes is conducted in terms of efficacy and accuracy of smoothing. Smoothing efficacy of traffic signals refers to closeness of the filtered signal to the original signal and is measured in terms of the MSE value of the signal. Smoothing accuracy is measured in terms of classifying the traffic signals into incident-free and incident-related in any two-dimensional projection of the three-dimensional traffic parameters space (volume-occupancy, volume-speed, or speed-occupancy plane). In this chapter, a sample set of simulated speed, volume, and occupancy data for a two-lane straight freeway at 30-second intervals for 30 minutes is used.

Haar, second order Daubechies, and second- and fourth-order Coifman wavelets are used to test three multi-resolution filtering schemes A, B, and C, shown in Figs. 5.4 to 5.6, respectively. In scheme A (Fig. 5.4), the unfiltered data signal is resolved into high resolution and low resolution components. The high resolution components are discarded as they correspond to traffic noise. The low resolution components are subjected to further decomposition to yield medium and low resolution components. After a total of three levels of decomposition, the six lowest resolution components are used to reconstruct the filtered signal through the process of inverse wavelet transform (Eq. 3.16). The first two levels of decomposition are the same in schemes B and C. In scheme B (Fig. 5.5), after three levels of decomposition the two lowest medium resolution components and the two lowest low resolution components are used to reconstruct the filtered signal. In scheme C (Fig. 5.6), a fourth level of decomposition is performed and the lowest medium resolution component and the three lowest low resolution components are used to reconstruct the filtered signal. Scheme A uses six wavelet coefficients while schemes B and C use four different wavelet coefficients.

Table 5.2 provides a comparison of smoothing efficacy of traffic signals in terms of MSE value of the signal for four different wavelets and three different filtering schemes. The average MSE values in Table 5.2 are the average of three MSE values calculated for speed, volume, and occupancy signals. First, different wavelets were tested with filtering scheme A which uses more wavelet coefficients than the other two schemes. As seen in Table 5.2, the Haar wavelet produced the largest error followed by second-order Daubechies, second-order Coifman, and fourth-order Coifman. Based on these results, Haar and second-order Coifman wavelets were dropped for further studies. Next, second-order Daubechies and fourth order Coifman wavelets were tested with filtering schemes B and C. Table 5.2 shows that the fourth-order Coifman wavelet with scheme A generates the least error among all cases studied. Further, scheme C produces less error than scheme B and fourth-order Coifman wavelet produces less error than the Daubechies wavelet in all the filtering schemes. At this point, it is concluded that the Coifman wavelet is more effective than the Daubechies wavelet. In the rest of the chapter, the Coifman wavelet will be used and the accuracy of filtering schemes A and C will be investigated.

5.6   Training the Neural Network Classification Model

The Levenberg-Marquardt BP neural network classification model is trained using sample data generated from the TSIS/CORSIM software (TSIS 1999) developed to simulate traffic conditions microscopically for a range of parameters. The freeway traffic simulation module of the software, FRESIM, was used to generate 500 continuous minutes (8 hours and 20 minutes) of incident-free traffic data for a straight two-lane

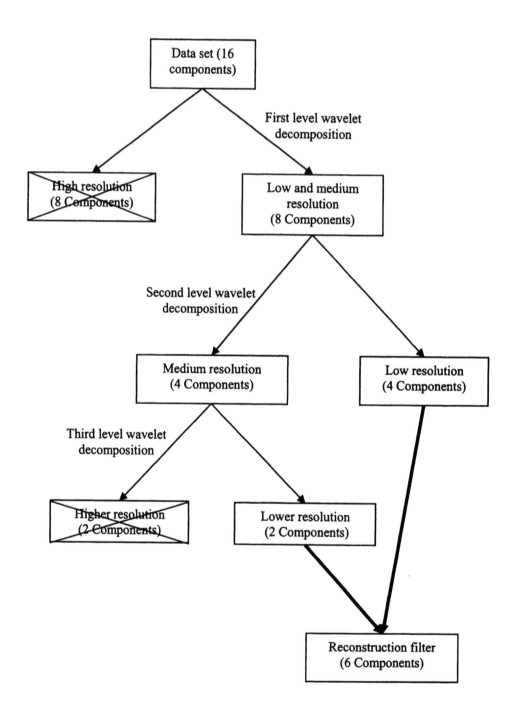

Figure 5.4 Wavelet-based filtering scheme A

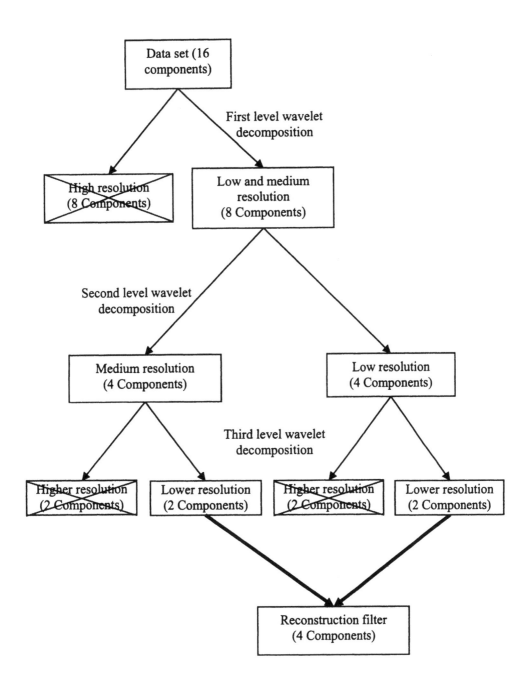

Figure 5.5 Wavelet-based filtering scheme B

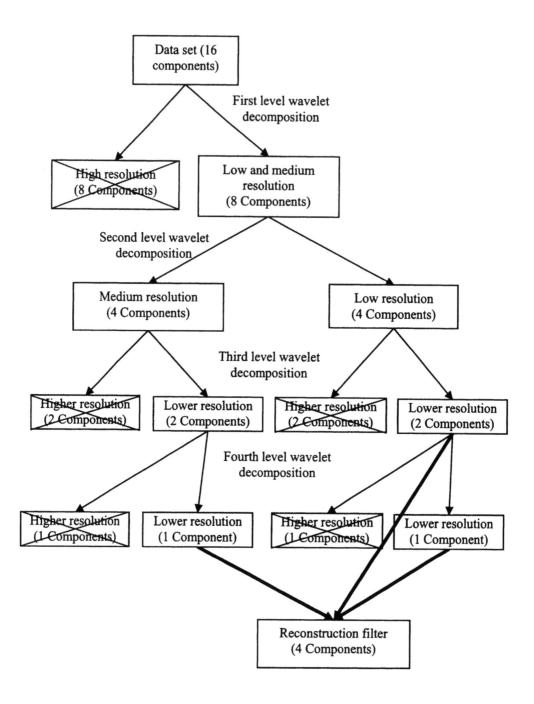

Figure 5.6 Wavelet-based filtering scheme C

freeway with no lane blockage, as well as 93 eight-minute discrete sequences of incident-related data sequences by varying the traffic demand and specifying lane blockages of different durations. The data consist of speed, volume, and occupancy values at intervals of 30 seconds. For the incident-free data, the traffic parameter values in each 30-second time period are combined with the values from the 7.5 minutes preceding it to form sequences of 8-minute duration with a time period of 30 seconds. Detection time cannot be less than the selected time period (30 seconds) for the traffic data sequence. Selecting a period greater than 30 seconds will increase the detection time which is not desirable. The duration of the traffic signal (8 minutes in this work) must be long enough to include recognizable patterns and produce accurate results.

Table 5.2 Comparison of smoothing efficacy of traffic signals in terms of mean squared error (MSE) value of the signal.

Filtering scheme	Wavelet type	Order	Average MSE
A	Haar	2	2987.625
A	Daubechies	2	2745.345
A	Coifman	2	2155.456
A	Coifman	4	1869.916
B	Daubechies	2	3622.909
B	Coifman	4	3579.061
C	Daubechies	2	3037.892
C	Coifman	4	2410.747

A total of 993 incident-free sequences and 93 incident-related sequences for each traffic parameter are used to train the network. The size of the original input space before clustering is $3 \times 16 = 48$ (16 measurements each for traffic speed, volume, and occupancy). The fourth-order Coifman discrete wavelet transform is applied to the original traffic sequences to filter out the normal random traffic fluctuations and noise in the data.

The filtered data are used in the clustering stage to reduce the classification parameters or the input to the neural network model to a single parameter by clustering the data into two predominant groups representing incident-free and incident-related data with an inconsistency coefficient of $Y_p = 1.3$ (Eq. 5.8). This threshold value was obtained by numerical experimentation. A lower value results in more than two predominant clusters. A higher value does not yield the required minimum two clusters. Clustering analysis is performed simultaneously on all objects or data points in the filtered traffic speed-volume-occupancy signal space. The Mahalanobis distance of each object from the incident-free cluster centroid is selected as the classification parameter. This reduces the input space of the Levenberg-Marquardt BP neural network to a 16×1 sequence and the number of input nodes to 1.

The desired output of the neural network is either the integer 0 representing no incident or 1 representing an incident. The training of the Levenberg-Marquardt BP neural network is initiated using random values for the weights of the links. The weights of the neural network are updated after each iteration using Eq. (5.20) and this process is repeated until the system error, Eq. (5.13), is reduced to a given tolerance (a value of 0.005 is used in this work). Figure 5.7 shows the convergence curves up to 60 iterations using (a) the unfiltered traffic speed, volume, and occupancy and (b) the filtered and clustered data based on the Mahalanobis distance classification parameter. It is observed that the former needs 527 iterations for convergence while the latter needs only 13 iterations for convergence.

5.7   Testing the Network
In this section, the model is tested using both simulated and real data.

5.7.1   Example 1: Simulated Data
Using the FRESIM module of the TSIS/CORSIM software, a test data set having the same number of traffic parameter data points as the training set (993 incident-free sequences and 93 incident-related sequences) is created in order to test the incident detection algorithm. The unfiltered data are filtered using the fourth-order Coifman wavelet with filtering scheme C (Fig. 5.6). Figures 5.8 to 5.10 show the unfiltered data and the effect of wavelet filtering in delineating the boundaries of the two groups of data – incident-free and incident-related. For clarity of representation, the three-dimensional traffic parameters input space is projected on three two-dimensional planes, volume-

occupancy (Fig. 5.8), volume-speed (Fig. 5.9), and speed-occupancy (Fig. 5.10). Figures 5.8 to 5.10 demonstrate the power of the Coifman wavelet in de-noising the data so that they can be classified more effectively by a pattern recognizer or classifier.

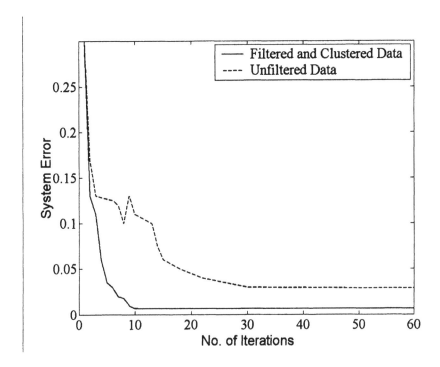

Figure 5.7 Convergence curves of the Levenberg-Marquardt BP neural network using (a) unfiltered data (dashed line) and (b) filtered and clustered data using fourth order Coifman wavelet (solid line)

Clustering analysis is performed on the complete three-dimensional filtered data space for the entire duration of the simulation. Clustering acts both as a rough classifier for the filtered data as well as a parameter reduction technique preceding the neural network classifier. The clustering analysis identifies individual data objects as incident-free or incident-related based on the cluster to which each object belongs. However, this classifier does not consider any patterns in the signal. When the unfiltered data are input directly into the cluster classifier based on the Mahalanobis distance without de-noising by the wavelet transform, the incident detection rate is a low 71.0%, the false alarm rate

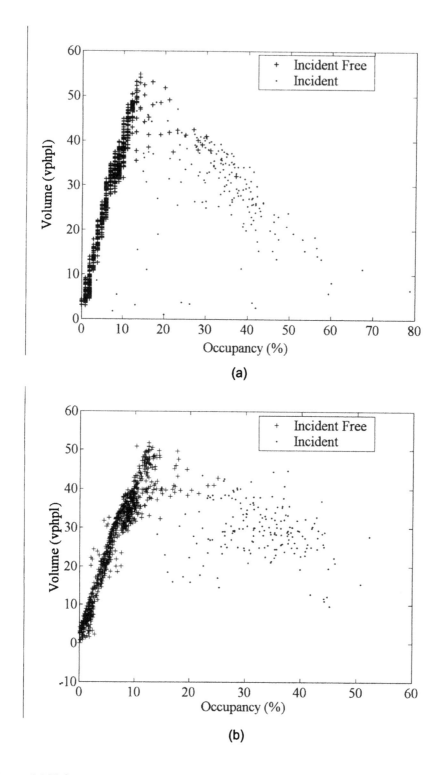

Figure 5.8 Volume-occupancy projection of three-dimensional traffic data space: (a) Unfiltered data, and (b) data filtered using fourth-order Coifman wavelet

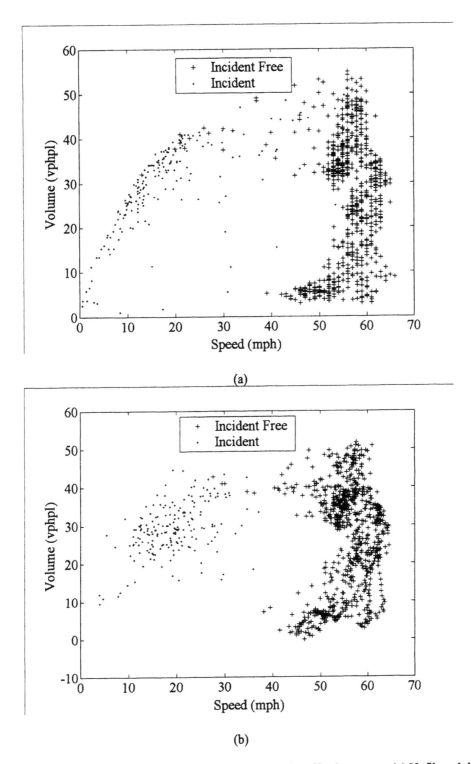

Figure 5.9 Volume-speed projection of three-dimensional traffic data space: (a) Unfiltered data, and (b) data filtered using fourth-order Coifman wavelet

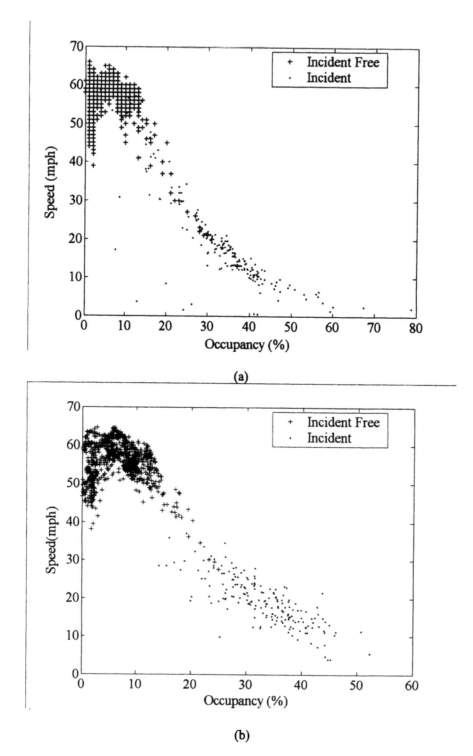

Figure 5.10 Speed-occupancy projection of three-dimensional traffic data space: (a) Unfiltered data, and (b) data filtered using fourth-order Coifman wavelet

is a high 9.3%, and the detection time is 69.1 seconds. The neural network classifier applied directly to the unfiltered data shows slightly better accuracy (79.6% incident detection and 9.0% false alarm rate) and lower detection time (52.9 seconds) as patterns in the signal corresponding to change in traffic flow characteristics are recognized faster. The accuracy and detection rate for the neural network classifier are low because the incident-free and incident-related groups are not distinct enough.

Table 5.3 presents a comparison of smoothing accuracy and computational efficiency of various classifying approaches using simulated data for Example 1. Results are substantially improved when data are filtered using the fourth-order Coifman wavelet. Between the two filtering schemes tested, scheme C shows greater smoothing accuracy and lower detection time than scheme A with both clustering and neural network classifiers (Table 5.3). Out of these, a combination of scheme C and the neural network classifier yields the best results with an incident detection rate of 90.3%, a false alarm rate of 4.2%, and a detection time of 43.5 seconds.

Table 5.3 Comparison of smoothing accuracy and computational efficiency of various classifying approaches using simulated data for Example 1

Wavelet filtering	Parameter reduction	Classifier	Incident detection (%)	False alarms (%)	Detection time (sec)
Adeli and Samant (2000)	LDA	ANN	97.78 (44/45)	1.11 (8/720)	38.1
-	-	Cluster	70.97 (66/93)	9.37 (93/993)	69.1
Coifman - A	-	Cluster	80.65 (75/93)	8.36 (83/993)	64.3
Coifman - C	-	Cluster	82.80 (77/93)	7.96 (79/993)	61.3
-	-	ANN	79.57 (74/93)	8.96 (89/993)	52.9
Coifman - A	-	ANN	87.10 (81/93)	4.43 (44/993)	45.3
Coifman - C	-	ANN	90.32 (84/93)	4.23 (42/993)	43.5
-	Cluster	ANN	81.72 (76/93)	2.11 (21/993)	49.3
Coifman - A	Cluster	ANN	96.77 (90/93)	0.81 (8/993)	38.2
Coifman - C	Cluster	ANN	98.92 (92/93)	0.30 (3/993)	33.8

Among all cases considered and summarized in Table 5.3, however, a sequential combination of fourth-order Coifman wavelet, filtering scheme C, clustering analysis, and the Levenberg-Marquardt BP neural network classifier provides the most accurate and efficient results with an incident detection rate of 98.9%, a false alarm rate of 0.3%, and detection time of 33.8 seconds. For comparison, Adeli and Samant (2000) use Daubechies second-order wavelet, linear discriminant analysis for clustering, and the adaptive conjugate gradient neural network model of Adeli and Hung (1994) and report an incident detection rate of 97.8%, a false alarm rate of 1.1%, and a detection time of 38.1 seconds.

5.7.2   Example 2: Real Data

In this example, real data from a single detector station obtained from the ARTIMIS project near Cincinnati, Ohio, are used to test the algorithm. The data set provided by the Ohio Department of Transportation had missing and inconsistent data. Only a small part of the data set could be extracted and used to test the algorithm. The extracted data cover a total period of 300 incident-free minutes (593 incident-free data sequences) and 11 incidents. The results obtained using fourth-order Coifman wavelet, filtering scheme C, clustering analysis, and the Levenberg-Marquardt BP neural network classifier are as follows: incident detection is 100% (11/11), false alarm rate is 0.3% (2/593), and detection time is 35.6 seconds.

5.8   Concluding Remarks

In this chapter, a freeway incident detection model was presented based on speed, volume, and occupancy data from a single detector station using a combination of wavelet de-noising, statistical clustering analysis, and neural network pattern recognition. It was shown that the fourth-order Coifman wavelet is superior to the Daubechies wavelet for denoising of traffic data. Clustering analysis is performed on the filtered data and the sequence of Mahalanobis distances of data samples from the centroid of the incident-free cluster is input to the Levenberg-Marquardt BP neural network.

The model was trained using simulated data obtained from the FRESIM module of the TSIS/CORSIM software. However, the model was tested for a straight freeway segment using both simulated and real data. The results show improved performance compared with previous models.

Chapter 6

MICRO-SIMULATION MODEL FOR TRAFFIC DELAY AND QUEUE LENGTH ESTIMATION IN FREEWAY WORK ZONES

6.1   Introduction

Little scientific work has been reported on the freeway work zone delay and queue length estimation problem specifically. Some research, however, has been done on queuing theory models based on the demand-capacity dynamics of traffic flow on freeways with special lanes such as car pool or priority vehicle lanes (Daganzo 1997; Daganzo *et al.* 1997) and breakdown of traffic flow behind bottlenecks on freeways due to entry and exit ramps, uphill gradients, and lane closures (Newell 1998 & 1999; Cassidy and Bertini 1999; Son 1999). It has been shown that the type of bottleneck does not affect the traffic flow characteristics (Treiber *et al.* 2000). Jiang (1999a) presents a deterministic model to compute delay in work zones as a sum of deceleration delay at the start of a work zone, reduced speed delay, acceleration delay at the end of a work zone, queue delay for congested traffic and waiting time delay for uncongested traffic. However, this model assumes fixed values for vehicle velocity and acceleration, and does not take traffic flow characteristics into consideration.

A widely used Microsoft Excel-based software named QuickZone Delay Estimation Program uses the macroscopic deterministic queuing theory to estimate queue lengths and travel times in a freeway work zone (MITRETEK 2000). QuickZone uses the difference between hourly and daily traffic demand and freeway capacity taking into account the seasonal demand factors to identify congestion and estimate queuing and travel times. It compares the user travel times for a freeway without construction with the corresponding values for the same freeway in the presence of a work zone to compute the

This chapter is based on the article: Ghosh-Dastidar S. and Adeli H. "Neural Network-Wavelet Micro-Simulation Model for Delay and Queue Length Estimation at Freeway Work Zones," ASCE-*Journal of Transportation Engineering* 2006, 132(4), 331-341, and is reproduced by the permission of the publisher, American Society of Civil Engineers.

additional travel time or delay due to the work zone. The major disadvantage is that the basic macroscopic model is too simplistic to generate accurate traffic conditions in the freeway, especially the sections upstream of the work zone that are influenced by lane closure. The macroscopic model used in QuickZone considers the entry traffic volume or traffic demand of vehicles at a given time but not the entry traffic speed and density conditions. Moreover, it does not take into account any traffic speed or density fluctuations due to individual driver behavior. Further, QuickZone does not actually model traffic flow due to lane closure but considers it indirectly only in the form of a reduction in freeway capacity for the work zone segment.

A different approach to computing traffic delay and queue length estimation is to simulate traffic flow dynamics on the freeway directly. The simulations are based on either microscopic or macroscopic models of traffic flow. A microscopic model uses a differential equation of motion for a single vehicle subject to various boundary conditions. A widely used freeway traffic modeling software, the FRESIM module of TSIS/CORSIM software, is based on a microscopic model. FRESIM microscopically simulates traffic flow around a lane blockage or closure (TSIS 1999; Bloomberg and Dale 2000) to yield values of traffic speed and density. The usefulness of this software is severely limited when applied to freeway work zones because it does not provide congestion characteristics such as length of queues and travel delay times.

Alternatively, a macroscopic model attempts to model the traffic flow as a whole and is based on a differential equation of continuity, which is a function of three parameters: traffic flow, speed, and density. Neural network (Adeli and Hung 1995; Adeli and Park 1998; Adeli 2001) and statistical techniques such as cluster and multivariate analyses have been used to model traffic flows (Neubert *et al.* 1999; Hasebe *et al.* 1999). Zhang *et al.* (1997) use the simple BP neural network to simulate a macroscopic freeway traffic flow model. Park *et al.* (1998) use a radial-basis function neural network to forecast freeway traffic flow. Suzuki *et al.* (2000) use a combination of the BP neural network and Kalman filter and a macroscopic model to estimate origin-destination (O-D) travel times and traffic flows. However, these studies have not been directly applied to the delay estimation problem. Adeli and coworkers have developed novel computational models for freeway traffic incident detection using neural network, wavelet, and fuzzy logic (Adeli and Karim 2000; Samant and Adeli 2001; Karim and Adeli 2002a&b;

Ghosh-Dastidar and Adeli 2003). In addition, Jiang and Adeli (2004a) present a hybrid wavelet packet-statistical autocorrelation function method for analysis of traffic flow time series and determining its self-similar, singular, and fractal properties to be discussed in Chapter 11.

Ghosh-Dastidar and Adeli (2003) developed a mesoscopic-wavelet model for simulating freeway traffic flow patterns and extracting congestion characteristics. A traffic speed-density relationship is introduced with a lane drop factor to take into account lane closures in freeway work zones. An approximate solution for this equation is found by space-time discretization. Patterns of multiple parameters are input to a congestion feature extraction algorithm. The high frequency fluctuations of the signal are not recognizable at normal resolutions. To overcome this problem, a multi-resolution wavelet filter is introduced in the proposed model to enhance traffic features and extract congestion characteristics from the traffic data. Fourth-order Coifman wavelets are used for filtering because of their good approximation for high-resolution scaling, as discussed in the previous chapter.

In this chapter, a neural network-wavelet micro-simulation model is presented to track the travel time of each individual vehicle for traffic delay and queue length estimation at work zones. The extracted congestion characteristics obtained from the mesoscopic-wavelet model are used in a Levenberg-Marquardt backpropagation (BP) neural network for classifying the traffic flow as free flow, transitional flow, and congested flow with stationary queue. The model incorporates the dynamics of a single vehicle in changing traffic flow conditions.

6.2   Mesoscopic-Wavelet Flow Model

The input parameters for the mesoscopic-wavelet freeway work zone traffic model include traffic flow parameters existing before the start of construction, that is, the original flow rate (q) measured in vehicles per hour (vph), space mean speed (v) in miles per hour (mph) or kilometers per hour (kph), and traffic density (ρ) measured in vehicles per mile (vpm) or per kilometer (vpk), as well as the pavement conditions proposed for the work zone, that is, the length of the work zone (L) in miles (or kilometers), total number of lanes (N), and number of closed lanes (N_L). In addition, the model requires inputs for duration of the work zone (D) in hours, time interval (T) in hours, freeway section length (Δ) in miles (or kilometers), and number of sections (N_S). The mesoscopic model uses three primary equations to generate values for the three traffic parameters (flow rate, space mean speed, and traffic density) over multiple freeway sections before and after the work zone in order to obtain a realistic flow pattern including cases with queue formation and dissipation.

Equation (6.1) represents the principle of conservation of vehicle flow which states that the number of vehicles entering any freeway section is the same as that leaving it at a given time period k:

$$\rho_i(k+1) = \rho_i(k) + (T/\lambda_i\Delta_i)[\lambda_{i-1}q_{i-1}(k) - \lambda_iq_i(k)] \tag{6.1}$$

where λ is the number of open lanes. The subscript represents the section number and the value in parentheses represents the time period number. For example, $\rho_i(k+1)$ represents the density in freeway section i at time $k + 1$. Equation (6.2) expresses the relationship among traffic flow, space mean speed, and density in a section i at a given time k as

$$q_i(k) = v_i(k)\rho_i(k) \tag{6.2}$$

Equation (6.3), representing the discretized speed-density relation, is obtained as (Adeli and Ghosh-Dastidar 2004)

$$v_i(k+1) = v_i(k) + [\overline{V}(\rho_i(k)) - v_i(k)]T/\tau_d + v_i(k)[v_{i-1}(k) - v_i(k)]T/\Delta_i$$
$$+ (\mu T/\tau_d)\{\{[\overline{V}(\rho_{i+1}(k)) - \overline{V}(\rho_i(k))]/[\rho_{i+1}(k) - \rho_i(k)]\}\{(\rho_{i+1}(k) - \rho_i(k))/(2\rho_i(k)\Delta_i)$$
$$+ [\rho_{i+1}(k) - 2\rho_i(k) + \rho_{i-1}(k)]/[12\rho_i(k)^2\Delta_i^2] - [\rho_{i+1}(k) - \rho_i(k)]^2/[2\rho_i(k)^3(\Delta_i)^2]\}$$
$$- [\eta \cdot \phi \cdot T(\lambda_i - \lambda_{i+1})\rho_i(k)v_i(k)^2]/(\Delta_i\lambda_i\lambda_i\rho_{cr}) \tag{6.3}$$

where η is the lane closure factor (equal to zero when there is no lane closure and one when there is lane closure), φ is a layout parameter which depends on the freeway configuration, μ is the adjustment parameter to control the traffic flow fluctuations due to contribution of the higher order terms, τ_d is the driver reaction time, and ρ_{cr} is the critical density factor at which flow changes to congested flow. The optimal velocity function, $\overline{V}(\rho_i(k))$, needed in Eq. (6.3) is obtained as (Adeli and Ghosh-Dastidar 2004)

$$\bar{V}(\rho_i(k)) = \tanh\{1/\rho_i(k) - [\rho_{i+1}(k) - \rho_i(k)]/[2\rho_i(k)^3\Delta_i]$$
$$- \{1/[6\rho_i(k)^4]\}[\rho_{i+1}(k) - 2\rho_i(k) + \rho_{i-1}(k)]/(2\Delta_i^2)$$
$$+ [\rho_{i+1}(k) - \rho_i(k)]^2/\Delta_i^2/[2\rho_i(k)^5] - 2\} - \tanh(-2) \tag{6.4}$$

The data obtained using Eqs. (6.1), (6.2), and (6.3) consist of values for traffic flow, space mean speed, and density for any given freeway section at specified intervals T for duration of the work zone, D. All three equations are needed for modeling the traffic flow accurately. However, once the values of the three traffic parameters are obtained for the entire freeway segment, the values for the density and the space mean speed are sufficient to model congestion information accurately.

The discrete wavelet transform using fourth-order Coifman wavelets is applied individually to each one of the traffic space mean speed sequences for different freeway sections over the specified duration. The original sequences contain D/T data points and the first-level wavelet transform after downsampling results in $D/(2T)$ high resolution components and $D/(2T)$ low resolution components. The low resolution components are subjected to second-level wavelet transform to yield $D/(4T)$ medium resolution components and $D/(4T)$ lowest resolution components. Since the purpose is to magnify the high resolution information and compare it with the low resolution information, none of these components is discarded. The $D/(4T)$ lowest resolution components are used to reconstruct the trend of the signal. On the other hand, the $D/(2T)$ high resolution components, combined with the $D/(4T)$ medium resolution components, represent traffic fluctuations or noise and are used to reconstruct the high frequency temporal fluctuations in the signal.

6.3   Identification of Traffic Flow Phase

The resulting pattern recognition problem is solved using a multi-layer feed-forward network trained by the Levenberg-Marquardt BP algorithm (Hagan *et al.* 1996; Mathworks 2000c). The architecture of the neural network for identification of the traffic flow phase is shown in Fig. 6.1. It consists of an input layer with three nodes representing traffic space mean speed trend, space mean speed fluctuation, and density. A network with a single hidden layer failed to produce satisfactory results because of the complexity of the pattern recognition problem. Consequently, the number of hidden layers was increased to two. The number of nodes in each hidden layer is chosen to be six after concluding that fewer nodes would not yield satisfactory convergence based on numerical experimentation. The output layer has a single node representing the traffic flow phase.

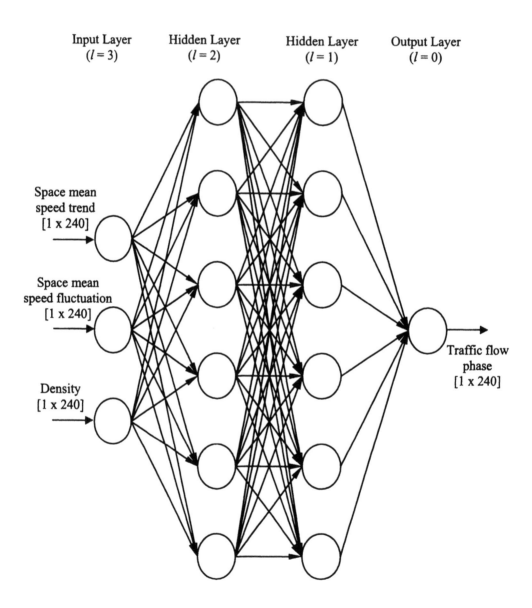

Figure 6.1 Architecture of neural network model for traffic flow phase identification

The 3×1 vector of the input variables for the nth training instance is $\mathbf{I}(n)$. The output for the nth training set is $d(n)$. The layers are numbered from right to left starting with $l = 0$ for the output layer and ending with $l = 3$ for the input layer. The total input to the jth node in a hidden or output layer l, is mathematically expressed as (Bose and Liang 1996)

$$i_j^l = \sum_i y_i^{l+1} w_{ij}^l , l = 0, 1, \text{ or } 2 \qquad (5.11) \text{ (Repeated)}$$

where y_i^{l+1} is the output of the ith node in layer $l+1$ and w_{ij}^l is the weight of the link connecting the ith node in layer $l+1$ to the jth node in layer l. The output of the jth node in layer l (= 0, 1, or 2) is computed using the sigmoid function, Eq. (5. 12).

The total error function for M training instances is expressed in vector form by Eq. (5.14). This error is propagated backward through the network and the weights of the links are updated until the network converges within a given tolerance. The Levenberg-Marquardt BP algorithm presented in Chapter 5 is used in this chapter.

6.4   Micro-Simulation Model for Queue Length and User Delay Estimation

6.4.1   Queue Length Estimation

User travel time on the freeway depends on whether the traffic flow is free flow, transitional flow or congested flow with stationary queue. The queue length is a measure of congestion and needs to be estimated accurately for computing the user travel time. A freeway segment under consideration is divided into a number of equal-length sections. Traffic parameter values for all sections of the freeway segment are obtained from the mesoscopic model (Adeli and Ghosh-Dastidar 2004). The proposed micro-simulation model assumes that a traffic flow phase holds for the entire freeway section under consideration for the entire time interval. Accordingly, the freeway section length selected for the mesoscopic model should be small enough to obtain an accurate (high resolution) estimate but not too small so as to avoid additional unnecessary computational costs.

The stationary queue length $L_i(k)$ in miles (or kilometers) at time k (expressed in terms of the time interval T) for any freeway section i having congested traffic flow with stationary queue (as identified by the neural network) is expressed as

$$L_i(k) = \Delta_i \qquad\qquad (6.5)$$

where Δ_i is the length of the freeway section i. The number of vehicles in the queue, $Q_i(k)$, is given by

$$Q_i(k) = \rho_i(k)\Delta_i \qquad\qquad (6.6)$$

where $\rho_i(k)$ is the traffic density in section i at time k. For queue length estimation, the maximum number of contiguous freeway sections having congested flow with stationary queue at any specific time is used as the total number of queued sections; the estimated queue length is calculated as equal to this number multiplied by the freeway section length. For computing the total user travel time, however, the queue length and duration of the queue for every queued section of the freeway segment, including non-contiguous queued sections, are used.

6.4.2   User Travel Time and Delay Estimation

Computation of the user travel time in the micro-simulation model is summarized in Fig. 6.2. For a micro-simulation approach considering the dynamics of a single vehicle in changing traffic flow conditions, the user travel time at time k is defined as the time taken by the vehicle entering the freeway segment at time k to traverse all the sections of the freeway segment.

Free and Transitional Flow

For free flow and transitional flow in any freeway section i, the vehicle travels at the space mean speed values obtained using the mesoscopic model (Adeli and Ghosh-Dastidar 2004) for a given time interval. The distance $x(k)$ traveled by the vehicle in the kth time interval is computed as

$$x(k) = v_i(k)T \qquad\qquad (6.7)$$

where $v_i(k)$ is the traffic space mean speed at time k in freeway section i. The total distance traveled by the vehicle during the time interval under consideration, X, is computed by incrementally adding the values of $x(k)$ obtained from Eq. (6.7). The distance X identifies the position of the vehicle on the freeway segment. The distance traveled in the next time interval $k+1$ is computed using the parameter values for the freeway section in which the vehicle is located. The computation is repeated for subsequent time intervals until the total distance traveled by the vehicle equals the total length of the freeway segment or the total time exceeds the duration of simulation (Fig. 6.2). If multiple freeway segments are traversed in a particular time interval, then the values of i and X are updated accordingly.

Congested Flow with Stationary Queue

Applying the same procedure used for free and transitional flow for a vehicle in a stationary queue results in the vehicle remaining virtually at the same spot throughout the duration of the queue because the mesoscopic-wavelet model yields near zero traffic space mean speed values. That model, by itself, cannot estimate the time it takes for the vehicle to clear the queue. In reality, a vehicle in a congested flow with stationary queue is in a stop-and-go motion and resumes free flow velocity when it clears the queue. Therefore, if the neural network model detects congested flow with stationary queue for the freeway section under consideration at a given time, the number of contiguous freeway sections having congested flow with stationary queue and the number of vehicles ahead of the vehicle under consideration in the queue are found. If the position of a vehicle located in a section p of the freeway is x_p with respect to the beginning of the same freeway section p and N_Q is the number of queued sections following section p, the total queue length in front of the vehicle under consideration is found from

$$L_T(k) = \sum_{i=p}^{p+N_Q} L_i(k) - x_p(k) \tag{6.8}$$

The total number of vehicles ahead in the queue, $Q_T(k)$, is computed as the sum of the number of vehicles in the freeway sections following the section under consideration, p, and the number of vehicles in section p ahead of the vehicle under consideration. For the latter term, a linear interpolation is used over the length of section p. The result is

$$Q_T(k) = \sum_{i=p}^{p+N_Q} Q_i(k) - [x_p(k)/\Delta_p]Q_p(k) \tag{6.9}$$

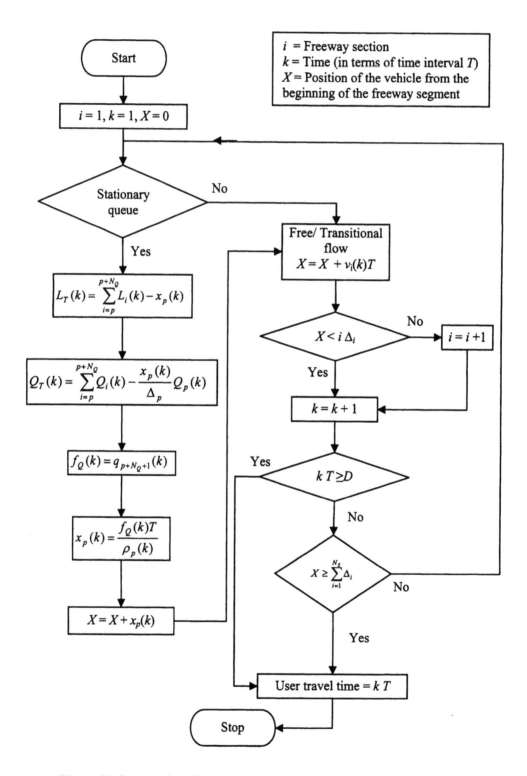

Figure 6.2 Computation of the user travel time in the micro-simulation model

The queue discharge rate (f_Q) defined as the traffic flow from the front end of the queue, is obtained from the flow values generated by the mesoscopic-wavelet model (Eq. 6.2) as follows:

$$f_Q(k) = q_{p+N_Q+1}(k) \tag{6.10}$$

where $q_{p+N_Q+1}(k)$ is the traffic flow at time k in freeway section $p+N_Q+1$. The number of vehicles leaving the queue at time k is $f_Q(k) T$, which is used to compute the change in position of the vehicle for the subsequent time interval as

$$x_p(k+1) = f_Q(k)T/\rho_p(k) \tag{6.11}$$

Equations (6.8) to (6.11) are used repeatedly at every time interval k as long as the vehicle is in the stationary queue (Fig. 6.2). When the vehicle clears the stationary queue the algorithm control is transferred to the free/transitional flow, as noted in Fig. 6.2. The algorithm control will be transferred back to the stationary queue and Eqs. (6.8) to (6.11) will be reused when the vehicle again enters a section having congested flow with stationary queue (Fig. 6.2).

The user delay time is simply the difference between the user travel time generated by the model for the freeway segment with and without lane closure due to a work zone.

6.5   Training the Neural Network

In order to train the Levenberg-Marquardt BP neural network classifier, training instances for different entry flow conditions with known start and end times for different phases of traffic flow are required. Traffic density and raw space mean speed data are created using the FRESIM module of the TSIS/CORSIM simulation software (TSIS 1999; Bloomberg and Dale 2000) for a straight segment of a freeway with no entry and exit ramps for the duration of the work zone. Space mean speed trend and fluctuations are obtained by applying the Coifman wavelet filter to the raw space mean speed data as described in the previous chapter. Even though the traffic flow model for FRESIM is different from the proposed mesoscopic model, the general shapes of the patterns are similar and therefore sufficient for training the classifier in this work.

For network training purposes, a simulation period of 60 hours is considered for a straight freeway section. This period is divided into 240 15-minute time intervals. As such, the input for each training instance consists of three 1×240 sequences corresponding to traffic density, space mean speed trend, and space mean speed fluctuations for the duration of simulation. The desired output of the neural network is a 1×240 sequence of the integers 0 (representing free flow), 1 (representing transitional flow) and 2 (representing congested flow with stationary queues). The training of the Levenberg-Marquadt BP neural network is initiated using random values for the weights of the links. The weights of the neural network are updated after each iteration, using all the training instances simultaneously. This process is repeated till the change in system error [defined by Eq. (5.14)] is reduced to 0.001.

Figure 6.3 shows the convergence curves using a) raw space mean speed and density data and b) space mean speed trend and fluctuations data obtained from the Coifman wavelet filter. It is observed that the latter needs 47 iterations for convergence whereas the former requires 99 iterations for convergence. Figure 6.3 clearly shows that wavelet filtering smoothens and speeds up the training convergence. It was also found that at least 400 training instances are required for neural network training in order to obtain accurate results.

The neural network recognizes transitional flow when the following facts are observed simultaneously: a) fluctuations in the space mean speed signal, b) a sharp drop in the trend of the space mean speed signal, and c) a sharp increase in the density signal. This information is used to classify the traffic flow on a freeway section at any specific time into free flow, transitional flow, and congested flow with stationary queues. The results are used for evaluation of the start and end times for each phase of the traffic flow and estimation of the stationary queue length.

6.6   Application and Examples

6.6.1   Constant Parameters in the Neural Network-Wavelet Micro-Simulation Model

The values of the following constant parameters used in the mesoscopic-wavelet model are based on previous research on traffic flow models. The driver reaction time, τ_d, in response to changing conditions is assumed to have a constant value of 36 seconds. The adjustment parameter μ in Eq. (6.3) is set to 13.52 mile^2/hr (35 km^2/hr) and the layout parameter φ in Eq. (6.3) for a straight and level freeway configuration is set to 0.8

(Papageorgiou *et al.* 1990). The critical value of density ρ_{cr} at which flow changes to congested flow, is fixed as 47 vpmpl (29.20 vpkpl) (Kerner 1999).

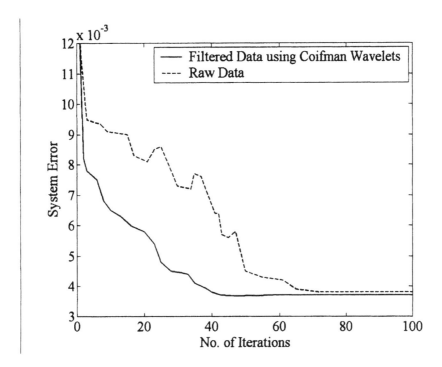

Figure 6.3 Convergence curves using a) raw space mean speed and density data (dashed line) and b) space mean speed trend and fluctuations data obtained from Coifman wavelet filter (solid line)

In order to avoid division by zero, the minimum space mean speed v_{min} is assumed to be 1 mph (1.61 kph) and the minimum density ρ_{min} is assumed to be 1 vpmpl (0.62 vpkpl). The upper boundary conditions are set by assuming that the space mean speed cannot exceed v_{max} = 80 mph (128.75 kph), density cannot exceed ρ_{max} = 193 vpmpl (119.92 vpkpl) (Kerner 1999) and the flow in a given freeway section cannot exceed the freeway capacity depending on the number of lane closures (Table 6.1) as defined by the Highway Capacity Manual (HCM 2000).

6.6.2   Common Freeway and Traffic Parameters

The micro-simulation model is applied to five examples of freeways with two and three lanes and one lane closure with varying entry flow or demand patterns. The common freeway and traffic parameters for all the examples are described in this section.

Table 6.1 Work zone capacities (HCM 2000)

Total no. of lanes	No. of lane closures	Average capacity (vphpl)
Any	0	2300
2	1	1550
3	1	1860

The freeway segment is straight and level with no entry or exit ramps. The length L of the work zone is assumed to be equal to 0.5 miles (0.80 kms). In order to obtain accurate results, the length of the freeway (Δ) considered for discretization should be of the order of twenty times the length of the work zone ($\Delta/L \approx 20$). In the examples presented here, the number of sections is assumed to be $N_S = 21$, each having a length of $\Delta = 0.5$ miles (0.80 kms). The total length of the freeway considered is thus 10.5 miles (16.90 kms). In all the examples, the work zone section is assumed to be the central section. The duration of simulation is the same as the duration D of the work zone which is assumed to be 60 hours for all the examples. The mesoscopic-wavelet model is employed to generate traffic flow, space mean speed, and density values using two different values of time intervals, $T = 15$ min, resulting in a 1×240 row vector, and $T = 5$ min, resulting in a 1×720 row vector for each parameter for any given freeway section (based on a 60-hour simulation). To maintain dimensional consistency and improve numerical performance, traffic space mean speed is normalized over average free flow space mean speed, $v_{avg} = 55$ mph (88.51 kph) and density is normalized over average free flow density, $\rho_{avg} = 30$ vpmpl (18.64 vpkpl).

6.6.3   Traffic Patterns

Three different entry flow (demand) pattern scenarios are tested using two- and three-lane freeways with one lane closure. For all the demand patterns, the data used as initial input to the model are the entry space mean speed and density signifying the conditions for the entry flow. Demand pattern A simulates two high-traffic demands over short periods separated by very low traffic demand during the remaining duration of simulation as shown in Fig. 6.4a. Demand pattern B simulates very high traffic demand for a moderate time period and about average traffic demand for the remaining duration of simulation, as shown in Fig. 6.4b. Patterns A and B are tested for 2-lane and 3-lane freeways with one lane closure to illustrate the effects of varying demand peaks and their duration and magnitude.

Demand pattern C (Fig. 6.5a) is generated using extracted parameter values from real data supplied by North Carolina Department of Transportation (NCDOT) for a 2-lane freeway with one lane closure. The numbers of vehicles passing through the detector station A3101 (Fig. 6.5b) are recorded every hour for a total of 60 hours from Wednesday (03/07/2001) 12:00 A.M. to Friday (03/09/2001) 11:00 A.M. Since the recorded data are hourly, they are interpolated to obtain the data for 15-minute intervals for the sake of comparison with patterns A and B.

Example 1: Two-Lane Freeway with One Lane Closure Using Demand Pattern A

The mesoscopic-wavelet model is used to generate the values of traffic flow parameters over the length of the freeway segment with and without construction for the duration of the work zone. Figures 6.6 to 6.8 show surface plots of normalized traffic space mean speed, flow, and density for Example 1, respectively. For the case of no construction, there is a short queue in the beginning of the segment (with a length of 25 vehicles as noted in Table 6.2) because the traffic flow exceeds the freeway capacity (Fig. 6.5a) which causes the traffic speed to approach zero as noted in Fig. 6.6a. After the queue dissipates the space mean speed approaches its maximum value (Fig. 6.6a) and the traffic flow (Fig. 6.7a) and density (Fig. 6.8a) become small. The presence of a work zone affects traffic flow by causing congestion in the upstream freeway sections. When there is a work zone, the same short queue at the beginning of the freeway segment is observed. Further, a second longer queue is formed in the work zone section and the section immediately before the work zone section (sections 10 and 11 in Figs. 6.6b, 6.7b, and 6.8b) resulting in a maximum queue length of 122 vehicles (Table 6.2). The computational model confirms that a work zone acts as a flow regulator for the downstream freeway sections by maintaining high space mean speed (Fig. 6.6b) and low density (Fig. 6.8b).

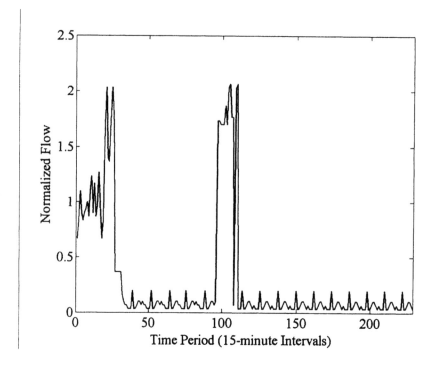

(a)

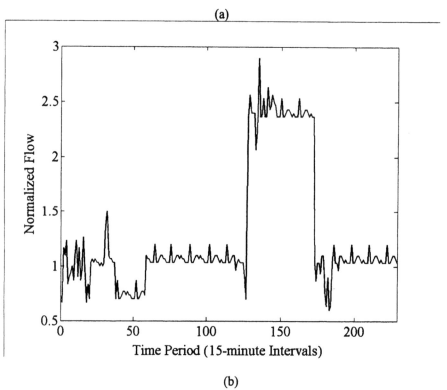

(b)

Figure 6.4 Traffic demand patterns (a) A and (b) B for duration of 60 hours

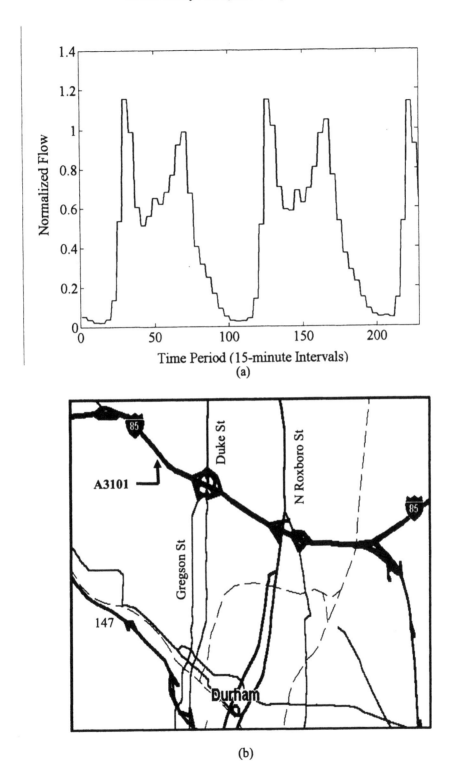

Figure 6.5 (a) Traffic demand pattern C obtained from NCDOT detector station A3101 on I-85 freeway for duration of 60 hours, and (b) location of NCDOT detector station A3101 on I-85, 0.3 mile west of Gregson Street (four-lane divided, grass median, urban area)

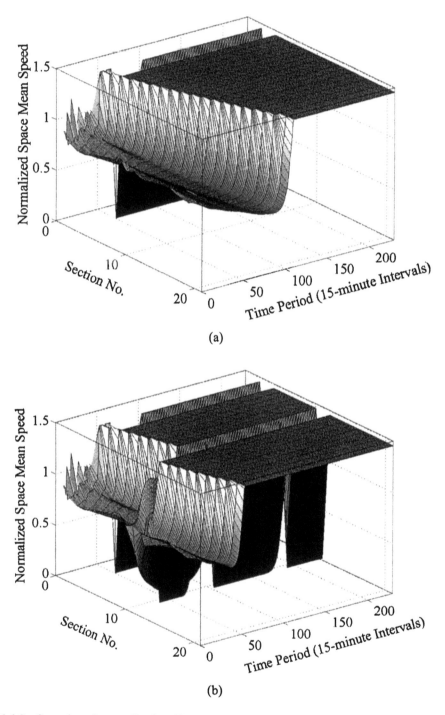

Figure 6.6 Surface plot of normalized traffic space mean speed across the freeway segment for the duration of the work zone for Example 1: (a) Without construction, and (b) with construction. See insert following Chapter 15.

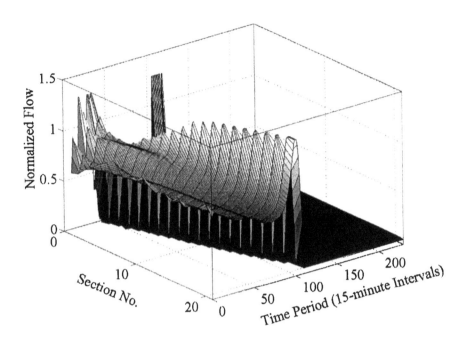

(a)

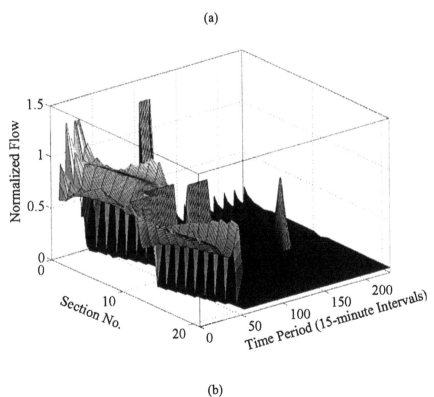

(b)

Figure 6.7 Surface plot of normalized traffic flow across the freeway segment for the duration of the work zone for Example 1: (a) Without construction, and (b) with construction. See insert following Chapter 15.

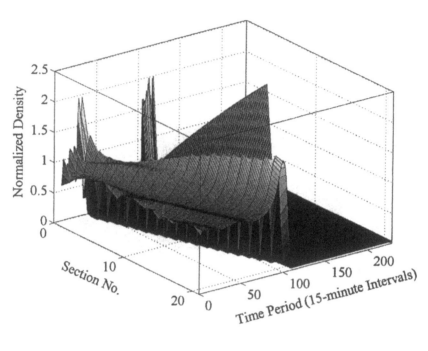

(a)

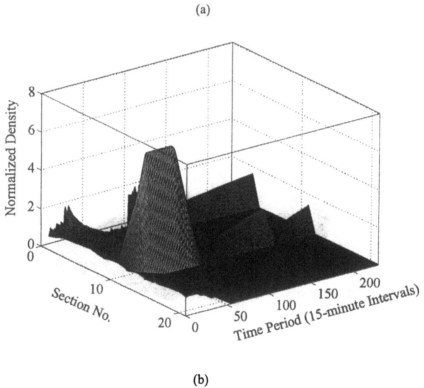

(b)

Figure 6.8 Surface plot of traffic density across the freeway segment for the duration of the work zone for Example 1: (a) Without construction, and (b) with construction. See insert following Chapter 15.

Table 6.2 Delay estimation results from the model using 15-minute time intervals

Example	Traffic pattern	Total no. of lanes	No. of lane closures	Maximum no. of vehicles in queue	Maximum travel time (minutes)	Maximum delay (minutes)
1	A	2	0	25	15	15
	A	2	1	122	30	
2	A	3	0	24	15	15
	A	3	1	78	30	
3	B	2	0	272	75	60
	B	2	1	292	75	
4	B	3	0	292	75	60
	B	3	1	305	90	
5	C	2	0	6	15	15
	C	2	1	57	30	

To identify congestion characteristics accurately, the raw traffic space mean speed data obtained from the mesoscopic-wavelet model are separated into space mean speed trend and fluctuations using Coifman wavelets. Similar to the training data, the input data for testing the trained Levenberg-Marquardt BP neural network consists of three 1×240 sequences corresponding to traffic density, space mean speed trend, and space mean speed fluctuations for the duration of simulation. The neural network model outputs the traffic flow phase at every time period in each freeway section.

In freeway sections 10 and 11, no stationary queues are detected when there is no lane closure. However, the presence of a work zone results in a maximum stationary queue length of 26 vehicles in freeway section 10 (Fig. 6.9a) and 96 vehicles in freeway section 11 (Fig. 6.9b) simultaneously. In such a situation, it is concluded that the queue is continuous over the two freeway sections. Additionally, in Fig. 6.9b, a second shorter queue of 27 vehicles is observed in freeway section 11 at the 155th time interval. This shorter queue is due to the second period of high entry traffic flow in Fig. 6.4a (at the 100th time period). This period of high traffic flow is of a shorter duration which results in faster dissipation of congested traffic and, therefore, a shorter queue. The high entry traffic flow is in the entry freeway section (section 1) whereas the queue is formed in

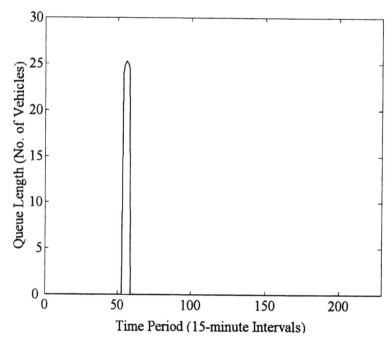

(a)

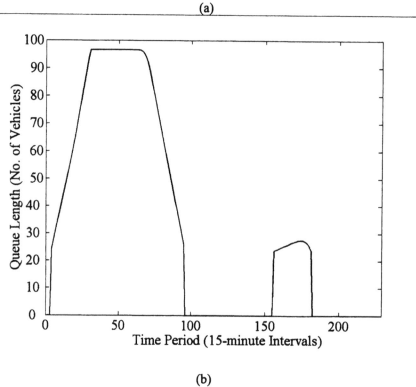

(b)

Figure 6.9 Queue lengths in the presence of the work zone for Example 1 in freeway section (a) 10 and (b) 11

section 11 which leads to a difference in time between the two occurrences. The travel time over the duration of the simulation ($D = 60$ hours) is estimated using the micro-simulation model for the freeway with and without lane closure; their difference is the user delay. The maximum user delay is estimated as 15 minutes when the maximum travel time without a work zone is 15 minutes and with a work zone is 30 minutes. The maximum queue lengths in the entire freeway segment with and without work zone are 122 and 25 vehicles, respectively (Table 6.2).

To increase the accuracy of estimation at the cost of computational performance, the time interval is reduced to 5 minutes. Repeating the entire process yields the same values for user travel time and delay, but the maximum queue lengths in the entire freeway segment with and without a work zone change to 133 and 27 vehicles, respectively (Table 6.3).

Table 6.3 Delay estimation results from the model using 5-minute time intervals

Example	Traffic pattern	Total no. of lanes	No. of lane closures	Maximum no. of vehicles in queue	Maximum travel time (minutes)	Maximum delay (minutes)
1	A	2	0	27	15	15
	A	2	1	133	30	
2	A	3	0	24	15	10
	A	3	1	75	25	
3	B	2	0	276	65	55
	B	2	1	301	70	
4	B	3	0	298	70	45
	B	3	1	311	75	
5	C	2	0	3	10	10
	C	2	1	61	20	

Example 2: Three-Lane Freeway with One Lane Closure for Demand Pattern A

In the case of a freeway with three lanes in either direction with one lane closure due to construction, the maximum travel time with and without a work zone is 30 and 15 minutes, respectively. The maximum user delay in the presence of the work zone is 15 minutes. The maximum queue lengths in the entire freeway segment with and without work zones are 78 and 24 vehicles, respectively (Table 6.2). Repeating the entire process with a time interval of 5 minutes, the maximum travel time with and without a work zone is 25 and 15 minutes, respectively, the maximum user delay in the presence of the work zone is 10 minutes, and the maximum queue lengths in the entire freeway segment with and without a work zone change to 75 and 24 vehicles, respectively (Table 6.3). The results are slightly lower compared with those for Example 1. Although the total number of vehicles accessing the freeway segment increases (since flow values are expressed as number of vehicles per hour per lane), an increased number of lanes allows greater flexibility for changing lanes during free flow, which leads to longer regions of free and transitional flow.

Example 3: Two-Lane Freeway with One Lane Closure for Demand Pattern B

The values generated for the traffic parameters for demand pattern B applied to the same configuration as Example 1 yield significantly different results especially when no work zone is involved. Queue formation is detected in some freeway sections when the demand exceeds the capacity without any specific pattern. However, if a freeway section has a lane closure, the sections downstream of the work zone neither form queues nor reach capacity flows even though they are congested when there are no lane closures. Very long queues spanning multiple freeway sections are generated due to the high demand flow associated with pattern B as noted in Tables 6.2 and 6.3.

Example 4: Three-Lane Freeway with One Lane Closure for Demand Pattern B

This example yields queue length and delay estimation values larger than those for Example 3 because the advantages of extra lanes, described for Example 2, are offset by the increase in traffic demand on the freeway segment (Tables 6.2 and 6.3).

Example 5: Two-Lane Freeway with One Lane Closure for Demand Pattern C

The data obtained from the NCDOT construction project on I-85 are used in this example. The original data consist of vehicle counts at the detector station from which the traffic flow and density are obtained. The space mean speed values are computed using Eq. (6.3). Following the same steps as in the other examples, the queue length and travel delay values are obtained and summarized in Tables 6.2 and 6.3. The queue lengths, travel times, and user delay times in this example are low because the demand pattern C extracted from the NCDOT data has low demand values (the normalized values are mostly below one). In spite of the low demand, a queue is observed in sections of the freeway (Tables 6.2 and 6.3) even without a lane closure because the traffic density used for initialization of the model exceeds the critical density at certain times. It is also noted that stationary queues are not formed when the critical density is exceeded for a very short duration.

6.7   Comparison with QuickZone

The proposed model was compared with QuickZone using all five examples. QuickZone yielded no traffic congestion or delays in any of the five examples. Since demand patterns A and B were specifically created to yield congested traffic flow in specific freeway sections, it is concluded that QuickZone does not yield accurate results for the examples presented in this chapter. This was also verified using simulations from FRESIM, which yielded congested traffic flow in some sections. The low accuracy of QuickZone for the examples presented in this chapter may be attributed to the nature of the underlying queuing theory which does not account for speed reduction due to high traffic density in the uncongested sections. Similar problems with QuickZone have also been reported by other researchers (e.g., Benekohal *et al.* 2003).

6.8   Concluding Remarks

In this chapter, a neural network-wavelet micro-simulation model was presented and applied to a straight freeway with two and three lanes in either direction with one lane closure for construction. The results from various examples show the expected patterns in traffic parameter values due to change in the demand pattern or lane closure configuration. When the traffic demand exceeds the freeway capacity, traffic flow changes from free flow to congested flow, and queue formation starts with the work zone section extending upstream. On the other hand, a phase change from congested flow to free flow occurs, when very low demand results in the dissipation of the stationary queue.

The model described in this chapter demonstrates substantially improved efficiency as the time required for simulation is much less compared to the FRESIM module of the TSIS/CORSIM software package. Additionally, the presented model directly yields values for queue lengths and user delay which can be used to study their impacts on construction planning and management.

The proposed model is particularly suitable for use in real-time traffic management. If the current traffic speed and density in any freeway segment and the demand pattern are known, the model can predict the effects of an immediate lane closure. Alternatively, if seasonal and daily demand patterns are known, the model can be used as a useful project management tool. The model can be applied to entirely different freeway configurations provided that the neural network is trained using field data for those configurations. The simulations in this chapter consider the work zone to be present in only one freeway section. The position and length of the work zone can be varied along the freeway segment to observe changes in traffic behavior and estimate the corresponding delay and queue lengths. Such a study can be used to estimate an optimum work zone length for minimum traffic disruption. The model is general enough to handle traffic parameter changes when entry and exit ramps are present in the freeway. Greater accuracy is achieved for the queue length estimation by incorporating a higher degree of spatial resolution (reducing the freeway section lengths). Similarly, accuracy of the travel time or delay estimation depends on the temporal resolution (time interval) of the model.

Since the model is designed to accept both short-term (with a duration of less than a day) as well as long-term (with a duration of more than a day), it can be used to simulate flow during the occurrence of an incident. The methodology presented in this chapter can be extended to match the generated traffic flow patterns with actual flow patterns to pinpoint the location of the incident and even the occurrence of isolated incidents that cannot be detected easily by other algorithms.

Chapter 7

TRAFFIC DELAY ESTIMATION AND COST OPTIMIZATION IN FREEWAY WORK ZONES

7.1   Introduction

Freeway work zones result in congestion and traffic delays leading to increased driver frustration, increased traffic accidents, and increased road user delay cost. The traffic delay costs to users have been mathematically modeled and evaluated based on simplifying assumptions. Since the freeway work zone segment length has a significant impact on both the agency and user costs, efforts have been made to find the optimum freeway work zone segment length so as to minimize the costs to users and freeway agencies.

McCoy and Mennenga (1998) developed a simple model to find the optimum work zone segment length for minimum work zone costs in a rural four-lane freeway with one lane closure. Based on the average daily traffic (ADT), it takes into account the construction cost, user delay cost, vehicle operating cost, and accident cost. A Microsoft Excel-based model has been developed for predicting the work zone delay, named QuickZone Delay Estimation Program (MITRETEK 2000) based on the deterministic queuing model for each network link in the work zone. The hourly estimation in QuickZone takes into account expected time-of-day utilization and seasonal variation in travel demand. QuickZone, however, does not have any optimization capability for finding the optimum work zone segment length or starting time of the project.

Chien and Schonfeld (2001) present a simplified and useful model for estimating the delay cost using the average daily traffic and finding the optimum work zone segment length in a four-lane freeway with one lane closure. They assume that if the work zone capacity is more than the ADT, no queue is formed. However, since the traffic flow varies within a day, this assumption does not hold at least during part of the day. Furthermore, the starting time of the work zone in a day (work during the day versus evening) and seasonal demand have significant effects on user delays and work zone costs. Chien and Schonfeld (2001), however, have tackled a problem of great practical significance in managing freeway work zones, which is to find the optimum work zone segment length.

The previous chapter presents a neural network-wavelet micro-simulation model to track the travel time of each individual vehicle for traffic delay and queue length estimation at work zones. This chapter presents a macroscopic computational model for estimating traffic delays in freeway work zones based on the flow theory using neural network and optimization techniques.

7.2   Optimization Models for Freeway Work Zones

7.2.1   Assumptions

The macroscopic computational model uses hourly traffic flow and takes into account the following factors: 1) number of lane closures (N_l), 2) length of the work zone segment (l), 3) anticipated hourly traffic flow of the freeway approaching the work zone, 4) starting time of the work zone (time of the day in hours), 5) darkness, 6) seasonal variation in travel demand, and 7) duration of the work zone in hours (D).

The following assumptions are made to formulate the problem:

1) All the vehicles travel at the same speed of V_w through the work zone, and at the same speed of V_a approaching and leaving the work zone.

2) The road user delay cost is represented by an average cost per vehicle hour c_{vh} expressed in dollars per vehicle hour.

3) The anticipated hourly traffic flow approaching the work zone in vehicles per hour (vph) at time t of day (measured in hours), f_t, is known. An intersection close to the work zone or a residential street in an urban area creates traffic diversion and affects the anticipated hourly traffic flow approaching the work zone. The model includes the effect of an intersection indirectly as long as the anticipated hourly traffic flow includes this effect as a percentage of diverted traffic.

4) The freeway work zone capacity, c_w, is assumed to be constant for any given number of lane closures. Also, the freeway capacity outside the work zone, c_0, is assumed to be constant.

5) The agency or maintenance cost (C_M) for maintaining a work zone segment is a linear function of the work zone segment length (l_w) and is expressed in the following form:

$$C_M = c_1 + (N_l c_2)l_w \tag{7.1}$$

where c_1 represents the fixed cost independent of the work zone segment length and c_2 represents the average additional maintenance cost per work zone kilometer per lane.

The time period required to complete the maintenance for the work zone, D_w, is a linear function of the work zone segment length and is expressed in the following form:

$$D_w = d_1 + (N_l d_2)l_w \tag{7.2}$$

where d_1 represents the setup time independent of work zone segment length and d_2 represents the additional maintenance time per work zone kilometer per lane.

The cost and time linearity assumptions 5 and 6 are also made by Chien and Schonfeld (2001). However, in this work we have included an additional parameter, that is, the number of lane closures in the formulations.

7.2.2 Traffic Delay Model

A deterministic delay method is developed to estimate the number of vehicles per hour in a queue. The user delay time consists of the queue delay time upstream of the work zone (t_q) and the moving delay time through the work zone (t_m). The total user delay time, t_d, during the duration, D_w, of the construction at the work zone is

$$t_d = t_q + t_m \tag{7.3}$$

It should be pointed out that these quantities are computed for all the road users and therefore expressed in terms of vehicle hours.

Within a specific time period Δt (in hours), if the anticipated hourly traffic flow approaching the work zone ($\alpha_s f_{\Delta t}$) exceeds the work zone capacity (c_w), a queue forms. That is, a queue forms when $\alpha_s f_{\Delta t} > c_w$, where α_s is the seasonal demand factor used to adjust the short-term traffic flow for seasonal variations. For example, the Ohio Department of Transportation (ODOT) specifies a value for each day of the week and for every month of the year depending on a classification of highways. For annual average daily traffic (AADT) used for the whole year, the seasonal demand factor is equal to one. For various days of different months, ODOT specifies a value in the range of 0.76 and

1.72 (http://www.dot.state.oh.us/ techservsite). When the real-time traffic flow measurement is used, $\alpha_s = 1.0$. The number of vehicles in a queue within the specific period Δt, $Q_{\Delta t}$, is equal to

$$Q_{\Delta t} = \alpha_s f_{\Delta t} - c_w \qquad (7.4)$$

and the cumulative number of vehicles $T_{t+\Delta t}$ in a queue at time $t + \Delta t$ is

$$T_{t+\Delta t} = \sum_{t=t_i}^{t+\Delta t} Q_{\Delta t} \qquad (7.5)$$

where t_i represents the starting time at the work zone in hours ranging from 1 to 24.

When $\alpha_s f_{\Delta t} < c_w$, the queue delay time is zero and the existing queue starts to disappear. In that case

$$Q_{\Delta t} = 0 \qquad (7.6)$$

and

$$T_{t+\Delta t} = \max\{T_t - s, 0\} \qquad (7.7)$$

where s represents the queue reduction. This parameter is formulated differently depending on whether the work zone has a long duration (more than one day) or a short duration (less than one day).

When the work zone duration is long-term (defined as work zones with duration of more than one day), the queue reduction factor is (Fig. 7.1a)

$$s = c_w - \alpha_s f_{\Delta t} \qquad (7.8)$$

When the work zone duration is short-term (Fig. 7.1b), the queue reduction factor is

$$s = c_0 - \alpha_s f_{\Delta t} \qquad (7.9)$$

where c_0 represents the freeway capacity in the absence of any work zone.

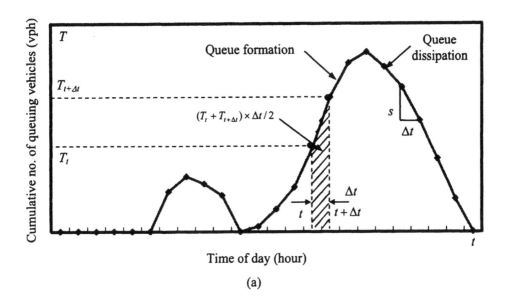

(a)

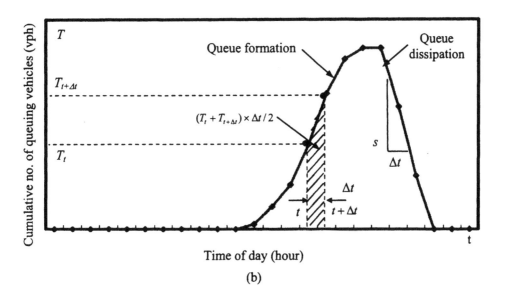

(b)

Figure 7.1 Mathematical model for work zone delay: (a) Long-term work zone delay, and (b) short-term work zone delay

The queue delay time, t_q, over the work zone duration, D_w, is obtained as

$$t_q = \sum_{t=t_i}^{t_i+D_w-1}[(T_t + T_{t+\Delta t})\Delta t / 2] \tag{7.10}$$

The shaded area in Figs. 7.1a and 7.1b represents the value of the function inside the parentheses in Eq. (7.10). The total area under all *queue waves* during the work zone duration represents the work zone queue delay, t_q. A queue wave is the hill-shaped curve representing the variation of the cumulative number of queuing vehicles over time from the start of the formation of one queue to the total dissipation of that queue. For example, there are two queue waves in Fig. 7.1a and only one in Fig. 7.1b. The queue reduction factor over the time period Δt, s, is also shown in Figs. 7.1a and 7.1b.

The moving delay time, t_m, is expressed as a function of the difference between the travel times on a freeway with and without a work zone. Within a given period Δt, if the anticipated hourly traffic flow approaching the work zone exceeds the work zone capacity ($\alpha_s f_{\Delta t} > c_w$), the maximum traffic flow through the work zone is c_w. Then, the moving delay time, Δt_m, over the given period, Δt, is expressed as

$$\Delta t_m = (1/V_w - 1/V_a)c_w \Delta t \tag{7.11}$$

When the anticipated hourly traffic flow approaching the work zone is less than the work zone capacity ($\alpha_s f_{\Delta t} < c_w$), the moving delay time Δt_m over the given period Δt, becomes

$$\Delta t_m = (1/V_w - 1/V_a)(\alpha_s f_{\Delta t})\Delta t \tag{7.12}$$

Thus, the total moving delay time during the work zone duration is

$$t_m = \sum_{t=t_i}^{t_i+D_w-1}\Delta t_m \tag{7.13}$$

where Δt_m is substituted from Eq. (7.11) or (7.12) depending on whether within any given time period the anticipated hourly traffic flow approaching the work zone exceeds the work zone capacity.

7.2.3   Cost Optimization Model

The freeway work zone cost is defined as the sum of three components: the user delay cost (C_d), the accident cost (C_a), and the work zone maintenance cost including the setup and removal cost (C_m)

$$C_w = C_d + C_a + C_m \tag{7.14}$$

All these three components are defined in dollars per length (kilometer) of the work zone.

The user delay cost per work zone kilometer per lane is the total user delay time, t_d, multiplied by the average cost per vehicle hour c_{vh} divided by the work zone segment length l_w and the number of lane closures in the work zone, N_l:

$$C_d = (c_{vh}t_d)/l_w N_l \tag{7.15}$$

The traffic accidents considered in this chapter are those occurring in the work zone and queue areas. The accident cost C_a per work zone kilometer per lane incurred by the traffic flow passing through the work zone is determined from the number of accidents, n_a, per 100 million vehicle hours, multiplied by the product of the increased delay, t_d, and the average cost per accident, c_a, divided by the work zone segment length and the number of lane closures:

$$C_a = (\alpha_n n_a c_a t_d)/(10^8 l_w N_l) \tag{7.16}$$

In this equation, α_n is a factor to take into account the effect of darkness and working at night. Increasingly more work is performed at night to ameliorate the impact of construction on road users and reduce traffic disruptions. On the other hand, the evening construction results in reduced worker productivity at the work zone, increased construction costs for utilities and labor fees, and increased risk of traffic accidents. A darkness factor of greater than one $(\alpha_n > 1.0)$ is used for construction work at night. Its value is determined based on the previous experience as well as the management plan in practical applications. The accident cost used in our formulation and represented by Eq. (7.16) includes two new factors not considered in previous research (McCoy and Mennega 1998; Chien and Schonfeld 2001): number of lane closures and darkness factor.

The maintenance cost in the work zone includes the setup and removal costs for the work zone and the average construction cost per work zone kilometer per lane, as noted in Eq. (7.1). The average maintenance cost C_m per work zone kilometer per lane is the total maintenance cost, C_M (defined by Eq. 7.1), divided by the work zone segment length and the number of lane closures modified by the darkness factor α_n:

$$C_m = \alpha_n c_1/(l_w N_l) + \alpha_n c_2 \qquad (7.17)$$

Substituting Eqs. (7.15) to (7.17) into Eq. (7.14) yields the work zone cost function C_W per work zone kilometer per lane.

Thus, the freeway work zone cost optimization model is expressed as follows: Minimize

$$C_w = C_d + C_a + C_m = [\alpha_n c_1 + (\alpha_n n_a c_a + 10^8 c_{vh})t_d/10^8]/(l_w N_l) + \alpha_n c_2 \qquad (7.18)$$

Subject to the following constraints:

$$t_d \geq 0 \qquad (7.19)$$

$$l_w \geq l_{min} \qquad (7.20)$$

where t_d is the total user delay time as expressed by

$$t_d = t_q + t_m = \sum_{t=t_i}^{t_i+D_w-1}[(T_t + T_{t+\Delta t})\Delta t/2] + \sum_{t=t_i}^{t_i+D_w-1}\Delta t_m = \sum_{t=t_i}^{t_i+D_w-1}[(T_t + T_{t+\Delta t})\Delta t/2 + \Delta t_m] \qquad (7.21)$$

and l_{min} is the minimum work zone segment length based on practical considerations. For example, a minimum work zone segment length of 0.1 km is chosen in the examples presented in this chapter.

There are two variables in the optimization formulation presented in this section: the work zone segment length (l_w), a real variable, and the starting time of the work zone in hours (t_i), an integer variable. The computational model presented in this section is general and can be used for both short-term work zones (with duration of less than one day) and long-term work zones (with duration of more than one day). In the following sections, we present an approach for solving this mixed real variable-integer nonlinear programming problem for short-term work zones.

7.2.4 Work Zone Cost Function

For practical reasons, the work zone segment length is chosen in a pre-selected increment of β kilometers (or miles), for example, $\beta = 0.05$ km or 0.1 km. The starting time of a short-term work zone can take an integer value between 1 and 24 for the twenty-four hours of a day. The maximum work zone segment length for short-term work zones is obtained from Eq. (7.2) by using the maximum work zone duration of 24 hours for the work duration, D_w:

$$l_{max} = (24 - d_1)/(d_2 N_l) \tag{7.22}$$

If the work zone segment length increment of $\beta = 0.05$ km is chosen, the number of possible work zone segment lengths becomes

$$n = (l_{max} - l_{min})/\beta = (l_{max} - l_{min})/0.05 \tag{7.23}$$

In the freeway work zone cost formulation presented in this chapter, the starting time of work zone, t_i, affects the total work zone cost. The user delays in the work zone vary depending on the starting time of the work zone. For short-term work zones, for any given work zone segment length l_i, there are 24 possible starting times for the construction work, corresponding to the 24 hours in a day, and the total work zone cost is obtained by substituting Eq. (7.21) into Eq. (7.18):

$$
\begin{aligned}
C_w &= \frac{1}{l_i N_l} \left\{ \alpha_n c_1 + \frac{\alpha_n n_a c_a + 10^8 c_{vh}}{10^8} \sum_{t=t_i}^{t_i + D_w - 1} [(T_t + T_{t+\Delta t})/2 + \Delta t_m] \right\} + \alpha_n c_2 \\
&= \frac{\alpha_n n_a c_a + 10^8 c_{vh}}{10^8 l_i N_l} \sum_{t=t_i}^{t_i + D_w - 1} [(T_t + T_{t+\Delta t})/2 + \Delta t_m] + [\alpha_n c_2 + \alpha_n c_1/(l_i N_l)]
\end{aligned}
\tag{7.24}
$$

7.3 Boltzmann Neural Network with Simulated Annealing for Cost Optimization

A combined Boltzmann neural network-simulated annealing algorithm is developed to solve the mixed real variable-integer cost optimization problem for short-term work zones. The goal is to find the global optimum solution for the work zone segment length and starting time. For a description of Boltzmann neural network and simulated annealing, see Section 2.4.

The architecture of the Boltzmann-simulated annealing neural network for solving the freeway work zone cost optimization problem is presented in Fig. 7.2. The network consists of three layers: input layer, hidden layer, and output layer. Unlike the conventional Botlzmann machine, the neural network created in this work has a set of *storage* nodes in addition to the standard Boltzmann network nodes. The inputs to the neural network are the hourly traffic flows approaching the work zone (f_i, i = 1, 24). These values are used to calculate the 24 vectors of work zone information $C_I(i,n)$ (i =1, 24) assigned to the storage nodes. Each vector contains the values for nine quantities: the work zone segment length (l_w), starting time (t_i), queue delay time (t_q), moving delay time (t_m), user delay time (t_d), user delay cost (C_d), accident cost (C_a), maintenance cost (C_m), and the total work zone cost (C_w). The Boltzmann network nodes in the input layer do not change in the process of training, and therefore are all assigned a value of one (i.e., x_i = 1, i = 1, 24).

The number of nodes in the hidden layer is equal to the number of possible work zone segment lengths, n, as determined by Eq. (7.23). Similar to the input layer, the local minimum work zone information for any given work zone segment length is stored in the vector $C_H(j)$ (j = 1, n), which contains the values for the same nine quantities mentioned in the previous paragraph. Again, the Boltzmann network nodes in the hidden layer do not change in the process of training, and are randomly assigned values of x_j = -1 or +1 (j = 1, n) in order to be different from the nodes in the input layer.

For training the Boltzmann network, we define an energy function in the following form:

$$E = \sum_{j=1}^{n} w_{i,j} x_i x_j \qquad (7.25)$$

where w_{ij} represents the weight of the link connecting the Boltzmann input node i to node j in the hidden layer. It is defined as

$$w_{ij} = (C_w)_{ij} \bigg/ \sqrt{\sum_{i=1}^{24} (C_w)_{ij}^2} \ , i = 1, 24, j = 1, n \qquad (7.26)$$

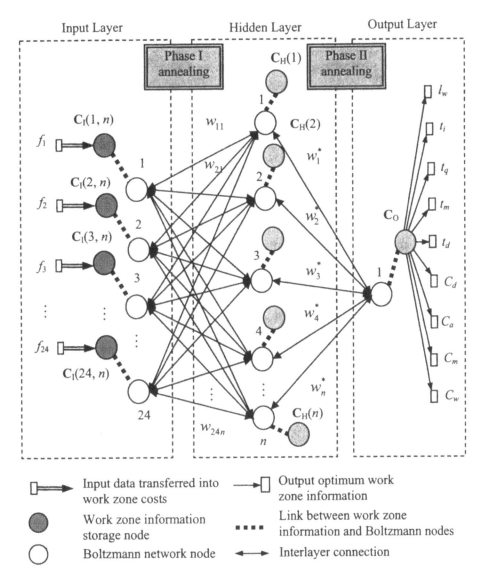

f_i = Hourly traffic flow approaching work zone at time i (i = 1, 24); $C_I(i, n)$ = vector to store the work zone information in the input layer (i = 1, 24); $C_H(j)$ = vector to store the local minimum work zone information in the hidden layer (j = 1, n); C_o = vector to store the optimum work zone information in the output layer; w_{ij} = weight of the link connecting the Boltzmann input node i to node j in the hidden layer; w_j^* = weight of the link connecting the Boltzmann hidden node j to the output node; n = number of possible work zone segment lengths

Figure 7.2 Architecture of the Boltzmann-simulated annealing neural network for short-term freeway work zone cost optimization problem

in which $(C_w)_{ij}$ represents the total work zone cost at the input node i corresponding to the jth work zone segment length.

To find the global optimal solution for the work zone cost optimization problem we need to find the global minimum solution for the energy function defined by Eq. (7.25). This is achieved by using the simulated annealing in two phases (Fig. 7.2). In the first phase, simulated annealing is applied between the input layer and the hidden layer yielding the local minimum total work zone cost solutions corresponding to various work zone segment lengths. In the second phase, simulated annealing is used between the hidden layer and the output layer to obtain the global optimum solution for the work zone cost optimization problem.

The flow diagram for the Boltzmann-simulated annealing algorithm for the work zone cost optimization problem is shown in Fig. 7.3. In phase one of the simulated annealing, the energy function (Eq. 7.25) is initially evaluated by summing over all hidden nodes but using the values of the weights associated with only one randomly selected input, i, for any hidden node, j. One input node x_i and one hidden node x_j are selected randomly and the value of the selected hidden node x_k is changed from -1 to 1 or from 1 to -1. Because other hidden nodes j ($j \neq k$) are not selected for updating the energy function at this step, the resulting change of energy becomes

$$\Delta E = w_{l,k} x_l x_k (t+1) - w_{i,k} x_i x_k (t) = w_{l,k} x_k (t+1) - w_{i,k} x_k (t) \tag{7.27}$$

where $x_k(t+1)$ and $x_k(t)$ represent the values of the selected k^{th} hidden node in the new and last steps, respectively, and x_l is the selected l^{th} input node. If the energy change, ΔE, is negative the change is accepted and the weight of the link connecting the selected k^{th} Boltzmann hidden node to the node in the output layer, w_k^*, is set to the weight of the link connecting the selected input node i to the selected hidden node k, w_{ik}:

$$w_k^* = w_{ik} \tag{7.28}$$

If the energy change, ΔE, is positive the change is accepted with a probability of

$$p = 1/(1 + e^{-\Delta E / \tau_w}) \tag{7.29}$$

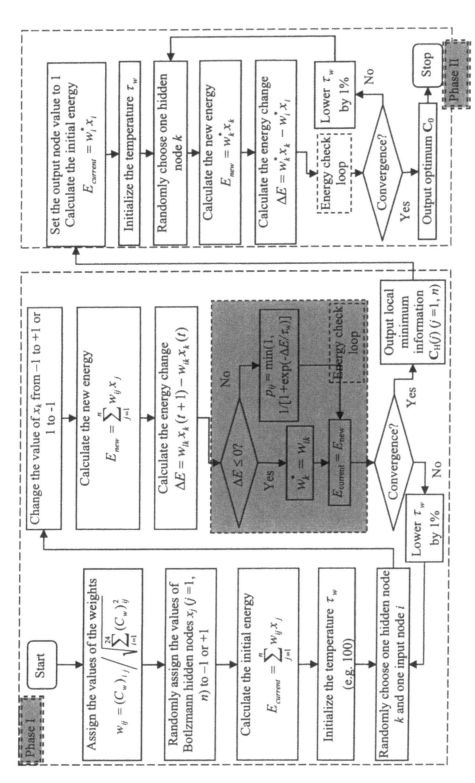

Figure 7.3 Flow chart of hybrid Boltzmann neural network-simulated annealing model used to solve the mixed real-integer nonlinear programming problem

where the parameter, τ_w, is the so-called temperature parameter in the simulated annealing algorithm (see Section 2.4). The initial temperature is set to some high value (e.g., 100°C in this work) and is reduced in subsequent iterations by a certain percentage of the previous value (e.g., 1% in this work). A local optimum work zone cost solution is found when the system reaches an equilibrium point at a temperature τ_w when the probability p approaches one. This solution is represented as the weights of the links connecting the nodes in the hidden layer and the output node.

In phase two of the simulated annealing, a similar process is performed between the hidden layer and the output layer yielding the global minimum work zone cost solution with the corresponding global optimum work zone segment length and starting time.

7.4   Numerical Examples

7.4.1   Example 1: Four-Lane Freeway

The data for this example, summarized in Table 7.1, are chosen to be the same as those of Chien and Schonfeld (2001) for the sake of comparison, with the addition of values for darkness and seasonal demand factors. This example is a four-lane freeway with one-lane closure. Chien and Schonfeld (2001) use the average daily traffic only. In this work, the anticipated hourly traffic flow approaching the work zone is used.

Example 1A (ADT = 1000 vph)

The anticipated hourly traffic flows approaching the work zone for this example for the duration of one day are given in Table 7.2, with an average daily traffic (ADT) of 1000 vph, the same value used in Chien and Schonfeld (2001). The hourly and cumulative numbers of vehicles in a queue calculated by the proposed computational model are presented in Table 7.2. The cumulative number of queuing vehicles as a function of the time of day is also displayed graphically in Fig. 7.4. Two queue waves are observed in this figure.

Chien and Schonfeld (2001) report an optimum work zone segment length of $l_w = 1.4$ km (corresponding to minimum work zone cost) and duration of $D_w = 10.4$ hours for the example data presented in Table 7.1 with the ADT of 1,000 vph. Using the same work zone length of 1.4 km and duration of 10.4 hours, the traffic delay estimation model presented in this chapter yields a maximum queue delay time (t_q) of 18,624 vehicle-hours and maximum moving delay time (t_m) of 149 vehicle-hours when the maintenance work is started at 8 A.M. In contrast, when the maintenance work is started at hour 19 (7 P.M.) in the evening, the queue delay is zero and the moving delay time is reduced to 56 vehicle-hours. Chien and Schonfeld (2001) report a queue delay time of 0 and moving delay time of 141.5 vehicle-hours (Table 7.3). The current investigation indicates that the starting time of the work zone affects the user delay time significantly. This factor is absent in the recently published work zone delay estimate models but is taken into account in the proposed model.

Table 7.1 Input data for Example 1 (chosen to be the same as those of Chien and Schonfeld, 2001, for the sake of comparison with the addition of values for darkness and seasonal demand factors)

Var	Description	Values
c_0	Freeway capacity in the absence of the work zone	2,600 vph
c_w	Work zone capacity	1,200 vph
V_a	Average approaching speed	88.0 km/h
V_w	Average work zone speed	48.0 km/h
n_a	Number of accidents per 100 million vehicle hour	40 acc/100mvh
c_a	Average accident cost	142,000 $/acc
c_{vh}	Average vehicle delay cost per hour	12.0 $/vph
c_1	Fixed set up cost	1,000 $/zone
c_2	Average maintenance cost per work zone kilometer per lane	80,000 $/km
d_1	Fixed setup time	2 h/zone
d_2	Average maintenance time per kilometer	6 h/km
N_L	Number of lane closures in the work zone	1
N_o	Number of open lanes in the work zone	1
α_n	Darkness factor (cost increase ratio for night work)	2.0
α_s	Seasonal demand factor	1.0

Assuming that the work zone duration is less than one day (short-term work zone), a maximum work zone segment length of $l_{max} = 3.65$ km is obtained from Eq. (7.26). The maximum number of possible work zone segment lengths from Eq. (7.27) is then $n = 72$. A darkness factor of $\alpha_n = 2.0$ is assumed for all the examples in this chapter. Figure 7.5 shows the variation of the work zone costs versus the work zone segment length. It should be pointed out that the data for each given work zone segment length correspond

Table 7.2 Queue delay results obtained from the computational model for Example 1 (four-lane freeway with one lane closure)

Time (Hour of day)	Anticipated traffic flow approaching the work zone, f_t (vph)		Number of vehicles in the queue per hour, Q (vph)		Cumulative number of vehicles in the queue per hour, T (vph)	
	Example 1A	Example 1B	Example 1A	Example 1B	Example 1A	Example 1B
1	180	360	0	0	0	0
2	50	100	0	0	0	0
3	117	234	0	0	0	0
4	420	840	0	0	0	0
5	833	1,681	0	481	0	481
6	1,145	2,290	0	1,090	0	1,571
7	2,161	4,322	961	3,122	961	4,693
8	821	1,642	0	442	582	5,135
9	1,020	2,075	0	875	402	6,010
10	930	1,660	0	460	132	6,470
11	910	1,831	0	631	0	7,101
12	1,320	2,651	120	1,451	120	8,552
13	1,620	3,242	420	2,042	540	10,594
14	1,728	3,456	528	2,256	1,068	12,850
15	2,154	4,325	954	3,125	2,022	15,975
16	2,420	4,840	1,220	3,640	3,242	19,615
17	2,021	4,142	821	2,942	4,063	22,557
18	1,460	2,920	260	1,720	4,323	24,277
19	850	1,700	0	500	3,973	24,777
20	700	1,425	0	225	3,473	25,002
21	400	800	0	0	2,673	24,602
22	280	560	0	0	1,753	23,962
23	240	480	0	0	793	23,242
24	210	420	0	0	0	22,462

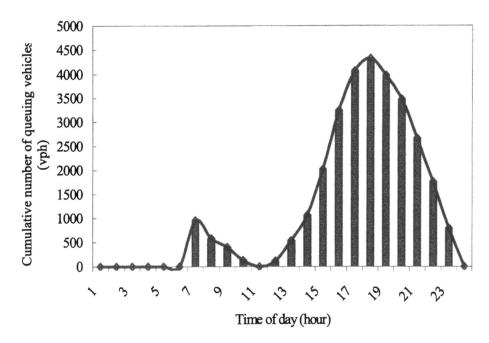

Figure 7.4 Cumulative numbers of queuing vehicles for Example 1A
(four-lane freeway with one lane closure, ADT = 1,000 vph)

Table 7.3 Traffic delay estimate results (unit: vehicle-hours)

Work zone traffic delay model	Example 1A ADT = 1,000 vph, l_w = 1.4 km			Example 1B ADT = 2,000 vph, l_w = 0.34 km		
	Chien & Schonfeld (2001)	Proposed model		Chien & Schonfeld (2001)	Proposed model	
		Max	Min		Max	Min
Queue delay	0	18,624	0	44,804	38,579	0
Moving delay	141.5	149	56	66.1	35	4

to a local minimum solution for that particular work zone segment length for various starting times. The optimum work zone segment length corresponding to the minimum total work zone cost is the global optimum solution. In this example, it is found that the maintenance cost is a significant factor in the total work zone cost. The global optimum starting time is found to be 8 A.M. and the global optimum work zone segment length is 0.35 km resulting in the global minimum total work zone cost of 83,147.55 $/km. Figure 7.6 shows the variation of work zone costs versus the starting time of the

128 Intelligent Infrastructure

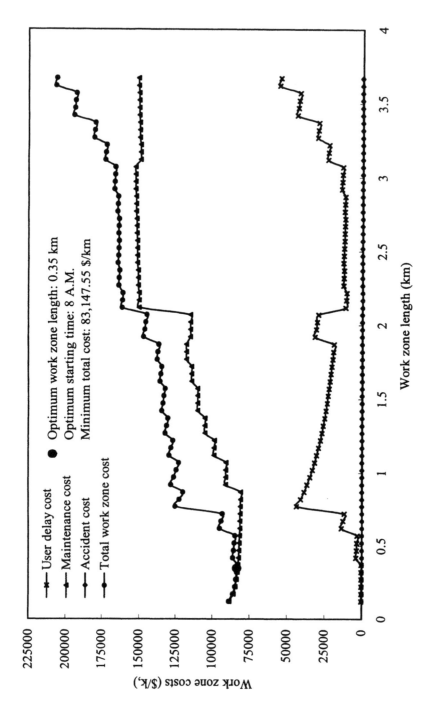

Figure 7.5 Variation of work zone costs versus work zone segment length for Example 1A (four-lane freeway with one lane closure, ADT = 1,000 vph)

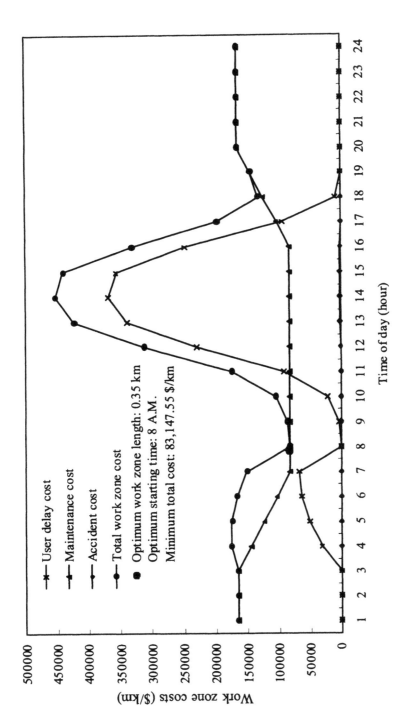

Figure 7.6 Variation of work zone costs versus starting time of the day for global optimum solution of 0.35 km for work zone segment length and 8 A.M. for starting time for Example 1A (four-lane freeway with one lane closure, ADT = 1,000 vph)

day for the global optimum solution. This figure demonstrates that selection of the starting time using the same work zone segment length of 0.35 km has a significant impact on the total work zone cost.

The work zone traffic delay estimation model presented in this chapter also allows one to choose the starting time of the work zone. This is a desirable feature as the optimum starting time provided by the model may not be acceptable for non-economical reasons. For instance, assume a starting time of 9 A.M. is selected for this example. Figure 7.7 shows the variation of work zone costs versus the work zone segment length. The model yields an optimum work zone segment length of 0.20 km resulting in minimum total construction cost of 84,941.92 \$/km. This cost is about 2% higher than the global optimum solution presented in Fig. 7.5. The proposed model can be used as an intelligent decision support system to quickly study the relation between the total work zone cost versus the work zone segment length and starting time.

Example 1B (ADT=2000 vph)

This example uses the same data as Example 1A with the exception of ADT = 2,000 vph. The anticipated hourly traffic flows approaching the work zone in a day with an ADT of 2,000 vph as well as the queue delay results obtained from the computational model are presented in Table 7.2.

Initially, for the sake of comparison, we use the same work zone segment length of 0.34 km and duration of 4 hours given in Chien and Schonfeld (2001). The proposed work zone traffic delay estimation model yields a maximum queue delay time t_q of 38,579 vehicle-hours and maximum moving delay time t_m of 35 vehicle-hours with a staring time of 3:00 A.M. However, if the maintenance work is performed at midnight the model yields the minimum queue delay of zero and the minimum moving delay time t_m of 4 vehicle-hours. Chien and Schonfeld (2001) report a queue delay of 44,804 vehicle-hours and a moving delay of 66.1 vehicle hours (Table 7.3) without considering the effects of the starting time of the work zone.

The new model yields a global optimum value of 1.05 km for the work zone segment length and a global optimum starting time of 21 o'clock (9:00 P.M.), resulting in a global minimum work zone cost of 162,310.22 \$/km for a work duration of 8 hours.

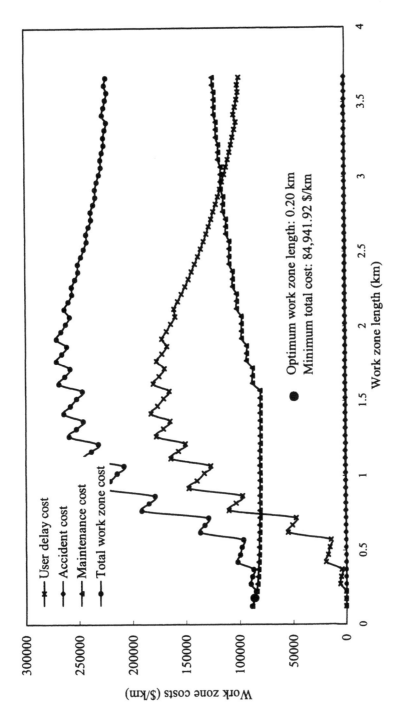

Figure 7.7 Variation of work zone costs versus work zone segment length at starting time 9 A.M. for Example 1A (four-lane freeway with one lane closure, ADT = 1,000 vph)

7.4.2   Example 2: Six-Lane Freeway

This example is created in this work to demonstrate the capability of the proposed work zone traffic delay estimation model to account for the number of lane closures. A six-lane freeway with three lanes in each direction is considered. Data in Table 7.1 are used in this example with the exception of values for freeway capacity (i.e., the capacity in the absence of a work zone) (a value of 5,400 vph is used in this example), work zone capacity, number of open lanes, and number of lane closures. The anticipated hourly traffic flows approaching the work zone in a day are presented in Table 7.4.

Example 2A: One-lane Closure

This example has only one lane closure. The work zone capacity is assumed to be 2,980 vph per the *Highway Capacity Manual* (HCM 1985). The hourly and cumulative numbers of vehicles in a queue calculated by the new computational model are presented in Table 7.4. Similar to Example 1, in this example the maintenance cost is a significant factor in the total work zone cost. The global optimum work zone segment length is 1.05 km and the global optimum starting time is 7 A.M. resulting in the global minimum total work zone cost of 85,107.26 $/km and duration of 8 hours. For one-lane closure, having the work done during the day is more economical than having it done during the night.

Example 2B: Two-lane Closures

This example has two lane closures. The work zone capacity is assumed to be 1,170 vph per the *Highway Capacity Manual* (HCM 1985). The hourly and cumulative numbers of vehicles in a queue calculated by the new computational model are presented in Table 7.4. Figure 7.8 shows the variation of the work zone costs versus the work zone segment length. The global optimum work zone segment length is 0.55 km and the global optimum starting time is 3:00 A.M. resulting in a global minimum total work zone cost of 15,900.70 $/km and duration of 5 hours. In this example, the user delay cost becomes the dominant work zone cost for work zone segment lengths of greater than about 5.7 km. For two-lane closure, having the work done during the night is more economical than having it done during the day. But, compared with Example 2A the cost is increased substantially because the darkness increases the maintenance cost, resulting in a considerable increase in the total work zone cost.

By comparing the results obtained for Examples 2A and 2B, it is concluded that having the work done during the day with a starting time of 7:00 A.M. with one lane closure is the most economical solution for the work zone project at hand.

Table 7.4 Queue delay results obtained from new computational model for Example 2 (six-lane freeway with one-lane closure or two-lane closures)

Time (Hour of day)	Anticipated traffic flow approaching the work zone, f_t (vph)	Number of vehicles in the queue per hour, Q (vph)		Cumulative number of vehicles in the queue per hour, T (vph)	
		Example 2A (one-lane closure)	Example 2B (two-lane closures)	Example 2A (one-lane closure)	Example 2B (two-lane closures)
1	682	0	0	0	0
2	431	0	0	0	0
3	304	0	0	0	0
4	323	0	0	0	0
5	312	0	0	0	0
6	580	0	0	0	0
7	1,934	0	764	0	764
8	2,986	6	1,816	6	2,580
9	2,666	0	1,496	0	4,076
10	3,067	87	1,897	87	5,973
11	2,681	0	1,511	0	7,484
12	3,035	55	1,865	55	9,349
13	2,887	0	1,717	0	11,066
14	2,761	0	1,591	0	12,657
15	3,133	153	1,963	153	14,620
16	3,503	523	2,333	676	16,953
17	3,586	606	2,416	1,282	19,369
18	4,027	1,047	2,857	2,329	22,226
19	2,609	0	1,439	1,958	23,665
20	1,895	0	725	873	24,390
21	1,591	0	421	0	24,811
22	1,492	0	322	0	25,133
23	1,423	0	253	0	25,386
24	833	0	0	0	25,049

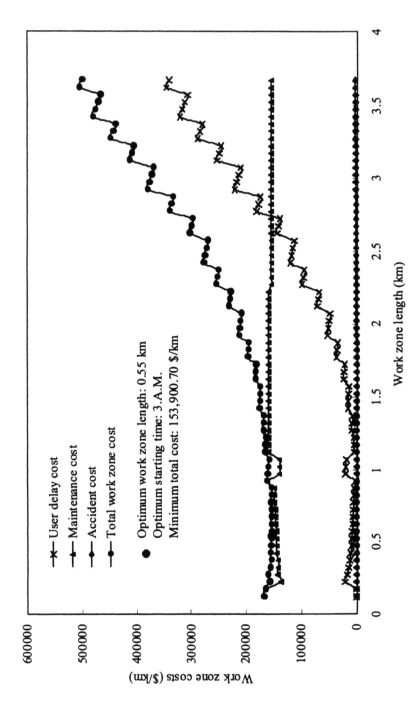

Figure 7.8 Variation of work zone costs versus work zone segment length for Example 2B (six-lane freeway with two-lane closures)

7.4.3   Example 3: Four-Lane Highway in North Carolina

In this example, actual traffic data measured in a work zone in a four-lane highway in the state of North Carolina with one lane closure are used. The hourly traffic flows approaching a work zone on route NC 147, 0.1 miles south of SR 1171, measured by the North Carolina Department of Transportation in a day (August 28, 2000) are presented in Table 7.5. Data in Table 7.1 are also used in this example except for values of freeway capacity in the absence of a work zone (a value of 2,400 vph is used) and work zone capacity (a value of 1,000 vph is used). The hourly and cumulative numbers of vehicles in a queue calculated by the new computational model are also presented in Table 7.5. The new model yields a global optimum value of 0.20 km for the work zone segment length and a global optimum starting time of 7 A.M., resulting in a global minimum work zone cost of 87,954.99 $/km for a work duration of 3 hours. Figure 7.9 shows the variation of work zone costs versus the starting time of the day.

7.5   Concluding Remarks

A new freeway work zone traffic delay estimation and total work zone cost optimization model was presented in this chapter. In contrast to the previous published works that are based on the average daily traffic flow, the new model is based on average hourly traffic flow. A total work zone cost function is defined as the sum of user delay, accident, and maintenance costs. The model takes into account the number of lane closures, the darkness factor, and the seasonal demand factor. The work zone traffic delay estimation and cost optimization model is applicable for both short-term (less than one day) and long-term (more than one day) work zones. The model yields the global optimum values for the work zone segment length and the starting time of the work zone. A Boltzmann-simulated annealing neural network model is developed to solve the resulting mixed real variable-integer short-term work zone cost optimization problem.

Numerical examples demonstrate that the starting time of the work zone has a significant impact on queue formation and total work zone cost. Thus, the proposed model based on the average hourly traffic flow allows the work zone traffic engineer to prepare a more effective traffic control plan for a given work zone based on detailed and accurate quantitative information in a systematic manner, resulting in substantial cost savings and minimum disruption of traffic for the traveling public. Using the proposed model, the work zone traffic engineer will be able to find the answer to important what-if questions, such as one-lane closure versus two-lane closure or selection of the starting time of the day systematically and quickly. The examples presented show how the transportation work zone engineer can observe the impact of the number of lane closures and the darkness. The model also incorporates a seasonal demand factor. That means the work zone engineer can use the model to find out the impact of seasonal demand on the user delay and total work zone costs.

Table 7.5 Queue delay results obtained from new computational model for Example 3

Time (hour of day)	Anticipated traffic flow approaching the work zone, f_t (vph)	Number of vehicles in the queue per hour, Q (vph)	Cumulative number of vehicles in the queue per hour, T (vph)
1	137	0	0
2	76	0	0
3	29	0	0
4	42	0	0
5	45	0	0
6	198	0	0
7	660	0	0
8	1,055	55	55
9	784	0	55
10	1,335	335	390
11	1,144	144	534
12	1,366	366	900
13	1,326	326	1,226
14	1,238	238	1,464
15	1,109	109	1,573
16	1,167	167	1,740
17	1,321	321	2,061
18	1,535	535	2,596
19	975	0	2,571
20	639	0	2,210
21	420	0	1,630
22	389	0	1,019
23	280	0	299
24	320	0	0

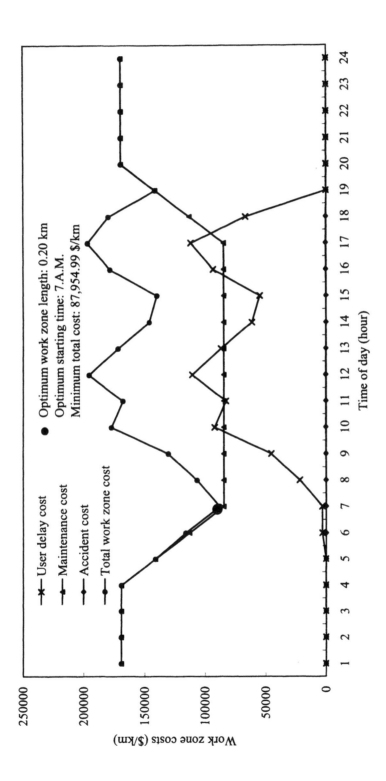

Figure 7.9 Variation of work zone costs versus starting time of the day for a global optimum solution of 0.20 km for work zone segment length and 7 A.M. for starting time for Example 3

Chapter 8

NEURO-FUZZY LOGIC MODEL FOR WORK ZONE CAPACITY ESTIMATION

8.1   Introduction

The work zone capacity in freeways is usually defined as the mean queue discharge flow rate at a freeway work zone *bottleneck* (any constricted location that restricts the flow of vehicles in a work zone) (HCM 2000). Work zone capacity has a significant impact on the congestion and traffic queue delays which result in increased driver frustration, increased traffic accidents, increased road user delay cost, and increased fuel consumption and vehicle emissions. Highway agencies often use the empirical and highly approximate method described in the *Highway Capacity Manual* (HCM) (HCM 2000) to determine the freeway work zone capacity with lane closures. The HCM provides a base capacity of 1,600 vehicles per hour per lane (vphpl) for short-term ideal unrestricted highway work zones. Guidelines are given on how to modify the base value to take into account percentage of trucks, work intensity, proximity of ramps, and lane widths. However, a large number of additional factors affect the freeway work zone capacity estimation which are neglected in the HCM guidelines.

The earlier field measurements and investigations can be traced to the work done at the Texas Transportation Institute (TTI) from the late 1970s to the mid-1980s. TTI's work provides the basis for the empirical freeway work zone capacity estimation guidelines included in the previous *Highway Capacity Manual* (HCM 1985). Recently, a few studies have been conducted for estimation of the work zone capacity based on measured field data (Krammes and Lopez 1994; Dixon and Hummer 1995; Dixon *et al.* 1997; Jiang 1999b; Al-Kaisy *et al.* 2000; Al-Kaisy and Hall 2001; Kim *et al.* 2001).

The work zone capacity cannot be described by any mathematical function because it is a complicated function of a large number of interacting variables. Karim and Adeli (2003a) present an adaptive computational model for estimating the work zone capacity and queue length and delay, taking into account the following factors: number of lanes, number of open lanes, work zone layout, work zone length, lane width, percentage of trucks, grade, speed, work intensity, darkness factor, and proximity of ramps. The model integrates judiciously the mathematical rigor of traffic flow theory with the adaptability of neural network analysis. In this chapter, a novel adaptive neuro-fuzzy logic model is presented for estimating the freeway work zone capacity. Seventeen different factors impacting the work zone capacity are included in the model. A neural network is employed to estimate the parameters associated with the bell-shaped Gaussian membership functions used in the fuzzy inference mechanism.

8.2   Factors Impacting the Work Zone Capacity

In this chapter, a computational model is presented for estimating the work zone capacity under a variety of possible work zone scenarios. A large number of factors impacting the work zone capacity are included as inputs to the model. They are 1) percentage of trucks (x_1), 2) pavement grade (x_2), 3) number of lanes (x_3), 4) number of lane closures (x_4), 5) lane width (x_5), 6) work zone layout (lane merging, lane shifting, and crossover) (x_6), 7) work intensity (work zone type) (x_7), 8) length of closure (x_8), 9) work zone speed (x_9), 10) interchange effects (proximity of ramps) (x_{10}), 11) work zone location (urban or rural) (x_{11}), 12) work zone duration (long-term or short-term) (x_{12}), 13) work time (daytime or night) (x_{13}), 14) work day (weekday or weekend) (x_{14}), 15) weather condition (sunny, rainy, or snowy) (x_{15}), 16) pavement conditions (dry, wet, or icy) (x_{16}), 17) driver composition (commuters or non-commuters such as tourists) (x_{17}), and 18) data collection locality (x_{18}).

Since heavy vehicles such as trucks occupy more space on the roadway and move slower than passenger cars, a higher percentage of trucks tends to reduce the work zone capacity. Krammes and Lopez (1994) study the work zone capacity and conclude that a high percentage of heavy trucks has a significant impact on the work zone capacity. The HCM (2000) suggests a heavy-vehicle adjustment factor given as a function of two parameters: proportion of heavy vehicles and the passenger-car equivalent for heavy vehicles. Kim et al. (2001) conduct a regression analysis of the interaction between the work zone grade and percentage of heavy trucks. The presence of grades may exacerbate the flow constriction in work zones, particularly in the presence of heavy vehicles.

Measurements made at freeway work zones in Texas (Dudek and Richards 1981; Krammes and Lopez 1994) and North Carolina freeways (Dixon and Hummer 1995;

Dixon *at el.* 1997) show clearly that the work zone capacity varies significantly with the number of freeway lanes as well as the number of lane closures. Among three commonly used work zone layouts, known as lane merging, lane shifting, and crossover, the lane shifting does not reduce the number of open lanes in the work zone but may affect the work zone capacity. The capacity value recommended for long-term work zones by HCM (2000) indicates that work zones with lane merging have a higher average capacity value than those with a crossover. A work zone with a lane width of less than the U.S. standard lane width of 12 feet may reduce the work zone capacity significantly. The HCM (2000) suggests a reduction factor up to 14% to account for the effect of lane width on work zone capacity.

Work zone capacity may decrease as the work intensity increases from the lightest (e.g., guardrail installation) to the heaviest (e.g., bridge repair). The intensity of work activities depends on a number of factors such as the type of work activities, the number of crews, the number and size of equipments, and the proximity of work activities to the open lane. The work zone intensity is classified into three levels (low, medium, or high) in Karim and Adeli (2003a) and six levels in Dudek and Richard (1981). The HCM (2000) suggests a modification of the base capacity value of the work zone to account for the intensity of work activity without actually providing any modification factors or guidelines. This important issue is left for the work zone engineer to decide subjectively based on his/her experience or professional judgment. Work intensity is a qualitative and subjective concept without any standard classification scheme. In this chapter, the work intensity is divided into six categories as defined in Table 8.1.

Table 8.1 Categories of work intensity in work zones used in the new model

Intensity level	Qualitative description	Work type examples
1	Lightest	Median barrier Installation or repair
2	Light	Pavement repair
3	Moderate	Resurfacing
4	Heavy	Stripping
5	Very heavy	Pavement marking
6	Heaviest	Bridge repair

The length of the closure affects the work zone capacity. Longer work zones often indicate more intensive work activity and display more traffic signs, causing travelers to drive more cautiously. A lower speed limit is often enforced in the work zone to improve safety, which decreases the work zone capacity.

Ramps proximity to the work zone, especially the entrance ramps inside the work zone activity area, can create traffic turbulence resulting in a reduction in the work zone capacity. The HCM (2000) suggests an adjustment for ramps without actually providing any modification factors. Again, the issue is left for the work zone engineer to decide subjectively based on his/her experience. Al-Kaisy *et al.* (2000) suggest that both duration of work zone and driver composition can affect the work zone capacity. The average capacity at long-term freeway work zones is greater than that at short-term work zones because the commuters and frequent travelers become familiar with the configuration of the long-term work zone.

The workday (weekdays or weekends) and work time (daytime or night) also affect the work zone capacity. In all likelihood commuters and regular travelers during the weekdays are more familiar with the configuration of the work zone and the traffic control plans in the affected areas (e.g., route diversion) than non-commuters (e.g., tourists) traveling during the weekends. Night construction on the one hand can help increase the work zone capacity by avoiding traffic congestion during peak hours, and on the other hand can decrease the work zone capacity because of the reduced travelers' attention (Al-Kaisy and Hall 2001).

Weather (snowy, rainy, or sunny) and pavement conditions will have a significant impact on the work zone capacity. The HCM (2000) suggests 10 to 20 percent capacity reductions due to bad weather conditions without providing any specific guidelines. Again, the issue is left for the work zone engineer to decide subjectively based on his/her experience or professional judgment. Similarly, a wet or icy pavement surface in the work zone forces travelers to lower their speed in the work zone, which reduces the work zone capacity. The HCM (2000) provides no guidelines in this regard.

Symbolically, work zone capacity can be expressed as a function of 18 variables defined in the previous paragraphs:

$$y = f(x_1, x_2, \ldots, x_{18})$$
(8.1)

The work zone capacity cannot be described by any mathematical function because it is a complicated and non-quantifiable function of a large number of interacting variables, some of which are linguistic.

8.3   Variable Quantification and Normalization

In the new work zone capacity estimation model, some of the variables are linguistic, such as work zone layout and weather conditions, some are binary-valued parameters such as the interchange effect representing the existence of ramps near or within the work zone, and others are numeric such as the work zone length. Spline-based nonlinear functions are used to quantify each linguistic as well as binary-valued variable mathematically. Spline-based nonlinear functions are also assigned to numeric variables in order to model the impact of their variations on the work zone capacity. These functions play another role, that is, to normalize the variables into the same range, 0 to 1. This normalization is desirable in the fuzzy inference mechanism developed in this work. The normalization prevents the undue domination of variables with large numerical values over the variables with small numerical values, thus improving the convergence of the network training. Compared with the conventional linear data normalization, the nonlinear normalization using spline-based functions represents the data variation more accurately.

Integer numbers are used to quantify the linguistic and binary-valued variables. Numbers 1, 2, and 3 are used to represent the three types of layouts (lane merging, lane shifting, and crossover), weather conditions (sunny, rainy, or heavy snowfall), and pavement conditions (dry, wet, or icy). Numbers 1 and 2 are used to represent the work zone location (urban or rural), work zone duration (short-term or long-term), work time (day or night), and day of week (weekday or weekend). Numbers 1 and 2 are also used to represent the binary-valued variables interchange effect (1 for no ramp and 2 for existence of ramp), driver composition (1 when it is not considered and 2 when it is considered), and pavement grade (1 when there is no grade and 2 for existence of grade). Numbers from 1 to 6 are used to represent the work intensity as defined in Table 8.1. Numbers from 1 to 7 are used to represent the seven localities where data are collected (data used in this chapter are collected from six different states and the city of Toronto).

The normalized spline-based nonlinear functions for variables 1 to 18 are shown in

Fig. 8.1. The value of each normalized function varies from 0 to 1. For variables x_1 (percentage of trucks), x_2 (pavement grade), x_3 (number of lanes), x_4 (number of lane closures), x_6 (work zone layout), x_7 (work intensity), x_{10} (interchange effects), x_{11} (work zone location), x_{12} (work zone duration), x_{13} (work time), x_{14} (work day), x_{15} (weather condition), x_{16} (pavement condition), x_{17} (driver composition), and x_{18} (data collection locality), an S-shaped spline-based nonlinear function is used, as defined by the following equation (Fig. 8.1):

$$q_i = \begin{cases} 2(x_i/b_i)^2 & \text{if } 0 \leq x_i \leq b_i/2 \\ 1 - 2[(b_i - x_i)/b_i]^2 & \text{if } b_i/2 < x_i < b_i \\ 1 & \text{if } x_i \geq b_i \end{cases} \qquad i = 1 \text{ to } 4, 6, 7, 10 \text{ to } 18 \qquad (8.2)$$

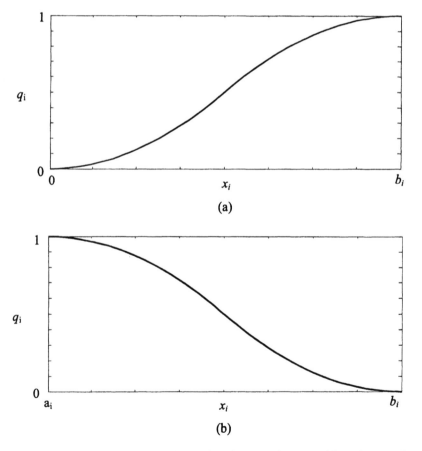

Figure 8.1 Spline-based nonlinear functions used to quantify and normalize the variables: (a) S-shaped, and (b) Z-shaped

where b_i is the upper bound for variable x_i. The upper bound is different for various input variables. The values of the upper bound used in this work for various variables are summarized in Table 8.2. When no data are available for any particular variable x_i, the value of that variable is entered as zero in the S-shaped spline function, Eq. (8.2), resulting in a corresponding value of zero for q_i.

For variables x_5 (lane width), x_8 (length of closure), and x_9 (work zone speed), a Z-shaped spline-based nonlinear function is used, as defined by the following equation (Figure 8.1b):

$$q_i = \begin{cases} 1 & \text{if } x_i \leq a_i \\ 1 - 2[(x_i - a_i)/(a_i - b_i)]^2 & \text{if } a_i < x_i \leq (a_i + b_i)/2 \\ 2[(b_i - x_i)/(a_i - b_i)]^2 & \text{if } (a_i + b_i)/2 < x_i < b_i \end{cases} \quad i = 5, 8, 9 \quad (8.3)$$

Table 8.2 Upper and lower limit values used in this work for various variables x_i

Variable	Description	Variable type	Upper bound	Lower bound
x_1	Percentage of trucks	Numeric	25%	—
x_2	Pavement grade	Binary	2	—
x_3	Number of lanes	Integer	6	—
x_4	Number of lane closures	Integer	5	—
x_5	Lane width	Numeric	12 ft	8 ft
x_6	Work zone layout	Linguistic/Integer	3	—
x_7	Work intensity	Linguistic/Integer	6	—
x_8	Length of closure	Numeric	5 km	0.5 km
x_9	Work zone speed	Numeric	60 km/hr	20 km/hr
x_{10}	Interchange effects	Binary	2	—
x_{11}	Work zone location	Binary	2	—
x_{12}	Work zone duration	Binary	2	—
x_{13}	Work time	Binary	2	—
x_{14}	Work day	Binary	2	—
x_{15}	Weather condition	Linguistic/Integer	3	—
x_{16}	Pavement condition	Linguistic/Integer	3	—
x_{17}	Driver composition	Binary	2	—
x_{18}	Data collection locality	Integer	7	—

where a_i and b_i are the lower and upper bounds for variable x_i. Note that for the three variables lane width, length of closure, and work zone speed, a Z-shaped spline function is used instead of an S-shaped spline function. Thus, in contrast to the S-shaped spline function, Eq. (8.2), the Z-shaped spline function, Eq. (8.3), is a function of a nonzero lower limit. The lower and upper bound values used in this work for variables x_5, x_8, and x_9 are summarized in Table 8.2. When no data are available for variable x_i (i = 5, 8, 9), the corresponding upper bound is entered in the Z-shaped spline function, Eq. (8.3), resulting in a value of zero for q_i.

The work zone capacity is also normalized to the range of zero to one using the following linear function:

$$C_n = (C - C_{min})/(C_{max} - C_{min}) \qquad (8.4)$$

where C_n is the normalized work zone capacity, and C_{min} and C_{max} are the minimum and maximum work zone capacity values for all the training data set, respectively.

8.4   Neuro-Fuzzy Model

An adaptive neuro-fuzzy model is developed for the nonlinear mapping of the inputs described earlier and the output, the freeway work zone capacity, incorporating fuzzy logic and neurocomputing concepts. Fuzzy logic is an effective approach for representing a) imprecision and b) linguistic variables (Adeli and Hung 1995; Adeli and Park 1998; Zadeh 1978). Neural network algorithms are powerful in providing solutions to complex pattern recognition problems where an analytical solution cannot be found (Adeli 2001; Adeli and Hung 1995; Adeli and Karim 2001). In the neuro-fuzzy model for work zone capacity estimation, a simple backpropagation neural network is employed to estimate the parameters associated with the membership functions used in the fuzzy inference mechanism.

8.4.1   Fuzzy Inference Mechanism

A fuzzy logic inference mechanism is employed using a set of IF-THEN fuzzy implication rules in the following form (Sugeno and Kang 1988):

IF $\mu_{i,1}:q_1$, AND $\mu_{i,2}:q_2$, AND, ..., AND $\mu_{i,18}:q_{18}$ THEN

$$C_i = C_n \sum_{j=1}^{18} [q_j \mu_{i,j}(q_j)] \quad i = 1, 2, ..., K \tag{8.5}$$

where $q_1,...,q_{18}$ are the 18 normalized input variables defined by Eqs. (8.2) and (8.3). The values of the variables are quantified and normalized to values between 0 and 1, employing the S-shaped and Z-shaped spline-based nonlinear functions described earlier. In Eq. (8.5), the variable $\mu_{i,j}$ is the membership function or the degree of membership of variable j in the i-th fuzzy implication rule, $\mu_{i,j} : q_j$ indicates the degree of membership of q_j is $\mu_{i,j}$, C_i is the value of the work zone capacity obtained from rule i, C_n is the normalized work zone capacity defined by Eq. (8.4), and K is the total number of fuzzy implication rules.

A bell-shaped Gaussian function is used for the membership function in the following form:

$$\mu_{i,j}(q_j) = \exp\left(-\frac{(q_j - c_{ij})^2}{2\sigma_{ij}^2}\right), i = 1, ..., K, j = 1, ..., 18 \tag{8.6}$$

where c_{ij} and σ_{ij} represent the center and the half-width of the membership function for the j-th variable and i-th fuzzy implication rule. The former determines the position of the function and the latter determines its shape.

The total number of fuzzy implication rules, K, is equal to the number of *clustering centers* for any given training data set. We use a *subtractive clustering* approach to determine the number of clusters and clustering centers. In this approach, it is assumed that each data point belongs to a potential cluster based on the minimum value of a predefined objective function. The approach is explained pictorially for a two-dimensional (two-variable) case in Fig. 8.2. An exponential data density measure is used as the objective function in the following form (Chiu 1994):

$$f(o_l) = \sum_{i=1}^{Q} \exp\left(-4\left\|\frac{\mathbf{q}_i - \mathbf{o}_l}{\mathbf{r}_c}\right\|^2\right), l = 1, ..., K \tag{8.7}$$

where $\|\mathbf{X}\| = \sqrt{\sum_{18} |X_i|^2}$ is the Euclidean distance, $\mathbf{q}_i$ is the 18×1 vector of the i-th input data set, $\mathbf{o}_l$ is the 18×1 vector of potential cluster data centers, and $\mathbf{r}_c$ is the 18×1 vector of predefined data cluster radii. Since all data have been normalized to the range 0 to 1, a constant value of r_c, in the range of 0 to 1, is chosen for all the radii corresponding to various variables. This is another advantage of normalizing the data. Without such normalization different values have to be chosen for various radii. At the beginning, Q is equal to the total number of training data sets, M. In other words, we start with Q cluster centers and compute Q different values for the data intensity measure or the objective function defined by Eq. (8.7) (Fig. 8.2a). The data set yielding the smallest objective function is selected as the first cluster and is excluded from further processing. Next, the subtractive clustering algorithm is applied to the remaining data points that do not belong to any cluster, or the *subtracted* set, and the second cluster is identified (Fig. 8.2b). This process is continued, finally resulting in K clusters. The selection of the data cluster radius (a value between 0 and 1) is a trial-and-error process. A smaller value of the cluster radius leads to a larger number of clusters requiring more computational resources for training the network and vice versa. The number of clusters should be just large enough to provide accurate results.

In Eq. (8.6), the initial values of the membership function centers for the i-th fuzzy implication rule, c_{i1} to c_{i18}, are set to the values of the clustering data centers $\mathbf{o}_1$ to $\mathbf{o}_{18}$, and the initial values of σ_{ij} are determined from

$$\sigma_{ij} = r_c(q_{j,\max} - q_{j,\min}), i = 1, ..., K, j = 1,...,18 \qquad (8.8)$$

where r_c, the cluster center's radius, represents the influence range of the variable, and $q_{j,\max}$ and $q_{j,\min}$ are the maximum and minimum values of the j-th variable among all training data sets, respectively. The standard deviations of q_j in the membership function, Eq. (8.6), may be used to estimate the initial values of σ_{ij}. In this work, however, Equation (8.8) is used instead in order to speed up the training convergence of the model.

For the fuzzy inference mechanism, any fuzzy implication rule i performs an AND operation, that is, multiplying the degrees of membership function of all the variables and finding the following output:

$$w_i = \prod_{j=1}^{18} \mu_{i,j}(q_j), i = 1, ..., K \qquad (8.9)$$

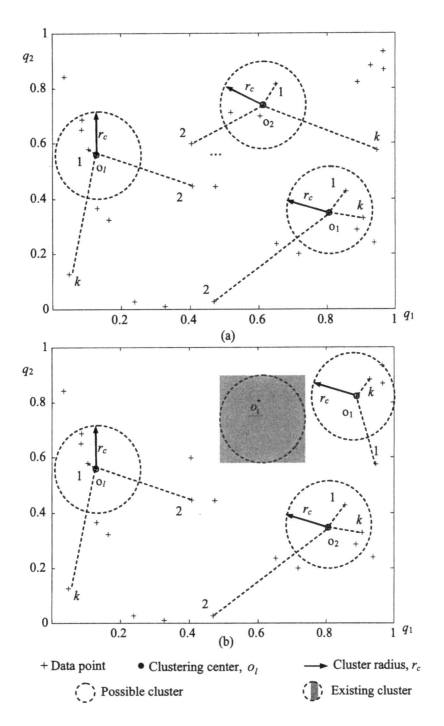

+ Data point   ● Clustering center, o_l → Cluster radius, r_c

⌜⌝ Possible cluster   ▓ Existing cluster

Figure 8.2 Illustration of subtractive clustering for a two-dimensional (two-variable) case in searching for (a) first cluster and (b) second cluster

This output represents the *firing strength* of the *i*-th fuzzy implication rule. The estimated work zone capacity, $\hat{C}$, is obtained from the fuzzy inference mechanism as the aggregation (or summation) of the outputs of K fuzzy implication rules as follows:

$$\hat{C} = \sum_{i=1}^{K} C_i \left(w_i \Big/ \sum_{i=1}^{K} w_i \right) \tag{8.10}$$

8.4.2   Topology of the Neuro-Fuzzy Model

Figure 8.3 shows the topology of the neuro-fuzzy inference model for estimating the work zone capacity. It consists of an input layer, a fuzzy implication layer, and an output layer. The input layer has 18 nodes representing the 17 variables defined in the previous section and an 18th node to indicate the data collection locality. The values of the variables in the input layer are quantified and normalized to values between 0 and 1, employing the S-shaped and Z-shaped spline-based nonlinear functions described earlier.

The fuzzy implication layer consists of two sub-layers. The first (left) sub-layer represents the fuzzy membership function layer, where every node represents one Gaussian membership function ($\mu_{i,j}$). The number of nodes in this sub-layer is equal to 18K. Each membership function is used to map one normalized input variable to one cluster. The second (right) sub-layer consists of K nodes representing K fuzzy AND operations. The output layer has only one node that performs the fuzzy aggregation (summation) operation. The output of this node is the estimated work zone capacity.

8.4.3   Training the Neuro-Fuzzy Model

The fuzzy inference mechanism presented in a previous section requires estimation of the parameters c_{ij} and σ_{ij} associated with the membership functions. For this problem, the parameters are adjusted using the backpropagation (BP) neural network-training algorithm.

A mean squared error function, $E(c_{ij}, \sigma_{ij})$, is defined as the average of the squared differences between the measured and estimated work zone capacity values over all the training data sets:

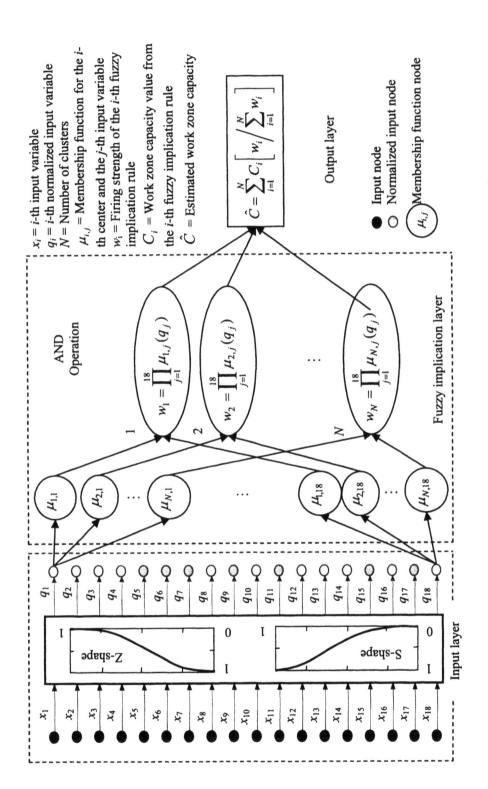

Figure 8.3 Topology of neuro-fuzzy model for estimating work zone capacity

$$E(c_{ij}, \sigma_{ij}) = \frac{1}{M} \sum_{m=1}^{M} \left| C_n^m - \hat{C}^m \right|^2 , \ i = 1, \ ..., \ K; \ j = 1,...,18 \qquad (8.11)$$

where C_n^m and $\hat{C}^m$ are the m-th normalized measured and estimated work zone capacity values, respectively, and M is the number of training data sets.

The parameters c_{ij} and σ_{ij} are updated after *all* the training data sets are applied. In other words, a training iteration or *epoch* is based on all the training data sets. The adjustment is performed by the following equation:

$$W_{ij,\text{new}} = W_{ij,\text{old}} + \sum_{m=1}^{M} \eta(C_n^m - \hat{C}^m)q_j , \ j = 1, \ ..., \ 18 \qquad (8.12)$$

where $W_{ij} = \{c_{ij}, \sigma_{ij}\}$ $(i = 1, \ ..., \ K)$ represents the parameter set of the membership function, $\mu_{i,j}$, for the j-th variable and i-th fuzzy implication rule, and η is the learning ratio.

The generalization capability of the neuro-fuzzy model demonstrates how accurately it can estimate the work zone capacity with a new data set. In training the network, one has to be cognizant of the *overgeneralization* problem, also known as *overfitting* problem in the statistics literature. In this work, rather than simply minimizing the error as defined by Eq. (8.11) we employ an *optimum generalization* strategy in order to avoid overgeneralization and achieve the most accurate results. This is done by dividing the available data sets for training into two groups: a training group and a checking group. The latter consists of only a fraction, in the order of 10 to 20%, of the total data sets available for training chosen randomly. The mean squared error term for the training set normally decreases with the number of iterations, as noted by convergence curve A in Fig. 8.4. In each iteration of the training stage, the trained network is used to estimate the work zone capacity for each set of the checking data sets, and their mean squared error is computed. The variation of this mean squared error with the iteration number of the training set is normally a concave curve with a minimum (curve B in Fig. 8.4). The iteration number corresponding to the minimum point on this curve is where the training of the network is stopped; the values of the membership function parameters obtained at this iteration provide the optimum generalization results.

As observed in Fig. 8.4, at iterations beyond the minimum point of curve B, the mean squared error in the checking set increases, indicating the overgeneralization of the network.

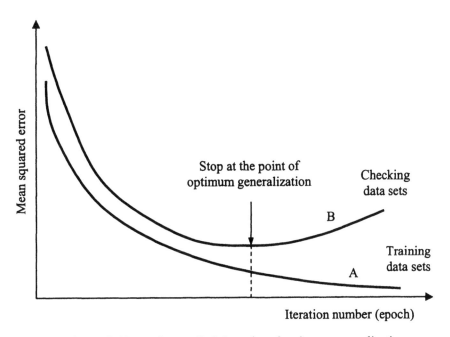

Figure 8.4 Procedure to find the point of optimum generalization

After training of the neuro-fuzzy work zone capacity estimation model using the optimum generalization strategy, a third group of data sets consisting of the testing data sets is used to evaluate the model.

8.5   Numerical Example

8.5.1   Data Collection

Data for training, checking, and testing of the model were collected from the existing literature and augmented by four data sets provided by the Ohio Department of Transportation. A total of 168 data sets were collected including 9 sets from the state of North Carolina (Dixon and Hummer 1995), 79 sets from Texas (Dudek and Rochards 1981; Krammes and Lopez 1994), 17 sets from California (Krammes and Lopez 1994),

12 sets from Indiana (Jiang 1999b), 12 sets from Maryland (Kim *et al.* 2001), 4 sets from Ohio, and 35 sets from Toronto, Canada (Al-Kaisy and Hall 2001), as summarized in Table 8.3. None of the data sets include all the 17 input variables used in the new computational model. The number of input variables provided ranged from four (number of lanes, number of lane closure, work zone intensity, and work zone duration) to fourteen (percentage of heavy trucks, grade of pavement, number of lanes, number of lane closures, work zone intensity, length of closure, work zone speed, proximity of ramps to work zone, work zone location, work zone duration, work time, work day, weather conditions, and driver composition). For those unavailable input variables, values of zero are obtained after variable quantification and normalization, as described earlier.

Table 8.3 Raw data for training, checking, and testing the model

State	Index	168 data sets		
		Training	Checking	Testing
California	1	14	2	1
Indiana	2	8	2	2
Maryland	3	8	2	2
North Carolina	4	6	2	1
Ohio	5	2	1	1
Texas	6	66	9	4
Toronto	7	29	3	3
Total		133	21	14

The 168 data sets summarized in Table 8.3 are divided into three parts: 133 sets are used for training, 21 sets are used for checking, and finally 14 sets are used for testing the neuro-fuzzy work zone capacity estimation model.

8.5.2 Training and Testing the Model

After trying several different values, a value of 0.1 was selected for the learning ratio, η, and a constant value of 0.3 for the cluster center's radii, r_c, for various variables. The number of clusters resulting from the subtractive clustering approach is 11, which is set to the number of fuzzy implication rules, K. In other words, every input node in the network topology presented in Fig. 8.3 has 11 membership functions. Convergence results for training the network based on $M = 133$ training data sets and 21 checking data sets are displayed in Fig. 8.5a.

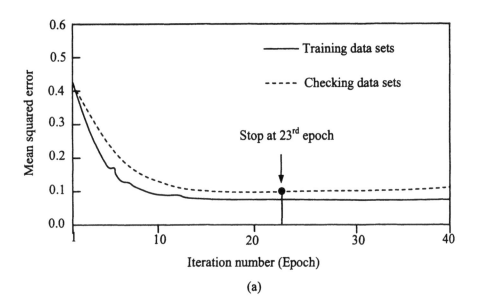

(a)

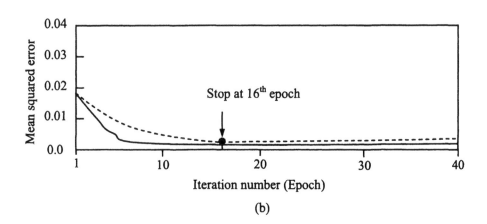

(b)

Figure 8.5 Convergence curves for training the new neuro-fuzzy model: (a) Raw data sets, and (b) improved data sets

Figures 8.6a and 8.6b show the normalized measured and estimated work zone capacity values for checking and testing data sets, respectively. There exist four and three *outliers* in the checking and testing data sets, respectively. In this work, an outlier is defined as any point with an error value 50% larger than the mean error for all the data points. There are three explanations for the existence of outliers: measurement error, inhomogeneity in the collected data (that is, data collected in one state may not be representative of data collected in another state), and lopsidedness of the data set (that is, some input variables are observed in a small number of data sets and other input variables are observed in a large number of data sets). Table 8.3 shows that a large number of training data sets are available from Texas (66), Toronto (29), and California (14), but only two training data sets are available from Ohio.

To improve the estimation accuracy of the new neuro-fuzzy model, the training data set are modified or denoised to make them more representative of actual conditions. First, the data sets yielding outliers are deleted (in the example presented in Fig. 8.6, the outliers are from the three localities with the most data, that is, Texas, Toronto, and California). Next, roughly the same numbers of training and checking data sets are chosen from various localities randomly and the network is retrained. If outliers are observed again after the second training, the outlier data sets are replaced with new remaining data sets from the same locality and the network is trained again. This process is continued until there is no outlier. For the example presented in Table 8.3, the final denoised data sets are presented in Table 8.4 which includes roughly the same number of data sets for each locality except Ohio, where only limited measured work zone data are available at the time of this writing. Convergence results for training the network based on the reduced and denoised training data sets are displayed in Fig. 8.5b, which indicates a substantial reduction in the error as well as a faster convergence. Figures 8.7a and 8.7b show the normalized measured and estimated work zone capacity values for checking and testing data sets, respectively, using the reduced data set. It is important to note that the mean squared values in Fig. 8.7 using the reduced data set is more than an order of magnitude smaller than the values in Fig. 8.6 using the raw data. This large reduction in the error indicates a significant improvement in the accuracy for estimating the work zone capacity.

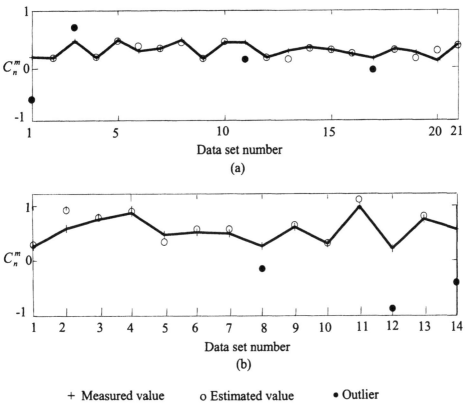

+ Measured value   o Estimated value   • Outlier

C_n^m = Normalized work zone capacity for the m-th data set

Figure 8.6 Normalized measured and estimated work zone capacity values for checking and testing data sets using the raw data (MSE = mean squared error): (a) checking data sets (MSE = 0.1669), and (b) testing data sets (MSE = 0.1872)

Table 8.4 Denoised data for training, checking, and testing the model

State	Index	67 data sets		
		Training	Checking	Testing
California	1	6	1	1
Indiana	2	8	2	2
Maryland	3	8	2	2
North Carolina	4	6	2	1
Ohio	5	2	1	1
Texas	6	8	2	2
Toronto	7	8	1	1
Total		46	11	10

After training of the neuro-fuzzy work zone capacity estimation network using the denoised data sets, the parameters of the Gaussian membership functions corresponding to 18 variables and 11 clusters are obtained and the corresponding membership functions are computed. As an example, Figure 8.8 shows the eleven different bell-shaped Gaussian membership functions for the normalized input variable q_1 (percentage of trucks).

8.5.3   Model Validation

In order to evaluate the accuracy of the new model, it is compared with two approximate empirical equations using the 10 sets of testing data given in Table 8.4. The input values for the 10 data sets are summarized in Table 8.5.

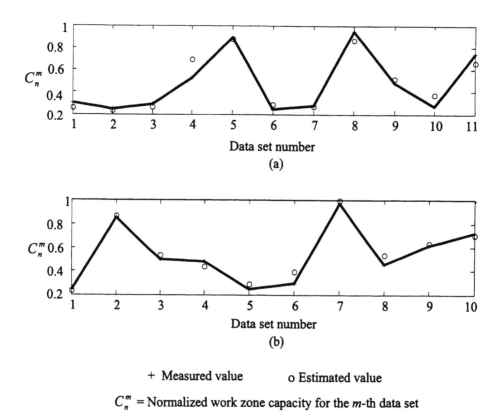

+ Measured value   o Estimated value

C_n^m = Normalized work zone capacity for the m-th data set

Figure 8.7 Normalized measured and estimated work zone capacity values for checking and testing data sets using the denoised data (MSE = mean squared error): (a) checking data sets (MSE = 0.00543), and (b) testing data sets (MSE = 0.00222)

Krammes and Lopez (1994) proposed the following empirical equation for estimating the work zone capacity:

$$C = (1600 \text{pcphpl} + I_a - R_a) \times H_a \times N_o \qquad (8.13)$$

where C = work zone capacity in vehicles per hour (vph), I_a = adjustment value for work intensity ranging from − 160 to 160 passenger cars per hour per lane (pcphpl), R_a = adjustment value for presence of ramps, H_a = adjustment factor for heavy vehicles given as a function of two parameters: proportion of heavy vehicles and passenger-car equivalent for heavy vehicles given in HCM (2000), and N_o = number of open lanes in the work zone. Equation (8.13) adjusts a single base capacity value of 1,600 pcphpl based on the effects of the intensity of the work activity, the percentage of heavy vehicles, and the presence of entrance ramps near the starting point of the lane closure. Similar to broad guidelines for work zone capacity estimation in HCM (2000), the values of various adjustment values are left for the work zone engineer to choose based on prior experience.

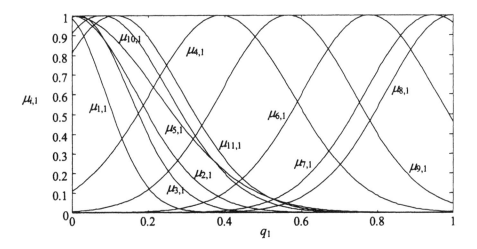

Figure 8.8 Bell-shaped Gaussian membership functions for the normalized input variable q_1 (percentage of trucks) after training the network with the denoised data

Table 8.5 Input values for 10 work zone scenarios used to test the neuro-fuzzy work zone capacity estimation model

Var.	x_1	x_2	x_3	x_4	x_5	x_6	x_7	x_8	x_9
Data	Trucks (%)	Grade (%)	No. of lanes	Lane closures	Lane width	Layout	Intensity	Length of closure (m)	Speed (mph)
1	-	-	3	1	-	-	3	-	-
2	32	-	2	1	-	M	6	11.7	-
3	10	-	2	1	-	C	2	11.7	-
4	8	5	4	1	-	-	1	0.18	30
5	8.5	0	4	2	12	-	6	2.2	21
6	-	-	3	1	-	M	-	0.6	-
7	26.2	-	2	1	-	-	6	-	-
8	-	-	4	1	-	-	1	-	-
9	-	-	5	3	-	-	3	-	-
10	-	3	3	1	-	-	-	-	-

M = Merging,   C = Crossover,   - Unavailable data

Table 8.5 (con'd)

Var.	x_{10}	x_{11}	x_{12}	x_{13}	x_{14}	x_{15}	x_{16}	x_{17}	x_{18}
Data	Ramp	Location	Work dur.	Work time	Work day	Weather cond.	Pave. cond.	Driver comp.	State
1	-	-	Long	-	-	-	-	-	California
2	-	Rural	Long	Day	Weekday	-	-	-	Indiana
3	-	Rural	Long	Day	Weekday	-	-	-	Indiana
4	Yes	Urban	Short	Day	Weekday	Sunny	-	0	Maryland
5	Yes	Urban	Short	Night	Weekday	Sunny	-	0	Maryland
6	No	-	Short	Day	Weekend	Sunny	-	-	Ohio
7	-	Rural	Long	Day	Weekday	-	-	-	N. Carolina
8	-	-	Short	-	-	-	-	-	Texas
9	-	-	Short	-	-	-	-	-	Texas
10	-	Urban	Short	Day	-	Sunny	Dry	1	Toronto

Kim *et al.* (2001) suggest the following empirical equation for work zone capacity based on multiple-variable regression analysis of 12 sets of measured work zone capacity values obtained in the state of Maryland:

$$C = 1857 - 168.1N_c - 37.0L_c - 9.0H_t + 92.7L_d - 34.3L_w - 106.1I_w - 2.3G_w \times H_t \quad (8.14)$$

where N_c = number of lane closures in the work zone, L_c = location of closed lanes (right = 1, otherwise = 0), H_t = percentage of heavy vehicles, L_d = lateral distance to the open lane, L_w = work zone length, I_w = work intensity, and G_w = work zone grade.

The work zone capacity estimates obtained by the new adaptive neuro-fuzzy model as well as two empirical Eqs. (8.13) and (8.14) are summarized in Table 8.6. The root mean square (RMS) error value obtained for the new neuro-fuzzy model, 127, is substantially lower than the RMS values 267 and 358 obtained for Kim et al. (2001) and Krammes and Lopez (1994) equations, respectively. The error percentage between the estimated and measured work zone capacity values ranges from 0.9 to 13.5 for the new neuro-fuzzy model (less than 10% with the exception of one), compared with 0.2 to 21.8 for the empirical equation presented by Kim *et al.* (2001), and 0.1 to 23.1 for that of Krammes and Lopez (1994).

Figure 8.9 shows a comparison of the estimated $(\hat{C})$ and measured work zone capacity (C) values in terms of number of open lanes. The solid line represents $\hat{C}_i = C_i$, that is, perfect correlation. The slight departure of the scattered dots representing the estimated work zone capacity from the ideal line indicates high estimation accuracy of the neuro-fuzzy model.

8.6   Concluding Remarks

A novel neuro-fuzzy freeway work zone capacity estimation model was presented in this chapter using fuzzy logic and neurocomputing concepts. A backpropagation neural network is employed to estimate the parameters associated with the bell-shaped Gaussian membership functions used in the fuzzy inference mechanism. The network has been trained using measured data obtained from six different states and the city of Toronto in Canada.

Comparisons with two empirical equations demonstrate that the new model in

Table 8.6 Comparison of work zone capacity estimates
obtained from the new neuro-fuzzy model with two empirical equations

	Data set number	Open lanes	Closed lanes	Measured values (C_i) (vph)	Krammes and Lopez (1994)		Kim et al. (2001)		Neuro-fuzzy model ($\hat{C}_i$) (vph)	
					Values (vph)	Error (%)	Values (vph)	Error (%)	Values (vph)	Error (%)
California	1	2	1	2,600	3,200	23.1	3,166	21.8	2,364.4	9.1
Indiana	2	1	1	1,308	1,307	0.1	1,295	1.0	1,395.7	6.7
	3	1	1	1,595	1,362	14.6	1,464	8.2	1,810.2	13.5
Maryland	4	3	1	5,205	4,545	12.7	4,695	9.8	5,342.6	2.6
	5	2	2	2,456	3,020	23.0	2,451	0.2	2,687.1	9.4
North Carolina	6	1	1	1,284	1,536	19.6	1,471	14.6	1,272.2	0.9
Ohio	7	2	1	3,318	3,200	3.6	3,378	1.8	3,414.8	2.9
Texas	8	3	1	4,590	4,800	4.6	5,067	10.4	4,644.5	1.2
	9	2	3	2,680	3,200	19.4	2,705	0.9	2,899.8	8.2
Toronto	10	2	1	3,904	3,200	18.0	3,378	13.5	3,779.4	3.2
Root mean square $\sqrt{\sum\limits_{i=1}^{10}(\hat{C}_i - C_i)^2 \Big/ 10}$					358		267		127	

general provides a more accurate estimate of the work zone capacity, especially when the data for factors impacting the work zone capacity are only partially available. The new model provides two important additional advantages over the existing empirical equations. First, it incorporates a large number of factors impacting the work zone capacity. Second, unlike the empirical equations, the new model does not require selection of various adjustment factors by the work zone engineers based on prior experience. The new model can be implemented into an intelligent decision support system a) to estimate the work zone capacity in a rational way, b) to perform scenario analysis, and c) to study the impact of various factors influencing the work zone capacity.

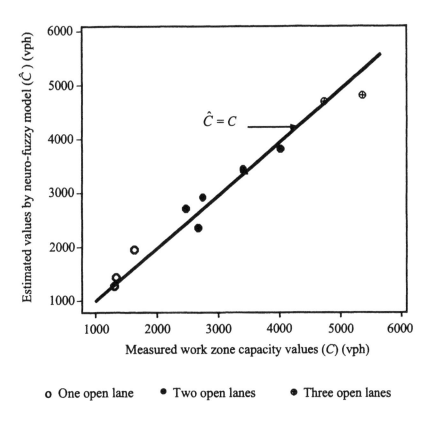

Figure 8.9 Comparison of the estimated $(\hat{C})$ and measured work zone capacity (C) values

Chapter 9

CLUSTERING NEURAL NETWORK MODELS FOR WORK ZONE CAPACITY ESTIMATION AND PARAMETRIC STUDY

9.1   Introduction

In Chapter 8, a neuro-fuzzy model was presented for estimating the work zone capacity taking into account seventeen different numeric and linguistic factors and data collection locality. A backpropagation neural network was employed to estimate the parameters associated with the bell-shaped Gaussian membership functions used in the fuzzy inference mechanism. An optimum generalization strategy was used in order to avoid overgeneralization and achieve accurate results.

In this chapter, the subtractive clustering approach described in Section 8.4.1 is judiciously integrated with the radial basis function (RBF) and backpropagation (BP) neural network models to create the clustering-RBF and clustering-BF neural network models (RBFNN and BFNN), respectively. The two clustering-neural network models are developed for estimating the work zone capacity in a freeway work zone as a function of eighteen different factors: 1) percentage of trucks (x_1), 2) pavement grade (vertical slope in the longitudinal plane) (x_2), 3) number of lanes (x_3), 4) number of lane closures (x_4), 5) lane width (x_5), 6) work zone layout (x_6), 7) work intensity (x_7), 8) work zone length (length of closure) (x_8), 9) work zone speed (x_9), 10) proximity of ramps (x_{10}), 11) work zone location (x_{11}), 12) work zone duration (x_{12}), 13) work time (x_{13}), 14) work day (x_{14}), 15) weather conditions (x_{15}), 16) pavement conditions (x_{16}), 17) driver composition (x_{17}), and 18) data collection locality (x_{18}). A detailed discussion of impact of these factors was presented in Section 8.2. The variables are quantified and normalized using the methods described in Section 8.3. Spline-based nonlinear functions are used to quantify each linguistic as well as binary-valued variable mathematically. Spline-based nonlinear functions are also assigned to numeric variables in order to model the impact of their variations on the work zone capacity.

The clustering-RBFNN model presented in this chapter is a modification of the fuzzy RBFNN model of Karim and Adeli (2003a). Work zone patterns are first grouped into similar clusters using a data clustering approach. Similarity of any new work zone pattern to the training patterns is measured by its proximity to the centers of the clusters. Karim and Adeli (2003a) use the fuzzy c-means algorithm (Adeli and Karim 2000) to find the cluster centers. In this work, the *subtractive clustering* approach described in Section 8.4.1 is used to determine the optimum number of clusters and clustering centers where it is assumed that each data point belongs to a potential cluster based on the minimum value of a predefined objective function. Subtractive clustering is an effective approach for grouping data into clusters and discovering structures in data (Chiu 1994; Yager and Filev 1994). The clustering-BPNN model is similar to the clustering-RBFNN model except that the neural network classifier in the former is the simple BP algorithm and in the latter it is the RBFNN algorithm.

9.2   Clustering-Neural Network Models

9.2.1   Topology of Neural Networks

The topology of the neural network models for estimating the work zone capacity is presented in Fig. 9.1. It consists of an input layer, a hidden layer, and an output layer. The input layer has 18 nodes representing the 18 variables defined in the previous section. The values of the variables in the input layer are normalized to values between 0 and 1 employing the S-shaped and Z-shaped spline-based nonlinear functions as explained in Section 8.3. The normalization prevents the undue domination of variables with large numerical values over the variables with small numerical values, thus improving the accuracy of estimating work zone capacity and accelerating the convergence of the network training. The normalized variables are denoted by q_1 to q_{18} in Fig. 9.1. The parameter w_{ij} represents the weight of the link connecting the input node i to node j in the hidden layer.

The number of nodes in the hidden layer, K, is equal to the number of cluster centers used to characterize and classify any given training data set. For the number of nodes in the hidden layer, instead of the trial-and-error approach commonly used in creating the neural network topology, the *subtractive clustering* method described in Section 8.4.1 is used. In Fig. 9.1, the variables in the hidden layer are denoted by p_1

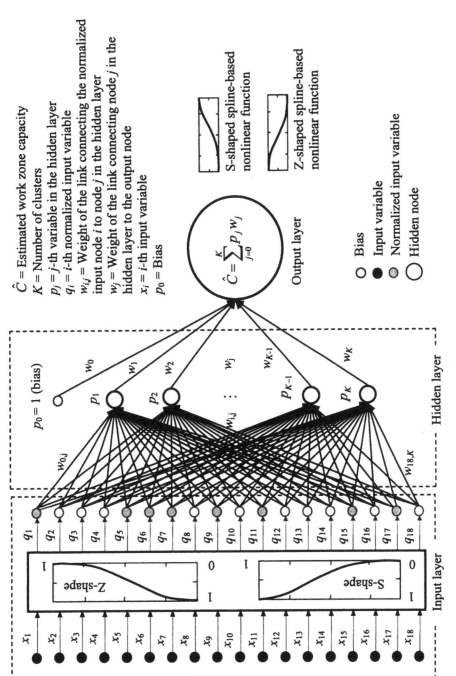

Figure 9.1 Topology of neural network models for estimating the work zone capacity

to p_K. A bias node with the value of one ($p_0 = 1$) is added to the hidden layer. Without the bias, the hyperplane separating the patterns is constrained to pass through the origin of the hyperspace defined by the inputs, which limits the adaptability of the neural network model. The output layer has only one node for the estimated work zone capacity. The estimated work zone capacity, $\hat{C}$, is obtained from the clustering-neural network model as the aggregation of the weighted outputs of $K + 1$ hidden nodes as follows:

$$\hat{C} = \sum_{j=0}^{K} w_j p_j \tag{9.1}$$

where the first term in the summation (for $j = 0$) represents the bias and w_j is the weight of the link connecting the j-th node in the hidden layer to the output node.

9.2.2 Clustering-RBFNN

Adeli and Karim (2000) used the fuzzy c-means clustering algorithm to improve the performance of RBFNN for another pattern recognition problem, the freeway traffic incident detection problem. Karim and Adeli (2003a) present a fuzzy RBFNN model for mapping eleven quantifiable and non-quantifiable factors influencing the work zone capacity. In this work, the Gaussian function is used as the basis in the hidden or the radial basis function (RBF) layer of the neural network model in the following form (Fig. 9.2):

$$p_j = \exp\left[\left\| \mathbf{q} - \mathbf{c}_j \right\|^2 / \left(2\sigma_j^2 \right) \right], j = 1, 2, ..., K \tag{9.2}$$

where $\left\| \mathbf{X} \right\| = \sqrt{\sum_{18} \left| \mathbf{X}_i \right|^2}$ is the Euclidean distance, p_j is the value of the j-th node in the hidden layer, $\mathbf{q}$ is the 18×1 vector of the normalized input variables, $\mathbf{c}_j$ ($j = 1, 2, ..., K$) is the 18×1 vector of the j-th clustering data center (which is determined by the subtractive clustering approach, as the optimum number of clusters), and K is the number of radial basis functions which is also equal to the optimum number of clusters.

In Eq. (9.2), the factor σ_j is the influencing range of the Gaussian function centered at $\mathbf{c}_j$, whose squared value in this chapter is approximated using the mean squared distance between cluster centers, as expressed by

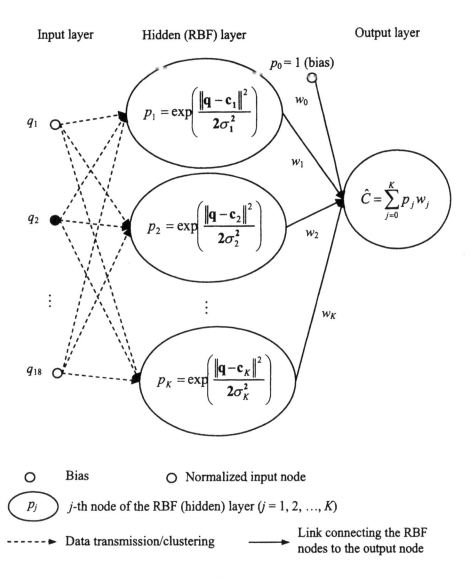

Figure 9.2 Architecture of clustering-RBFNN model for estimating freeway work zone capacity

$$\sigma_j^2 = \frac{1}{K} \sum_{i=1}^{M} \left\| \mathbf{c}_j - \mathbf{c}_i \right\|^2, j = 1,2,\ldots, K \tag{9.3}$$

where M is the total number of training data sets. The work zone capacity estimated by the clustering-RBFNN model is obtained as the aggregation of the weighted outputs of K + 1 hidden nodes from Eq. (9.1).

The weights of the links connecting the hidden nodes to the output nodes are updated by minimizing the mean squared error (MSE) of the normalized work zone capacity and using the gradient descent optimization algorithm (Rumelhart *et al.* 1986). Two stopping criteria are used for convergence of the clustering-RBFNN model. One is the acceptable mean squared error value (a value of 0.001 is used in this work) and the other is the maximum number of iterations (a value of 400 is used in this work).

In a conventional RBFNN, the weights of the links connecting the input layer to the hidden layer (i.e., the RBF parameters $\mathbf{c}_j$ defining the cluster centers) have to be updated in every iteration, similar to a standard multiple-layer feedforward neural network. In contrast, in the clustering-RBFNN model presented in this chapter, the centers of RBF clusters ($\mathbf{c}_j$) are determined in one step using the subtractive clustering approach described in Section 8.4.1, resulting in substantial speedup in the training convergence of the network and reduction of computer processing time for training the network.

9.2.3   Clustering-BPNN

The BP neural network (Hagan *et al.* 1996) has been popular because of its simplicity despite its slow convergence rate for complex pattern recognition problems (Adeli and Hung 1994). It is based on the gradient descent unconstrained optimization approach where weights are modified in a direction corresponding to the negative gradient of a backward-propagated error measure. The simple BP neural network algorithm is integrated with the subtractive clustering technique and used as an alternative approach for estimation of the work zone capacity. The output of the j-th hidden node in the BP neural network, p_j, is determined by the sigmoid activation function (Fig. 9.3):

$$p_j = 1/(1 + \exp(-X_j)), j = 1, 2, \ldots, K \tag{9.4}$$

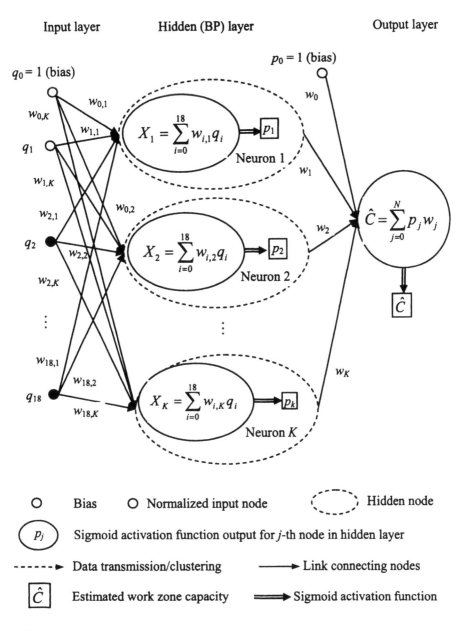

Figure 9.3 Architecture of clustering-BPNN model for estimating freeway work zone capacity

where $X_j = \sum_{i=0}^{18} w_{ij} q_i$ is the aggregation of the 18 weighted normalized input variables

plus the bias (for $i = 0$). The output value estimated by the clustering-BPNN model is obtained also using the sigmoid activation function as follows:

$$\hat{C} = 1/(1 + \exp(-X)) \qquad (9.5)$$

where $X = \sum_{j=0}^{K} w_j p_j$ is the aggregation of the weighted outputs of K nodes in the hidden

layer plus the bias (for $j = 0$).

Figure 9.3 shows the architecture of the clustering-BPNN model for the work zone capacity estimation. There are a number of differences between this model and the clustering-RBFNN model shown in Fig. 9.2. In the clustering-BPNN model: 1) weights of the links connecting the input layer to the hidden layer are required to be updated in each iteration of training the network, 2) aggregation is executed in both hidden and output layers, 3) a so-called *momentum* term is added to the weight modification equation or learning rule to help prevent the neural network getting trapped in a local minimum (Hagan *et al.* 1996), and 4) the *overgeneralization* problem is avoided by employing an *optimum generalization* strategy described in Section 8.4.3 for training the neural network. The resulting clustering-BPNN model requires more computation time for estimating the work zone capacity compared with the clustering-RBFNN model.

9.3   Training and Validating the Models
9.3.1   Training
Training of neural networks is performed similar to the approach described in Section 8.5.2. Convergence results for training the networks based on the entire 39 training data sets in Table 9.1 are displayed in Fig. 9.4. It is noted that RBFNN is a generalization network which does not have the overfitting problem, as described in Section 2.4. Therefore, the optimum generalization strategy is not needed for the clustering-RBFNN model (since there is only one training curve in Fig. 9.4b). In addition, the convergence rate for the clustering-RBFNN is substantially faster than the clustering-BPNN. On a 1.5-GHz Intel Pentium 4 processor, the CPU time for training the former is 0.25 seconds and the latter 1.42 seconds.

Table 9.1 Training, checking, and validation data set

State	Index	52 data sets		
		Training	Checking	Validation
Indiana	1	9	1	2
Maryland	2	9	1	2
North Carolina	3	7	1	1
Texas	4	7	1	2
Toronto	5	7	1	1
Total		39	5	8

9.3.2   Validation

Eight sets of validation data sets selected randomly from the collected data sets are used to validate the accuracy of the clustering-neural network models (Table 9.2). The input values for the 8 data sets are summarized in Table 9.2. There are two sets each from the states of Indiana, Maryland, and Texas, and one set each from North Carolina and Toronto. The work zone capacities estimated by three different models, the neuro-fuzzy logic described in Chapter 8, the clustering-BPNN, and the clustering-RBFNN models, are summarized in Table 9.3. The root mean square error (RMSE) values obtained for the three models are 229 vehicles per hour (vph), 215 vph, and 114 vph, respectively. As such, based on the limited training and validation data used, the clustering-RBFNN model provides the most accurate results. The error percentage for this model ranges from 0.1% to 8.7% (with one exception, the error is generally under 5%). For the other two approaches, the error is in general less than 10% with the exception of one case for each method.

The clustering-RBFNN model appears to have the attractive characteristics of training stability (the training results are not sensitive to the initial selections of the weights), accuracy, and quick convergence. In the next section, the clustering-RBFNN model is used to perform a parametric study of the main factors affecting the work zone capacity.

9.4   Parametric Studies of Work Zone Capacity

This study is done for an actual freeway work zone scenario with measured data provided in Dixon *et al.* (1997). The work zone site is a two-lane rural freeway on I-95 in North

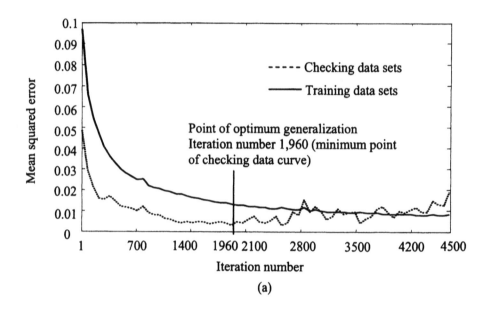

(a)

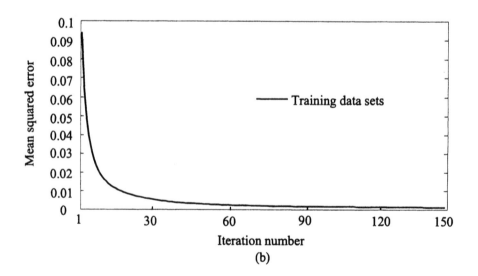

(b)

Figure 9.4 Convergence curves for training the clustering-neural network
models: (a) Clustering-BPNN, and (b) clustering-RBFNN

Table 9.2 Input values for 8 work zone scenarios used to validate three work zone capacity estimation models

Variable	x_1	x_2	x_3	x_4	x_5	x_6	x_7	x_8	x_9
Data set	Trucks (%)	Grade (%)	No. of lanes	No. of lane closures	Lane width	Layout	Work intensity	Length of closure (km)	Speed (km/h)
1	32	-	2	1	-	M	6	11.7	-
2	10	-	2	1	-	C	2	11.7	-
3	8	5	4	1	-	-	1	0.18	48
4	8.5	0	4	2	-	-	6	2.2	34
5	26.2	-	2	1	-	-	6	-	-
6	-	-	4	1	-	-	1	-	-
7	-	-	5	3	-	-	3	-	-
8	-	3	3	1	-	-	-	-	-

M = Merging,   C = Crossover,   - Unavailable data

Table 9.2 (Con'd)

Variable	x_{10}	x_{11}	x_{12}	x_{13}	x_{14}	x_{15}	x_{16}	x_{17}	x_{18}
Data set	Ramp	Location	Work dur.	Work time	Work day	Weather cond.	Pave. cond.	Driver comp.	State
1	-	Rural	Long	Day	Weekday	-	-	-	Indiana
2	-	Rural	Long	Day	Weekday	-	-	-	Indiana
3	Yes	Urban	Short	Day	Weekday	Sunny	-	0	Maryland
4	Yes	Urban	Short	Night	Weekday	Sunny	-	0	Maryland
5	-	Rural	Long	Day	Weekday	-	-	-	N. Carolina
6	-	-	Short	-	-	-	-	-	Texas
7	-	-	Short	-	-	-	-	-	Texas
8	-	Urban	Short	Day	-	Sunny	Dry	1	Toronto

Table 9.3 Comparisons of work zone capacity estimates by neuro-fuzzy logic, clustering-BPNN, and clustering-RBFNN models

State	Data set number	Open lanes	Closed lanes	Measured value (C_i) (vph)	Neuro-fuzzy logic ($\hat{C}_i$) (vph) (Chapter 8)		Clustering-BPNN ($\hat{C}_i$) (vph)		Clustering-RBFNN ($\hat{C}_i$) (vph)	
					Values (vph)	Error (%)	Values (vph)	Error (%)	Values (vph)	Error (%)
Indiana	1	1	1	1,308	1,320	0.9	1,326	1.4	1,287	1.6
	2	1	1	1,595	2,138	34.1	1,265	20.7	1,540	3.4
Maryland	3	3	1	5,205	5,343	2.6	4,982	4.3	5,211	0.1
	4	2	2	2,456	2,652	8.0	2,624	6.8	2,588	5.4
North Carolina	5	1	1	1,284	1,290	0.5	1,287	0.2	1,264	1.6
Texas	6	3	1	4,590	4,649	1.3	4,200	8.5	4,563	0.6
	7	2	3	2,680	2,900	8.2	2,779	3.7	2,914	8.7
Toronto	8	2	1	3,904	3,779	3.2	4,039	3.5	3,793	2.8
Root mean square error $= \sqrt{\sum_{i=1}^{8}(\hat{C}_i - C_i)^2/8}$					229		215		114	

Carolina with one lane closure (Fig. 9.5a). Dixon *et al.* (1997) provide values for only nine out of seventeen input variables used in the computational models created in this chapter, as summarized in Table 9.4. Data are not provided for pavement grade, lane width, work zone length, work zone speed limit, proximity to a ramp, weather and pavement conditions, and driver composition. Parametric studies presented in this chapter, however, are for eleven factors influencing the work zone capacity: work intensity, percentage of trucks, work zone configuration, layout, weather conditions, pavement conditions, work zone lane width, pavement grade, presence of ramps, work day, and work time. The impact of other factors is not investigated because insufficient data existed in the neural network training data set available to the authors.

9.4.1   Work Intensity

Work intensity in the parlance of the freeway work zone is a qualitative and subjective concept without any standard classification scheme. In this chapter, the work intensity is divided into six categories from the lightest to the heaviest, represented numerically by one to six, respectively, as summarized in Table 8.1. Keeping all other variables in the given work zone constant, the work zone capacities for six different work intensities are estimated using the clustering-RBFNN model. The results are summarized in Table 9.4 and displayed in Fig. 9.5b, which shows the work zone capacity reduces with an increase in the intensity of the work, as expected.

9.4.2   Percentage of Trucks

Keeping all other variables in the given work zone constant, the work zone capacities for nine different percentages of trucks, ranging from 8% to 30%, are estimated. The results are summarized in Table 9.4 and displayed in Fig. 9.5b, which shows the work zone capacity reduces with an increase in the percentage of trucks, as expected. The measured value provided by Dixon *et al.* (1997) for the truck percentage of 26.2 is 1,284 vehicles per hour per lane (vphpl). The clustering-RBFNN model provides the estimate of 1,265 vphpl with a small error of less than 2%.

9.4.3   Work Zone Configuration, Layout, and Weather/Pavement Conditions

Parametric studies of work zone configurations include the total number of lanes (2, 3 or 4), number of lane closures (1, 2 or 3), and work zone layout (i.e., merging, shifting, and

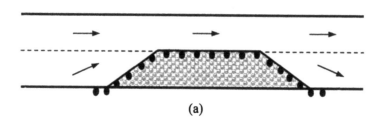

(a)

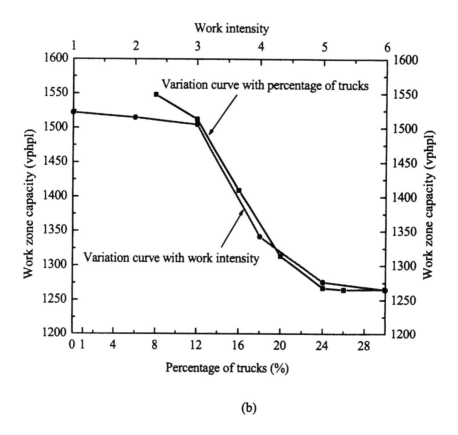

(b)

Figure 9.5 Variation of work zone capacity with work intensity and percentage of trucks:
(a) Work zone configuration, and (b) variation curves

Table 9.4 Work zone capacity variations
with work intensity and percentage of trucks

State: North Carolina	Location: Rural	Work duration: Long-term
Number of lanes: 2	Number of lane closures: 1	Truck percentage: 26.2
Work intensity: 6	Work time: day	Workday: weekday

Measured work zone capacity: 1,284 vphpl

Work intensity	1	2	3	4	5	6
Capacity (vphpl)	1,522	1,515	1,505	1,342	1,276	1,265
Percentage of trucks (%)	8	12	16	20	24	26.2	30
Capacity (vphpl)	1,548	1,513	1,409	1,314	1,268	1,265	1,264

Work intensity: 1 = Lightest, 2 = Light, 3 = Moderate, 4 = Heavy, 5 = Very heavy, 6 = Heaviest

crossover). Further, the influence of weather conditions (i.e., rainy or snowy) and pavement conditions (i.e., wet or icy) on the work zone capacity are also investigated. The work zone configurations are shown in Fig. 9.6 and their results are summarized in Table 9.5 and graphically shown in Fig. 9.7.

Three different work zone scenarios are studied. Scenario 1 is for a two-lane freeway with a one-lane closure, Scenario 2 is for a three-lane freeway with a two-lane closure, and Scenario 3 is for a four-lane freeway with a three-lane closure (Fig. 9.6). In all scenarios only one lane is open. The results are summarized in Table 9.5. For a single open lane, the work zone capacity reduces as the total number of lanes increases. Compared with a two-lane freeway, this reduction is only 1% for a three-lane freeway, but 9% for a four-lane freeway. This suggests that for a four-lane freeway a cost-benefit analysis should be performed for the option of keeping two lanes open versus maintaining just one lane open. The results of parametric studies indicate that the work zone capacity varies significantly with the number of freeway lanes as well as the number of lane closures which is consistent with the study on freeway work zones in Texas by Krammes and Lopez (1994).

The per-lane work zone capacity for the merging layout is about 14% more than that for the crossover layout and about 8% more than that for the shifting layout. The work zone capacity for the sunny weather (dry pavement) condition is about 6% more than that for the rainy weather (wet pavement) and about 10% more than that for the snowy weather condition.

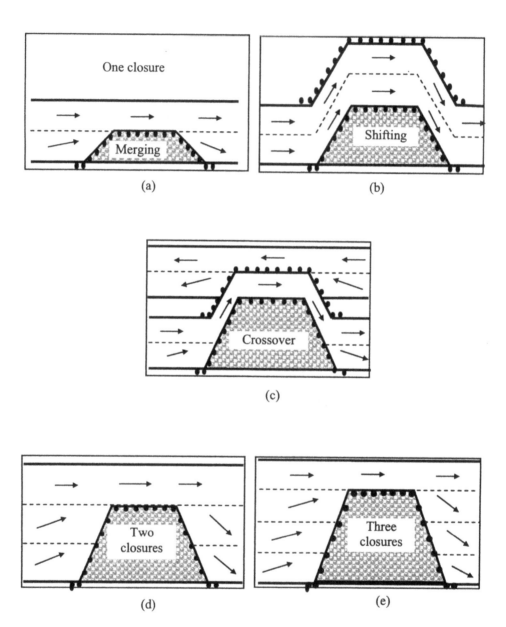

Figure 9.6 Work zone configuration and layout: (a) Merging with one lane closure in two-lane freeway; (b) shifting in two-lane freeway; (c) crossover; (d) two closures in three-lane freeway; and (e) three closures in four-lane freeway

Table 9.5 Variation of work zone capacities with influencing factors

Factors	Scenarios	No. of lanes	No. of lane closures	Estimated capacity (vphpl)
Work zone configuration	1	2	1	1,287
	2	3	2	1,274
	3	4	3	1,171
	Scenarios	**Layout**	**Estimated capacity (vphpl)**	
Work zone layout	1	Merging	1,287	
	2	Shifting	1,193	
	3	Crossover	1,112	
	Scenarios	**Weather condition**	**Pavement condition**	**Estimated capacity (vphpl)**
Weather/ Pavement	1	Sunny	Dry	1,287
	2	Rainy	Wet	1,213
	3	Snowy	Snowy/Icy	1,159

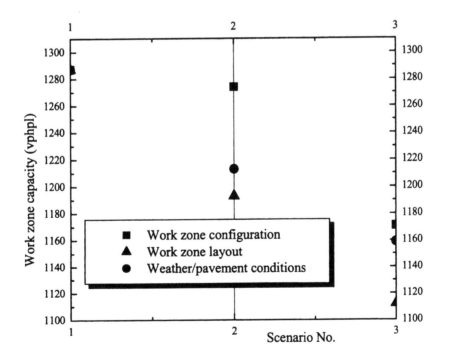

Figure 9.7 Variation of work zone capacities with work zone configuration, work zone layout, and weather/pavement conditions

9.4.4   Work Zone Lane Width and Pavement Grade

Keeping all other variables in the given work zone constant, the work zone capacities for seven different lane widths, ranging from 2.7 m (9 ft) to 3.6 m (12 ft) in increments of 0.15 m (0.5 ft), are estimated for two cases, in the presence and absence of the pavement grade. The results are shown in Table 9.6 and graphically in Fig. 9.8. In the presence of the pavement grade, the estimated work zone capacity ranges from 1,054 vphpl (for the smallest lane width of 2.7 m) to 1,342 vphpl (for the largest lane width of 3.6 m). In the absence of the pavement grade, the estimated work zone capacity ranges from 1,262 vphpl (for the smallest lane width of 2.7 m) to 1,862 vphpl (for the largest lane width of 3.6 m). The following observations are made. The work zone lane widths in the range of 3.3 m (11 ft) to 3.6 m (12 ft) (the U.S. standard lane width) do not affect the work zone capacity by any significant measure. As the work zone lane width reduces, the work zone capacity decreases significantly. The presence of the work zone pavement grade exacerbates the traffic flow constriction (e.g., speed) and affects drivers' behavior, resulting in a significant reduction in the work zone capacity in the range of 20% for a work zone lane width of 2.7 m (9 ft) to 39% for a width of 3.6 m (12 ft).

Table 9.6 Work zone capacities with lane width and pavement grade

Lane width (m)	Estimated work zone capacity (vphpl)	
	With pavement grade	Without pavement grade
2.70	1,054	1,262
2.85	1,132	1,422
3.00	1,225	1,615
3.15	1,294	1,761
3.30	1,327	1,830
3.45	1,339	1,856
3.60	1,342	1,862

9.4.5   Presence of Ramp

The neural network models take into account the effect of the presence of ramps on the work zone capacity. The presence of ramps is treated as a qualitative variable instead of a quantitative one. An example of ramp proximity to the work zone is illustrated in Fig. 9.9a. The work zone capacities estimated for a two-lane rural freeway on I-95 in North Carolina with one lane closure in the presence and absence of a ramp are summarized in Table 9.7 and shown in Fig. 9.9b. The presence of a ramp reduces the capacity by 12.6%.

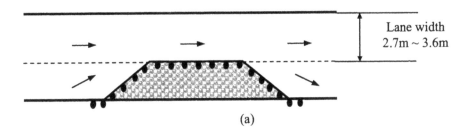

(a)

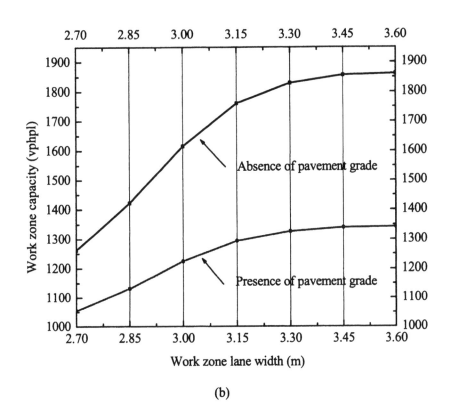

(b)

Figure 9.8 Variation of work zone capacities with lane width and pavement grade: (a) Configuration, and (b) variation curves

9.4.6   Work Day and Work Time

Work zone capacities for four combinations (i.e., four scenarios) of work day (weekday or weekend) and work time (daytime or night) are summarized in Table 9.7 and presented in Fig. 9.9b. Since in all likelihood commuters and regular travelers during the weekdays are more familiar with the configuration of the work zone and the traffic control plans in the affected areas (e.g., route diversion) than non-commuters (e.g., tourists) traveling during the weekends, the work zone capacity is somewhat larger during the weekday than during the weekend. The parametric study performed in this chapter can quantify this observation. The estimated capacities for the weekend are about 37% smaller than those for the weekday during both daytime and night.

The driver behavior and traffic characteristics differ during daytime and night time. Night construction can decrease the work zone capacity because of the reduced travelers' attention and inferior visibility during nighttime (Al-Kaisy and Hall 2001). Again, the models presented in this chapter can quantify this observation. The estimated work zone capacities for construction during the night are 10% to 11% smaller that those for construction during daytime.

Table 9.7 Work zone capacities with influencing factors

Group	Scenarios	Location	Ramp	Estimated capacity (vphpl)
Work zone Location/ramp	3	Rural	No	1,287
	4	Rural	Yes	1,143

Group	Scenarios	Work time	Workday	Estimated capacity (vphpl)
Work day/time	1	Daytime	Weekday	1,287
	2	Night	Weekday	1,164
	3	Daytime	Weekend	934
	4	Night	Weekend	847

9.5   Concluding Remarks

Validation results indicate that the work zone capacity can be estimated by clustering-neural network models in general with an error of less than 10%, even with limited data available to train the models. With additional data and training of the models the accuracy can be improved substantially. The computational models presented in this chapter are general. The parametric studies, however, are based on the adaptation of the work zone in a two-lane rural freeway on I-95 in North Carolina with one lane closure. There is no intention to offer generalized conclusions for every other work zone situation. However, the computational models provide a powerful tool to perform parametric studies for other work zone situations.

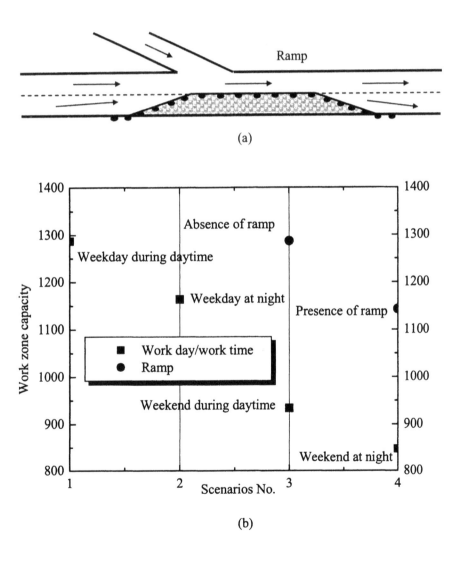

Figure 9.9 Variation of work zone capacities with workday and work time as well as work zone location and ramp: (a) Configuration, and (b) variation curves

The results of a parametric study of the factors impacting the work zone capacity can assist work zone engineers and highway agencies to create effective traffic management plans (TMPs) for work zones quantitatively and objectively. There is a definite need to collect additional data for various work zone conditions. Such data will have two significant applications. First, they can be used to further train the clustering-neural network models in order to improve the accuracy of work zone capacity estimation. Second, they can be used for more detailed sensitivity analysis.

Chapter 10

OBJECT-ORIENTED MODEL FOR WORK ZONE CAPACITY AND DELAY ESTIMATION

10.1 Introduction

Memmott and Dudek (1984) estimate the road user delay costs based on the average speed and average daily traffic (ADT) in freeway work zones. Using the deterministic queuing analysis approach and the conservation principle of traffic flow, they develop a computer program named QUEWZ for Queue and User Cost Evaluation of Work Zones to estimate the user costs and queue length for a work zone in the state of Texas. However, the work zone capacity is estimated from empirical speed–flow–density relationships independent of the work zone characteristics such as work zone layout and work intensity.

A Microsoft Excel-based model has also been developed for predicting the work zone delay, called QuickZone, based on the deterministic queuing model for each network link in the work zone (MITRETEK 2000). QuickZone estimates the hourly delay taking into account the expected time-of-day utilization and seasonal variation in travel demand. However, the existing computer models such as QUEWZ and QuickZone used to estimate queue delay upstream of the work zone have a number of shortcomings. They do not provide any model to estimate work zone capacity, which has a significant impact on the congestion and traffic queue delays. They cannot be used to perform parametric or scenario analysis for work zones with various characteristics such as work zone layout, number of closed lanes, work intensity, and work time.

To overcome these shortcomings, Adeli and coworkers have developed a number of computational models for accurate estimation of work zone capacity and traffic queue delays using computational intelligence approaches such as neurocomputing (Adeli and Hung 1995; Adeli and Park 1998; Adeli and Karim 2001), fuzzy logic, and case-based reasoning (CBR). Karim and Adeli (2002c) present a CBR model for freeway work zone traffic management. The model considers work zone layout, traffic demand, work characteristics, traffic control measures, and mobility impacts. A four-set case base schema or domain theory is developed to represent the cases based on the above characteristics of the problem. Three examples are presented to show the practical utility of the CBR system for work zone traffic management.

Chapter 7 presents a freeway work zone traffic delay and cost optimization model in terms of two variables: the length of the work zone segment and the starting time of the work zone using *average hourly traffic* data. The total work zone cost defined as the sum of user delay, accident, and maintenance costs is minimized. Number of lane closures, darkness factor, and seasonal variation in travel demand normally ignored in prior research are included. In order to find the global optimum solution, a Boltzmann-simulated annealing neural network is developed to solve the resulting mixed real variable-integer cost optimization problem for short-term work zones.

Chapter 8 presents an adaptive neuro-fuzzy logic model for estimation of the freeway work zone capacity. In Chapter 9, two neural network models, called clustering-RBFNN and clustering-BPNN models, are presented for estimating the work zone capacity in a freeway work zone as a function of eighteen different factors through judicious integration of the subtractive clustering approach with the radial basis function (RBF) and the backpropagation (BP) neural network models. The clustering-RBFNN model has the attractive characteristics of training stability, accuracy, and quick convergence.

In this chapter, an object-oriented (OO) model is presented for freeway work zone capacity and queue delay estimation. The model is implemented into a highly interactive software system, called *IntelliZone* (Intelligent decision support system for work zone traffic management) in an object-oriented programming (OOP) language. The integration of the modeling, control, and decision support features is described.

10.2 Function Architecture

Since 1991, Adeli and coworkers have advanced the use of the OO technology for development of flexible, maintainable, and reusable software systems for computer-aided engineering (CAE) applications (Yu and Adeli 1991 & 1993; Adeli and Yu 1993; Adeli and Kao 1996; Karim and Adeli 1999a&b). The object in the OO technology is a "*black box*" which abstracts a real world entity by encapsulating its characteristics (data and functionality). *Abstraction* means identifying the distinguishing characteristics of an object without having to process all the information about the object. *Encapsulation* is an OOP mechanism that combines data into codes and protects data and codes from outside interference and misuse. The additional two mechanisms provided by OOP languages (e.g., Visual C++), *inheritance* (the process by which the object of one class acquires the properties of another class) and *polymorphism* (a feature that allows one interface to be used for a general class of actions), allow easy extension and reusability of previously developed objects. The properly integrated application of polymorphism, encapsulation, and inheritance provides efficient development and management of a complicated software system.

Figure 10.1 shows the function diagram of *IntelliZone*. It consists of four interaction stages: input, analysis, output, and traffic management plan (TMP) stages. In the input stage, the user can select up to eighteen input parameters for work zone capacity estimation (noted in the left-upper box of Fig. 10.1). For traffic delay and queue estimation, up to four additional parameters may be identified in the input stage (noted in the left-lower box of Fig. 10.1). *IntelliZone* estimates the work zone capacity. Work zone capacity is also included in the list of input parameters for the sake of generality and to allow the user to input any predefined number (e.g., based on actual measurements at a particular work zone) or modify the estimated value provided by *IntelliZone*.

The primary computational model for work zone capacity estimation is the RBFNN presented in Chapter 9. Further, the simple BPNN and the fuzzy-RBFNN model of Karim and Adeli (2003a) are also provided as alternative approaches for estimation of work zone capacity. The queue delay and length estimation model is described in Chapter 7.

To improve the accuracy of estimation and accelerate the convergence speed of the neural network model, the values of the eighteen input variables for work zone capacity estimation are quantified and normalized to values between 0 and 1 employing the S-shaped and Z-shaped spline-based nonlinear functions described in Section 8.3. The normalization prevents the undue domination of variables with large numerical values over the variables with small numerical values, thus improving the convergence of the network training in estimating the work zone capacity. Compared with the conventional linear data normalization, the nonlinear normalization using spline-based functions represents the data variation more accurately.

IntelliZone provides three different types of output (Fig. 10.1). The convergence

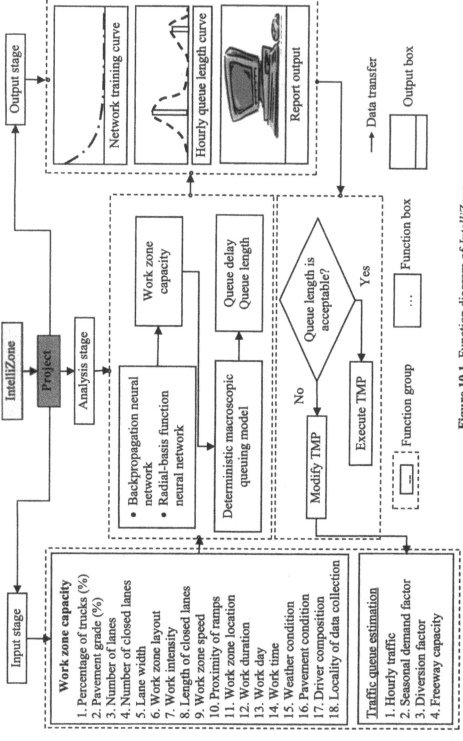

Figure 10.1 Function diagram of *IntelliZone*

results of neural network training can be viewed graphically. Similarly, the hourly queue length is presented graphically. A report output can be created where the work zone input and output information is summarized along with the results of the queue delay and length estimation. The hourly queue length plot assists the work zone engineer to modify the work zone traffic management plan and perform scenario analysis efficiently. For example, if the queue length within a given period of the day exceeds the acceptable limit, the work zone engineer can improve the work zone TMP by changing the work zone schedule (e.g., by changing the work time to avoid the traffic peak), or changing the work zone layout to increase the work zone capacity. Figure 10.2 shows how an effective TMP can be created for a particular work zone using *IntelliZone* interactively.

10.3   Application Architecture

The software system *IntelliZone* is based on the object-oriented software architecture using Visual C++ and the Microsoft Foundation Class (MFC) library. Design of a complex OOP software system usually requires the hierarchical use of multiple specialized *frameworks*. A *framework* is a collection of cooperating classes relevant to a specific domain (templates used to create multiple objects with similar features). Furthermore, the dependencies among the frameworks must be clearly delineated to avoid any conflicts. In this work, a layered approach (Baumer *et al.* 1997; Karim and Adeli 1999a&b) is used to separate the application functions and classes of *IntelliZone* from MFC classes, thus allowing for ease of development and maintenance.

The application architecture for constructing an object-oriented intelligent decision support system for work zone management is shown schematically in Fig. 10.3. Three levels of abstractions are modeled in layers. The outermost layer is the MFC *shell* layer. It is a framework created from the standard MFC library which encapsulates the most common functionality of the Windows Application Programming Interface (API) into an OO interface. The other two layers depend on and use the services of this shell layer. Typically, the shell layer provides commonly used data structures, mathematical functions, client/server middleware (low-level transaction management software), and request brokers (software that manages cooperation and communication among heterogeneous software components). The shell layer is closely connected to the Windows operating system, and its implementation in the form of a framework is

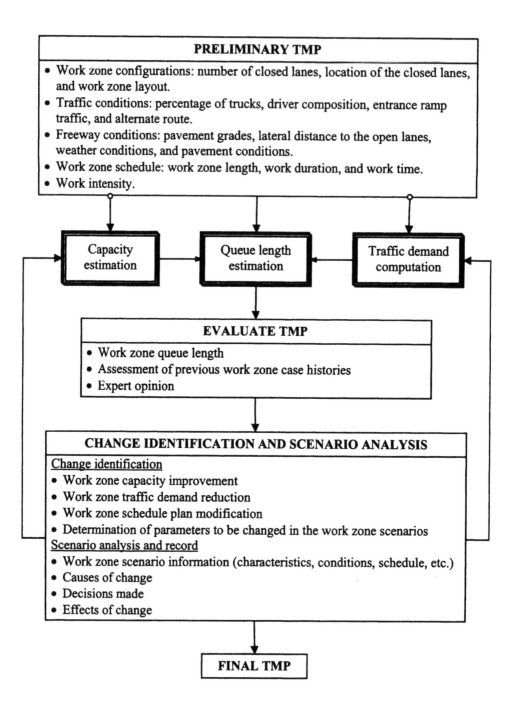

Figure 10.2 Intelligent decision support procedure for creating a freeway work zone TMP

available on various Windows operating systems.

Depending on the shell layer is the productivity layer (the middle layer in Fig. 10.3). In this work, it is subdivided into *database* and *user-interface* layers. The *database* layer provides an interface to the applications for data management, storage, and retrieval, including generating report document, processing data, and utilizing the existing databases. The frameworks in the *user-interface* layer aid in the design of user-friendly interactive interfaces. All input and output in the software are communicated through the user-interface layer. Depending on both the shell and productivity layers are the application domain layer. Generally, this layer contains algorithms and computational models for the solution of specific problems in the domain. It is usually subdivided to further categorize and generalize the application domain requirements. One or more frameworks may be used to implement this layer. In this work, four application frameworks are created in the domain layer to represent input variable quantification/normalization, traffic demand computation, work zone capacity estimation, and travelers' queue delay estimation.

Figure 10.4 shows the *IntelliZone* application architecture for freeway work zone traffic management in the form of a *package* diagram. This diagram shows the breakdown of the application into packages and their dependencies. Generally, a *package* is a collection of related software elements, which may be classes, components, or frameworks. In this work, the packages represent a collection of classes. In Fig. 10.4, the dashed-line arrows indicate the dependency of a package on another. A software dependency exists if any change in a package requires a change in the dependent package.

The high-level application domain layer is divided into two packages: an *Application* package and a *Domain* package. The Application package consists of the *Model* package and the *User-interface* package. The Model package in turn contains four packages for input variable quantification/normalization, work zone capacity estimation, traffic demand computation, and travelers' queue delay estimation. The *Domain* package contains two software packages, one for work zone scenario design and modification and the other for displaying queue length at various hours of the day. The Application and Domain software packages depend on the *Graphics* package, *File and database support* package, and *Miscellaneous support* package of the MFC library. The execution of

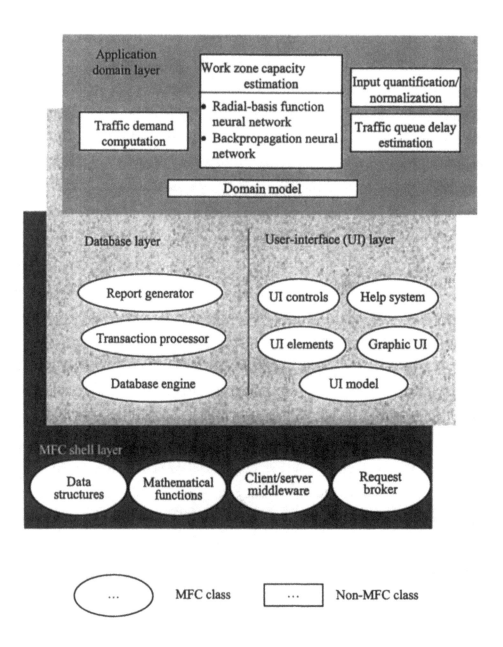

Figure 10.3 Schematic view of application architecture for constructing IntelliZone

packages is controlled by an MFC class, *CObject*, which has global dependency.

10.4   Class Diagram

IntelliZone is designed to run under all 32-bit Microsoft Windows environments such as Microsoft Windows 95/98/2000 and Windows NT 4.0 or the above version. Figure 10.5 shows the main classes for the controlling class, CObject, used in *IntelliZone*. Five control classes are used for overall work zone project management (*CWzProject*), interface windows for user-friendly data input/modification (*CDialog*), operation action for performing various computations (*CAction*), document management (*CDocment*), and graphical view (*CView*). These classes are directly derived from the MFC class *CObject* to take advantage of the services provided for object storage and retrieval. Subclasses are derived from the five control classes and inherit their properties. Functions of classes in Fig. 10.5 are described briefly in Appendix I. Figures 10.6 and 10.7 show the classes, their interrelationships, and the main methods used in each class for the work zone capacity estimation, and queue delay and length estimation parts of *IntelliZone*, respectively. Every box represents a class. The methods encapsulated in a class are listed in the box. The classes in boldface are derived directly from an MFC class and classes in italic represent non-MFC classes. Each one of the latter classes performs only a particular function and is therefore called a method class. The methods of classes in Fig. 10.6 and 10.7 are described briefly in Appendix II.

Figure 10.8 shows the information flows and exchanges among various user-interface windows. The sequence of user interactions with various interface windows is identified by shaded boxes from the bottom to top along the double-line arrows. The input/output information flow and exchanges are executed along the dashed-line arrows. A freeway work zone project is divided into several work zone segments based on the construction schedule. Work zone capacity and traffic queues are estimated for every segment. The work zone capacity and traffic queue are reestimated after any parameter change in any user-interface dialog.

10.5   User Interface for Capacity and Queue Estimation

IntelliZone provides an interactive user-friendly interface for training and using the neural network models to estimate the work zone capacity. Figure 10.9 shows the main

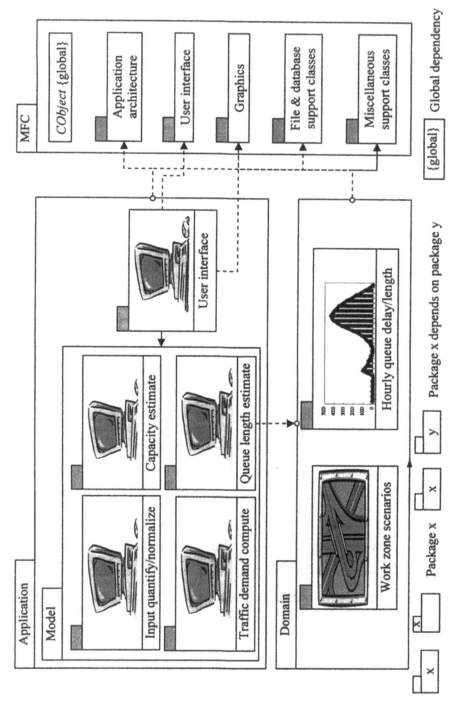

Figure 10.4 Package diagram of IntelliZone application architecture. See color insert following Chapter 15.

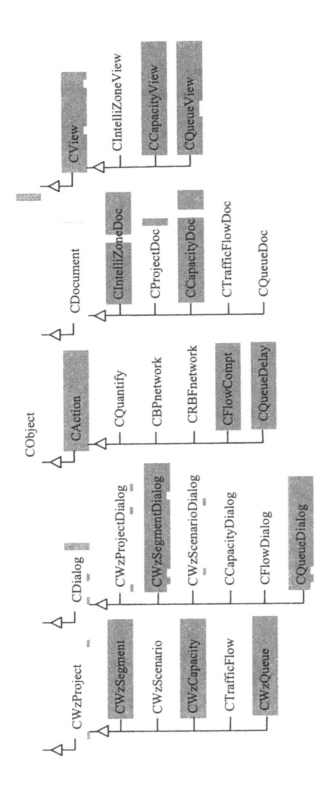

Figure 10.5 Main classes for the controlling class, CObject

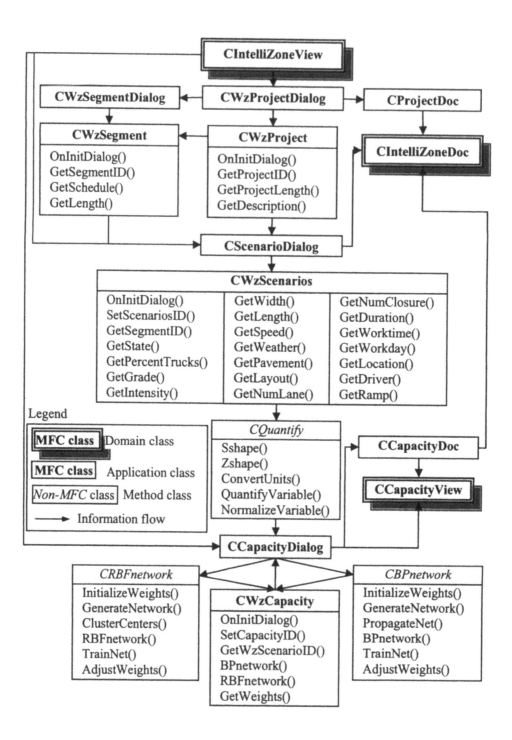

Figure 10.6 Classes, their interrelationships, and the main methods used in each class for the work zone capacity estimation

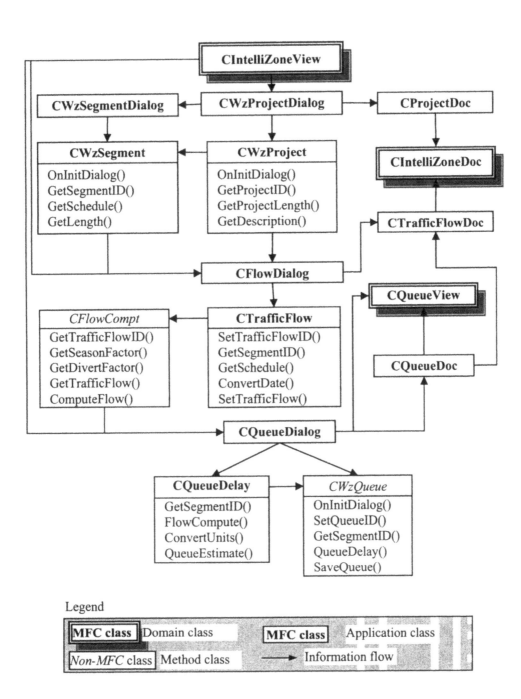

Figure 10.7 Classes, their interrelationships, and the main methods used in each class for the traffic queue estimation

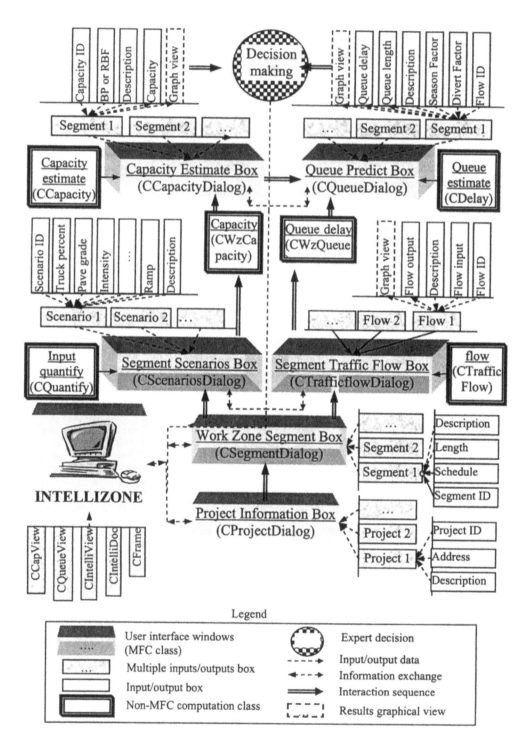

Figure 10.8 Information flows and exchanges among various user-interface windows

dialog box for inputting the values for up to seventeen input variables used in the capacity estimation model (the 18th variable, data collection locality, is inherited from the project information). The user provides numerical input values for 7 variables and chooses from a number of options for the remaining variables. For example, for the work zone intensity the user chooses from 6 different intensity levels. The user can input for up to 20 different scenarios. The data for various scenarios are summarized in a list box at the bottom of the dialog box. When a data item is not available the default value of N/A is used to indicate the lack of data.

The estimation result is shown in the three-window screen shown in Fig. 10.10. First, an option dialog is popped up asking for the capacity ID. Then, the results for training and testing of the network are displayed. The left window provides numerical results for the training and testing of the neural network. The upper-right window displays the testing results graphically along with the desired results for the data used to train the network. It shows the accuracy of the trained neural network model for estimating the work zone capacity. The lower-right window displays the training and checking curves. The lowest point in the checking curve (iteration number 23 in Fig. 10.10) represents the iteration number for the network training with the best generalization (see Section 8.4.3). The results of training at this iteration are used in the neural network model.

The entry and modification of input information for work zone queue estimation is handled by three dialog boxes. Figure 10.11 shows the dialog box for work zone queue delay and length estimation. The results can be obtained for any combination of work zone segment, traffic flow, and work zone capacity. In this dialog box the user is asked to enter the traffic flow and work zone capacity ID numbers (every traffic flow is associated with a given work zone segment ID number described in the previous dialog box), seasonal demand factor, diversion factor (e.g., for a 10% diversion, the factor is 0.9), and average length for vehicle occupancy. If a value of zero is entered for the last item, only queue delay in vehicles per hour per lane (vphpl) for every hour of the day and its maximum value during the day are presented in the box at the bottom of the dialog box. Otherwise, the total queue length in km or mi in every hour of the day as well as the maximum queue length during the day are presented.

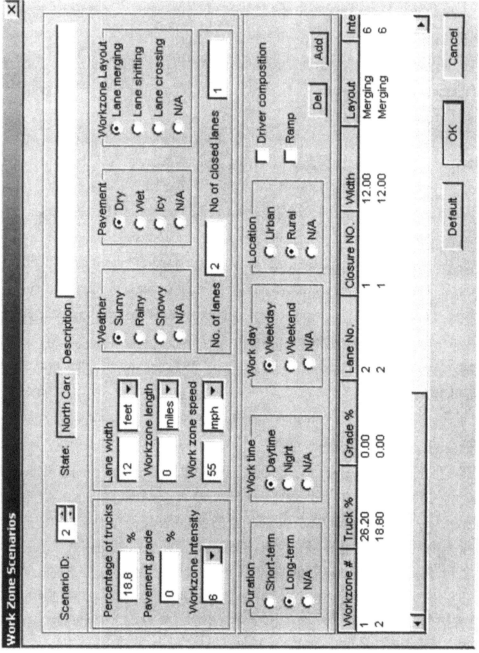

Figure 10.9 Work zone scenarios dialog box

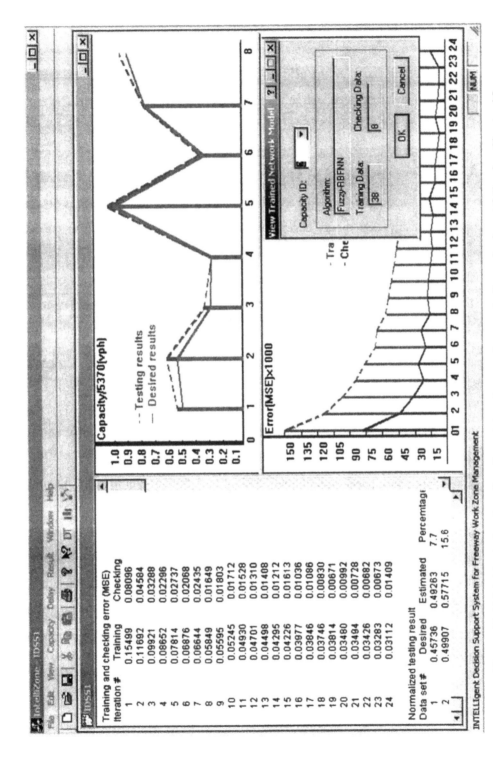

Figure 10.10 Multi-window interface for training the neural network. See color insert following Chapter 15.

If the existing flow data take into account the seasonal variation of the traffic the user (work zone engineer) shall enter a default value of one for the seasonal factor. Otherwise, the user will have the option to adjust the approaching traffic flow for seasonal variations by choosing a seasonal factor in the range of 0.5 to 2.0. The diversion factor is used to take into account the effect of an intersection close to the work zone or a residential street in an urban area. An intersection close to the work zone creates traffic diversion and affects the anticipated hourly traffic flow approaching the work zone. To take into account traffic diversion in the anticipated hourly traffic flows approaching the work zone the user has the option to adjust the approaching traffic flow by choosing a diversion factor in the range of 0.5 to 0.99. The work zone segment capacity is obtained from the capacity estimation model described in Section 9.2. The user is also provided with the option of overriding the computed work zone capacity in case a more accurate number is available based on actual measurements in the particular locality.

A three-window screen is used to display the estimation result shown in Fig. 10.12. First, a Graph Option dialog is popped up asking for the queue ID from the dialog box required in the queue delay estimation step. Then, the results for traffic flow and queue delay are displayed. The left window provides numerical values of the work zone traffic flow and queue delay (in vehicles per hour per lane) or length (in km or mi). The right-upper window displays the traffic flow graphically in the form of a bar diagram as a function of the hour of days. The maximum traffic flow in every day is noted in the display. The lower-right window displays the queue delay or length. The maximum queue value is also indicated in the display.

IntelliZone allows simultaneous execution of multiple work zone projects, each having different segments. The results for various projects are saved and may be displayed by toggling back and forth among various windows, as shown in the example of Fig. 10.12. The Result option in the menu bar of the introductory screen of *IntelliZone* (Fig. 10.9) creates a text file for input as well as outputs the results. It provides a report file for every project.

10.6 Illustrative Example

The data used for training the neural networks were obtained from California, Indiana, Maryland, North Carolina, and Texas. They are described in Section 8.5.1. As an illustrative example, *IntelliZone* is used to estimate the work zone capacity for an actual freeway work zone scenario with measured data provided by Dixon *et al.* (1997). The work zone site is a two-lane rural freeway on I-95 with one lane closure. Dixon *et al.* (1997) provide values for only nine out of seventeen input variables available in *IntelliZone*. The input values for the example are those used in Fig. 10.9. Data are not provided for pavement grade, lane width, work zone length, work zone speed limit, proximity to a ramp, weather and pavement conditions, and the driver composition. No values are used for pavement grade, work zone length, and the driver composition in this example.

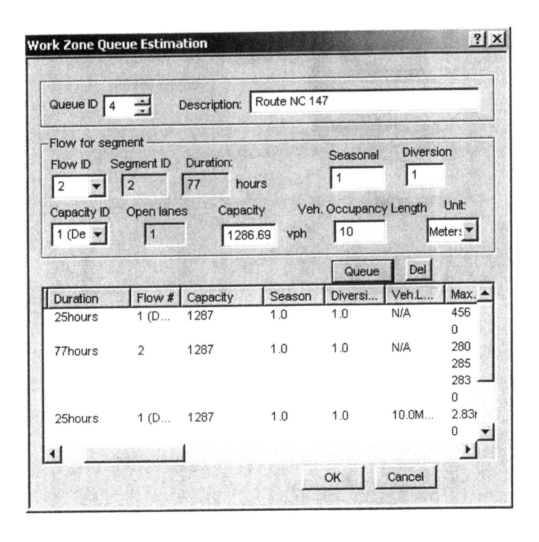

Figure 10.11 Work zone scenarios dialog box

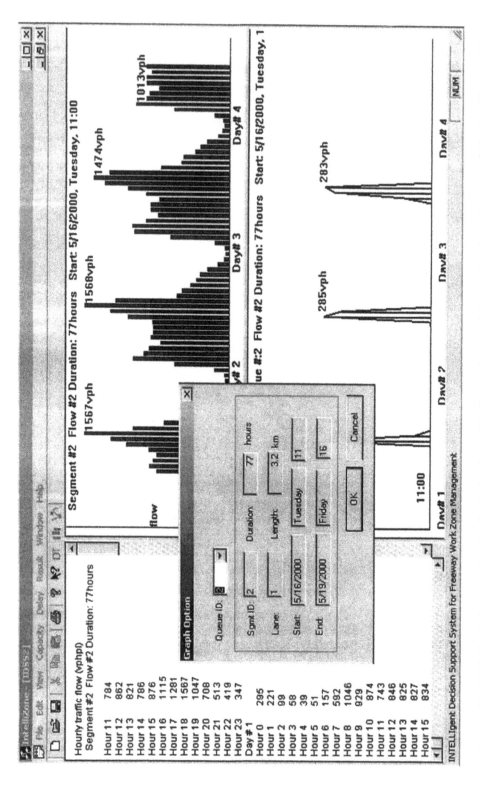

Figure 10.12 Multi-window interface for displaying the traffic flow and queue delay results. See color insert following Chapter 15.

Two different groups of scenario analysis are performed for this example. The results are summarized in Table 10.1. In group a) seven different values are used for the lane width in the work zone ranging from 9 ft to 12 ft in increments of 0.5 ft along with two different truck percentages. In this case, for the truck percentage of 26.2, the estimated work zone capacity ranges from 820 vphpl (for a lane width of 9 ft) to 1,312 vphpl (for a lane width of 12 ft). The measured value provided by Dixon *et al.* (1997) for the same truck percentage of 26.2 is 1284 vphpl. When the truck percentage is decreased to 18.8, the estimated work zone capacity ranges from 858 vphpl (for a lane width of 9 ft) to 1,339 vphpl (for a lane width of 12 ft). The measured value provided by Dixon *et al.* (1997) for the same truck percentage of 18.8 is 1,327 vphpl. A high truck percentage in the traffic flow reduces the work zone capacity value, as expected.

In group b), freeway work time (daytime or night) and workday (weekday and weekend) are considered while keeping the lane width constant at 12 ft and the truck percentages the same as those used in group a). The results are summarized in Table 10.1.

Table 10.1 Scenario analysis and influence of input variables
for work zone capacity determination

State: North Carolina	Location: Rural	Work duration: Long-term
Number of lanes: 2	Number of lane closures: 1	Truck percentage: 26.2 and 18.8
Work intensity: 6	Work time: day	Workday: weekday
Measured work zone capacity: 1,284 and 1,327 vphpl		Method: Clustering-RBF network

Group	Scenarios	Lane width (feet)	Estimated capacity (vphpl) 26.2% trucks	18.8% trucks
(a)	1	9.0	820	858
	2	9.5	943	986
	3	10.0	1,093	1,140
	4	10.5	1,206	1,257
	5	11.0	1,259	1,313
	6	11.5	1,280	1,334
	7	12.0	1,312	1,339

Group	Scenarios	Work time	Workday	Estimated capacity (vphpl)
(b)	1	Daytime	Weekday	1,265
	2	Night	Weekday	1,183
	3	Daytime	Weekend	1,008
	4	Night	Weekend	934

Traffic flow data measured in a work zone in a two-lane freeway in the state of North Carolina with one lane closure are employed to illustrate the use of *IntelliZone* for estimating the work zone queue delay or length. The hourly traffic flows approaching a work zone on route NC 147, 0.1 miles south of SR 1171, are stored in a text file (*flow.txt*) and used in the input dialog box. The period of data collection is approximately one year (year 2000). For the illustrative example, the project is divided into two work zone segments. A vehicle occupancy length of 10 ft is used. The work zone queue delays and lengths estimated by *IntelliZone* at various hours are shown in Fig. 10.11. Only part of the result can be seen in Fig. 10.11. To see the entire results the user has to scroll down and to the right in the list box. The estimated results are also shown graphically in Fig. 10.12.

10.7   Concluding Remarks

An OO model is presented for freeway work zone capacity and queue delay and length estimation. The model is implemented into an advanced intelligent decision support system, called *IntelliZone*, for effective management of work zones. *IntelliZone* has the following features and advantages:

- Integrated work zone capacity and queue estimation model.
- Capability of handling multiple-segment and multiple-traffic flow strategies.
- A mechanism to handle varying work zone scenarios.
- *IntelliZone*'s capacity estimation engine is based on pattern recognition and neural network models incorporating a large number of factors impacting the work zone capacity.
- *IntelliZone* provides a highly interactive user interface with all the tools necessary for scenario analysis and effective control of work zone traffic.
- *IntelliZone* provides a context-sensitive help facility readily available at any point of execution of the software.

Appendix I. Descriptions of Main Classes Shown in Figure 10.5

- *CAction*: Provides an interface for managing actions (input quantification/ normalization, traffic demand computation, and capacity and queue delay estimation).

- *CBPnetwork*: Encapsulates the backpropagation neural networks model.
- *CFlowDialog*: Provides an interface for inputting/modifying multi-flows for a work zone segment.
- *CCapacityDialog*: Provides an interface for estimating the work zone capacity.
- *CCapacityDoc*: Records the capacity estimation result for a work zone project.
- *CCapacityView*: Provides an interface for graphically viewing the neural network training results.
- *CIntelliZoneDoc*: Controls the presentation of results in the form of texts or documents.
- *CIntelliZoneView*: Controls the graphic presentation of results.
- *CProjectDoc*: Provides an interface for a work zone project application document.
- *CQuantify*: Quantifies and normalizes input variables for work zone capacity estimation.
- *CQueueDelay*: Encapsulates the work zone queue delay model.
- *CQueueDialog*: Provides an interface for inputting/modifying multi-queues by users.
- *CQueueDoc*: Records the queue result for a work zone project.
- *CQueueView*: Provides an interface for graphically viewing the queue results.
- *CRBFnetwork*: Encapsulates the radial basis function neural network model.
- *CScenarioDialog*: Provides an interface for abstracting multi-scenarios provided by the user.
- *CSegmentDialog*: Provides an interface for abstracting multi-segments provided by the user.
- *CTrafficFlow*: Encapsulates multi-flows for a work zone segment.
- *CTrafficFlowDoc*: Records the traffic flows for a work zone project.
- *CWzCapacity*: Provides an interface for estimating work zone capacity.
- *CWzProject*: Provides an interface for managing a work zone project.
- *CWzProjectDialog*: Provides an interface for abstracting a work zone project provided by the user.
- *CWzQueue*: Encapsulates multi-queues for a work zone project.

- *CWzScenario*: Abstracts a work zone segment scenario.
- *CWzScenarioDialog*: Provides an interface for abstracting multi-scenarios provided by the user.
- *CWzSegment*: Abstracts a segment of a work zone.
- *CWzSegmentDialog*: Provides an interface for abstracting multi-segments provided by the user.

Appendix II. Descriptions of Main Methods in Figures 10.6 and 10.7

- *AdjustWeights()*: Adjusts the weights of the links in the neural network.
- *BPnetwork()*: Executes the backpropagation neural network algorithm.
- *ClusterCenters()*: Creates cluster centers.
- *ComputeFlow()*: Modifies the traffic flow using seasonal and diversion factors.
- *ConvertDate()*: Converts the value of date to hour.
- *ConvertUnits()*: Converts the value of an input provided in SI units to a value in the U.S. customary system of units.
- *GenerateNetwork()*: Generates a network for the neural network model.
- *GetLength()*: Sets the value of the length provided by an input dialog box.
- *NormalizeVariable()*: Normalizes variables to the range of 0 to 1 by using a nonlinear normalization function.
- *OnInitDialog()*: Initializes a dialog.
- *PropagateNet()*: Propagates the errors from the hidden layer to the output layer in the backpropagation neural network model.
- *QuantifyVariable()*: Quantifies the linguistic variables.
- *QueueEstimate()*: Executes the traffic flow queue delay and length estimation.
- *RBFnetwork()*: Executes the radial basis function neural network algorithm.
- *Sshape()*: Uses the S-shaped spline-based nonlinear normalization function.
- *TrainNet()*: Trains the neural network model using the normalized training data.
- *Zshape()*: Uses the Z-shaped spline-based nonlinear normalization function.

Chapter 11

WAVELET ANALYSIS OF TRAFFIC FLOW TIME SERIES

11.1 Introduction

There has been a steady increase in both rural and urban freeway traffic in recent years resulting in congestion in many freeway systems. The freeway traffic congestion can no longer be dealt with simply by extending more highways for economical and environmental reasons (Kerner 1999). As a consequence, optimum use of the existing traffic network to manage the traffic congestion has increasingly become a more desirable alternative. Different types of freeway sections display various traffic flow characteristics. A number of researchers have investigated the characteristics of traffic flow in freeways (Disbro and Frame 1989; Dendrinos 1994; Smith and Demetsky 1997; Williams *et al.* 1998; Kerner 1999; Lee and Fambro 1999).

Intelligent Transportation Systems (ITS) play a significant role in optimizing the existing traffic network. The traffic information collected in ITS is broadcast through variable information display boards, or radio and GPS systems. Accurate and timely forecasting of traffic flow is of paramount importance in ITS in order to manage the traffic congestion in the freeway network effectively. A detailed understanding of the properties of traffic flow is essential for building a reliable forecasting model.

Actual observations of traffic flow in fields are typically time-series measurements of a scalar quantity at a fixed spatial point. Traffic flow often demonstrates a strong periodicity or seasonality (e.g., per day or per week). On the other hand, traffic flow also displays an atypical pattern near major recreational areas, shopping centers, or sports stadiums. Also, imprecision and noise often exist in the observed traffic flow due to errors in recording equipment, discontinuous records, sensitivity of equipment, and counting methods.

Wavelet-based signal processing is a powerful tool for analysis and synthesis of time series (Mallat 1999; Percieval and Walden 2000) as described in Chapter 3. Characteristics localization of time series in spatial (or time) and frequency (or scale) domains can be accomplished efficiently through wavelet decomposition. The power of wavelets for time series analysis stems from three features. First, wavelet analysis can determine the sharp transitions simultaneously in both frequency and time domains. Thus, wavelets can help identify nonlinear, chaotic, or fractal behavior displayed in any signal. Second, wavelet analysis allows for an effective representation of discontinuities in the chaotic time series. The wavelet representation of information in the time series allows for its hierarchical decomposition. In this way, the information can be analyzed in components of desired characteristics and at various levels of details. Third, when the information in time series is transformed into the wavelet domain less storage is required for its effective representation, resulting in computational efficiency for large time series.

The goal of this chapter is to identify important characteristics of traffic flow using wavelets and statistical autocorrelation function (ACF) analysis as described in Section 4.2. The ACF is used to judiciously choose the decomposition level of wavelet analysis. A hybrid wavelet packets-ACF method is proposed for analysis of traffic flow time series and determining its self-similar, singular, and fractal properties.

11.2 Discrete Wavelet Packet Transform-Based Denoising

A dilemma in the signal processing of traffic signals is that it is not possible to know by any measure of certainty whether and how the measured traffic flow is corrupted by noise. There exists no mathematical time series expression to represent the traffic flow dynamics or the noise in the traffic flow. The Fourier transform has been used in the past to reduce the noise in measured time series. This approach works reasonably well when the signal and the noise are located in different bands of the spectrum. It does not work when the time series is chaotic, which is the case for certain traffic flow situations (Disbro and Frame 1989; Dendrinos 1994). In such situations, the Fourier transform cannot be used effectively to separate the noise from the traffic flow. Another disadvantage of Fourier-based denoising methods is that abrupt changes in the frequency content of a signal are spread out over the entire spectrum. As a result, transient events cannot be properly isolated from the Fourier spectrum.

Adeli and coworkers developed wavelet-based denoising approaches for ITS applications and used them to create accurate and robust incident detection algorithms (Adeli and Karim 2000; Adeli and Samant 2000; Samant and Adeli 2000; Karim and

Adeli 2002b & 2003b). They demonstrate that the wavelet-based denoising approach can remove the low amplitude and high frequency noise effectively. The noise-free signal can then be retrieved with little loss of details through inversing wavelet transform. This denoising approach provides a powerful tool in eliminating undesirable fluctuations in observed data while at the same time preserving sharp transients

The discrete wavelet packet transform (DWPT) described in Section 3.3 provides more coefficients representing additional subtle details of the signal and therefore can be used to denoise the signal even more effectively than the conventional wavelet transform. Not only can the DWPT provide greater flexibility for detecting the oscillatory or periodic behavior and the fractal properties of time series, but also it can be used to denoise the signal even more effectively than the discrete wavelet transform (DWT).

Suppose that the traffic flow is contaminated by an additive white Gaussian noise, $e(t)$, as follows:

$$y(t) = x(t) + \sigma e(t) \qquad (11.1)$$

where $y(t)$ and $x(t)$ represent the noisy and denoised traffic flow, respectively, and σ represents the standard deviation of the noise.

The proposed DWPT-based denoising approach for the traffic flow involves three steps: DWPT-based decomposition of the noisy signal, thresholding the details coefficients, and reconstructing the denoised signal (Fig. 11.1). The DWPT-based decomposition and reconstruction of a time series signal were described in Section 3.3. In the thresholding step, two operations are performed to remove the noise from the details coefficients. First, the threshold parameter for de-noising at a given moment j, $\delta_j(t)$, is obtained by the wavelet coefficients penalization method (Barron *et al.* 1999):

$$\delta_j(t) = \sum_{i=0}^{t} [w_j(i)]^2 + 2\hat{\sigma}_j^2 t[\alpha + \log(N/t)] \qquad (11.2)$$

where the standard deviation $\hat{\sigma}_j$ is estimated from the wavelet coefficients of the j-th level decomposition and the second term is a penalization function, in which the parameter α, a real number in the range of one to five, is used to modify the penalization term (a value of $\alpha = 2$ is used in this work). Next, the denoised wavelet coefficients at a given moment, t, are obtained using soft-thresholding (Donoho 1995):

$$\overline{w}_j(t) = \begin{cases} \mathrm{sgn}[w_j(t)][|w_j(t)| - \delta_j(t)], & |w_j(t)| \ge \delta_j(t) \\ 0, & \text{otherwise} \end{cases} \qquad (11.3)$$

where sgn(w) is the signum function (it returns 1 if w is positive and -1 if w is negative).

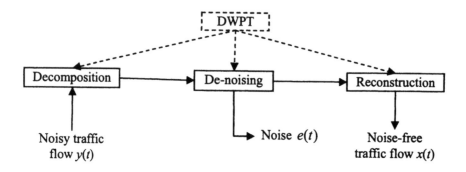

DWPT = Discrete wavelet packet transform

Figure 11.1 DWPT-based denoising procedure for traffic flow

11.3   Autocorrelation Function

In wavelet multiresolution analysis of a given time series, an important decision is selection of the decomposition level. To the best of the authors' knowledge, currently there is no theoretical criterion to select the decomposition level. A try-and-error process is most commonly used to select the level for wavelet decomposition (Percieval and Walden 2000). This approach, however, does not provide any rational basis for the selection and does not guarantee that the wavelet decomposition has effectively identified the desirable characteristics of the time series.

In this chapter, the statistical autocorrelation function (ACF) described in Section 4.2 is used for the selection of the decomposition level in wavelet multiresolution analysis of a given time series. The motivation for the use of ACF is its capability in characterizing *self-similarity* or periodicity (that is, the scale-invariance property of the time series). The ACF has been used widely to detect the trend and seasonality in a time series. In this study, the invariant characteristic is used to verify whether the wavelet decomposition level is sufficient or not. When the trend and/or seasonality shown in the ACF of the wavelet decomposition coefficients is the same as that of the original time series, we can reasonably conclude that the decomposition level used for DWPT provides sufficient accuracy.

Statistically, autocorrelation measures the degree of association between data in a time series separated by different time lags. The value of ACF for a time series y_i ($i = 1$, 2, ..., N) with N data points at any lag time index τ is estimated as follows:

$$\hat{\rho}(\tau) = \sum_{n=1}^{N-\tau}[y_{n+\tau} - \bar{y}][y_n - \bar{y}] \bigg/ \sum_{n=1}^{N-\tau}[y_n - \bar{y}]^2 \qquad (4.6)\ (\text{Repeated})$$

where $\bar{y}$ is the average of the time series. In this work, the time series y_i can be either the traffic flow or its wavelet decomposition coefficients. The value of ACF can range from -1 to 1.

11.4   Illustrative Example

11.4.1   Traffic Flow and ACF

Figure 11.2 illustrates the hourly average traffic flow along a two-lane freeway in the state of North Carolina with one lane closure obtained from the North Carolina Department of Transportation (NCDOT). It contains 1,824 hourly traffic flow data continuously recorded over a period of 76 days from October 1 to December 15, 2000 (that is, $N = 1,824$). Figure 11.2a shows the time series plot displaying a strong seasonal periodical pattern of 168 hours (one week) as expected. The three-dimensional graph of the time series with the plan axes of 24 hours and 76 days is shown in Fig. 11.2b. This graph shows the periodical pattern of 24 hours (one day) with two peak periods at around 8:00 A.M. and 6:00 P.M. every day. The 3-D graph also indicates an atypical pattern in the 61st day (November 30) from 10:00 A.M. (1450th data point) to 4:00 P.M (1456th data point) (this is not readily visible in Fig. 11.2a because of the small scale of the horizontal time axis).

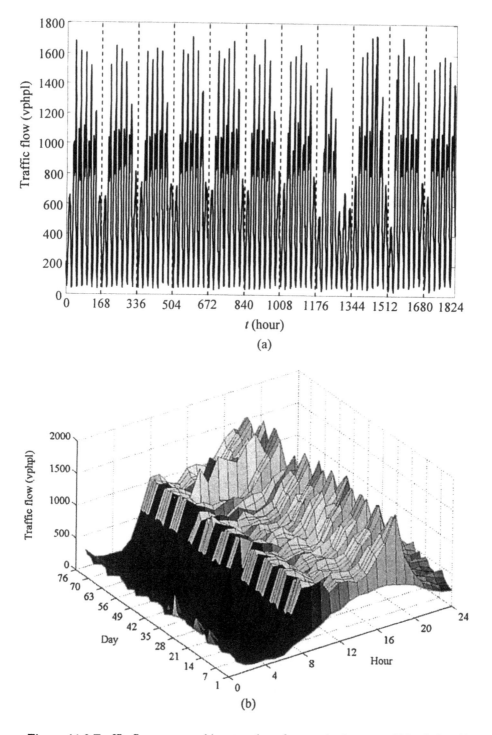

Figure 11.2 Traffic flow measured in a two-lane freeway in the state of North Carolina with one lane closure: (a) Time series, and (b) three-dimensional graph. See color insert following Chapter 15.

Figure 11.3 shows the histogram of the ACF for the traffic flow data of Fig. 11.2. Figure 11.3a displays the ACF values for a lag time of up to 600 hours, showing the periodicity over a long period of 25 days. Figure 11.3b displays the ACF values for a lag time of up to 32 hours, showing the periodicity over a short period of time. The ACF plots demonstrate strong seasonality and display both weekly and daily periodicity.

11.4.2 Wavelet Multiresolution ACF Analysis

Figure 11.4 presents a comparison of the DWT and continuous wavelet transform (CWT) of the first 512 (2^9) data points of the traffic flow series shown in Fig. 11.2 using Daubechies wavelet of order 4 (Fig. 1.1). Figures 11.4b and 11.4c show the grey-scaled graphs of multiresolution analysis for DWT and CWT, respectively. Five decomposition levels are used in the DWT analysis and 32 (equal to 2^5) scales in CWT. In Figs. 11.4b and 11.4c the bright (white) spots represent large traffic flows. The traffic flow decreases as the darkness increases. The grey-scaled graphs in Fig. 11.4 indicate scales and hours at which the traffic flow is large or low.

Figure 11.4b consists of a number of discrete grey-scaled blocks. It cannot indicate the self-similarity properties in the traffic flow. On the contrary, the CWT coefficients in Fig. 11.4c indicate continuous and subtle information about the traffic flow signals. The CWT multiresolution coefficients map provides a powerful tool for identifying self-similarity and fractal patterns in the traffic flow.

Figure 11.5 shows the DWT multiresolution analysis coefficients of the first 512 points of the traffic flow series shown in Fig. 11.2 using the Daubechies wavelet of order 4. The DWT coefficients are organized into six series. Five of these series, denoted by D_1 to D_5, are the wavelet coefficients representing the details of five levels of decomposition. They are equivalent to the CWT coefficients with scales of 2 to $2^5 = 32$ with the dyadic factor of two. It should be noted that in Fig. 11.5 for the sake of visibility of the illustration the scale of the vertical axis is chosen differently for different levels of details. The details identify subtle characteristics of the traffic flow at different levels. For example, D_1 and D_2 show the high-frequency and low-amplitude components, roughly reflecting the long-term periodicity of 168 hours (one week), but D_4 characterizes a contour with the periodicity of 24 hours (one day). The sixth series, denoted by A_5, represents the scaling coefficients and the decomposition approximation. Since the

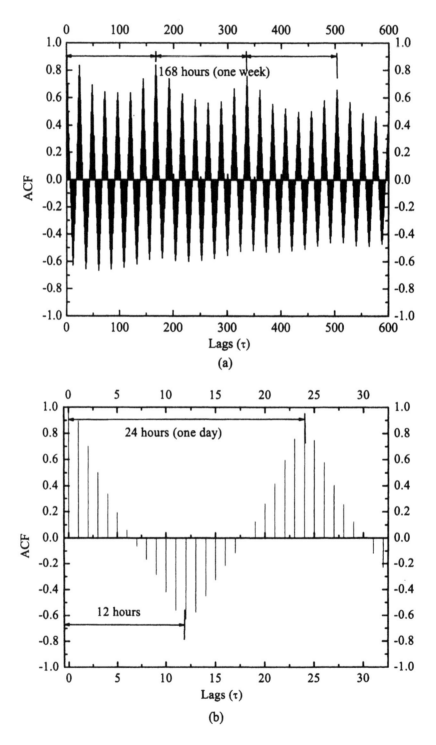

Figure 11.3 Histogram of ACF for the traffic flow data of Figure 11.2: (a) Large lags
($\tau_{max} = 600$), and (b) small lags ($\tau_{max} = 32$)

average of adjacent elements is used to approximate the traffic flow in the wavelet analysis, the scaling coefficient A_5 has the same sample mean as the traffic flow. The approximation series, A_5, reflects the long-term periodicity of the traffic flow (one week). Based on Eq. (3.16), the traffic flow can be numerically represented by

$$y(t) = A_5 + D_5 + D_4 + D_3 + D_2 + D_1 \qquad (11.4)$$

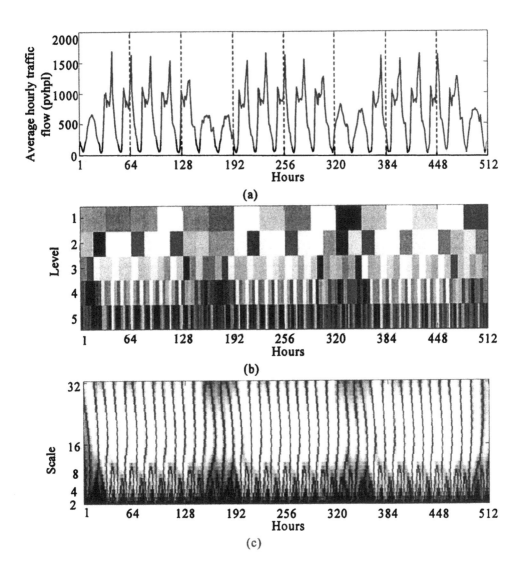

Figure 11.4 DWT and CWT of the first 512 data points of the traffic flow series shown in Figure 11.2 using the Daubechies wavelet of order 4: (a) Traffic flow; (b) absolute coefficients from DWT, and (c) absolute coefficients from CWT. See insert following Chapter 15.

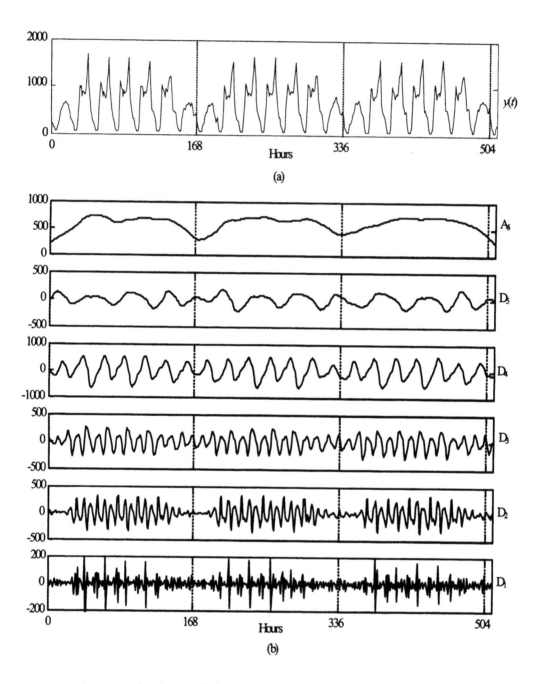

(a)

(b)

A = approximation coefficients; D_j = detail coefficients (j = 1, ..., 5);

$y(t)$ = traffic flow signal = $A_5 + D_5 + D_4 + D_3 + D_2 + D_1$

Figure 11.5 DWT multiresolution analysis coefficients of traffic flow series with five-level details and 5th approximation: (a) Traffic flow, and (b) decomposition coefficients

Figure 11.6 shows the autocorrelation functions for detail coefficients of the 7-level DWT for the first 512 points of the traffic data (Fig. 11.5a) using the Daubechies wavelet of order 4 and a maximum time lag of 50 hours ($\tau_{max} = 50$). This figure is used to determine what level of wavelet decomposition is necessary in order to be able to identify short-term periodicity. Short-term periodicity can be observed in the ACF of D_1 to D_4 within the maximum time lag of 50 hours. The period of periodicity is 24 hours, the same as that seen in the original flow data of Fig. 11.2b. It is concluded that four levels of wavelet decomposition are necessary and sufficient to capture and represent the short-term characteristics and periodicity in the traffic flow.

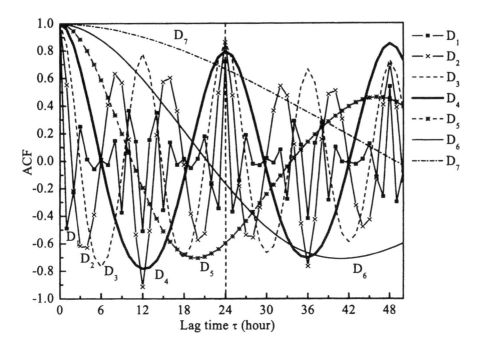

$D_j = j$-th level Detail coefficients ($j = 1, 2,..., 7$)

Figure 11.6 Autocorrelation functions for detail coefficients of the 7-level DWT using the Daubechies wavelet of order 4 ($\tau_{max} = 50$)

The DWT approximation coefficients for decomposition levels 4 to 7 are shown in Fig. 11.7a. This figure shows that the approximation coefficients can identify the long-term periodicity of the traffic flow when the decomposition level is 4 or 5. The approximation coefficients for decomposition levels 6 and 7 do not display any periodicity and consequently are not necessary for accurate traffic flow signal representation. The ACFs for approximations A_4, A_5, A_6, and A_7 with a maximum time lag of $\tau_{max} = 350$ hr are shown in Fig. 11.7b. The long-term periodicity or self-similarity of the traffic flow with a period of 168 hours (one week) is clearly identified in the ACF of A_5 and A_6 (Fig. 11.7b). It is concluded that five levels of wavelet decomposition are necessary and sufficient to capture and represent the long-term characteristics and periodicity in the traffic flow.

Based on the ACF analysis of the DWT multiresolution coefficients (details in Fig. 11.6 and approximations in Fig. 11.7), we can conclude that 5 decomposition levels capture both short-term and long-term periodicity and therefore can adequately characterize the details of the given traffic flow. The same analysis was performed using the DWPT. Results similar to those presented in Figs. 11.5 to 11.7 are obtained but not presented in the chapter for the sake of brevity. However, it is found that three decomposition levels of DWPT can adequately characterize the details of the given traffic flow as opposed to five decomposition levels necessary for DWT. DWPT-based results for denoising and singularity identification using only three decomposition levels are presented in the next section.

11.4.3 Denoising and Singularity Identification

Figures 11.8b and 11.8c show the denoised average hourly traffic flow for the first 512 points of the original traffic flow shown in Fig. 11.8a using the Daubechies wavelet of order 4 and DWT and DWPT methods, respectively. Considering the details of traffic flow in part of Figs. 11.8a to 11.8c enlarged and enclosed in a rectangular box, it is observed that the DWPT-based de-noising approach smoothes the traffic flow more effectively than the DWT-based de-noising approach while at the same time capturing the subtle features in the signal.

The DWPT is used to identify singularity in the traffic flow time series in addition to denoising the data. The three-level wavelet packet decomposition using the

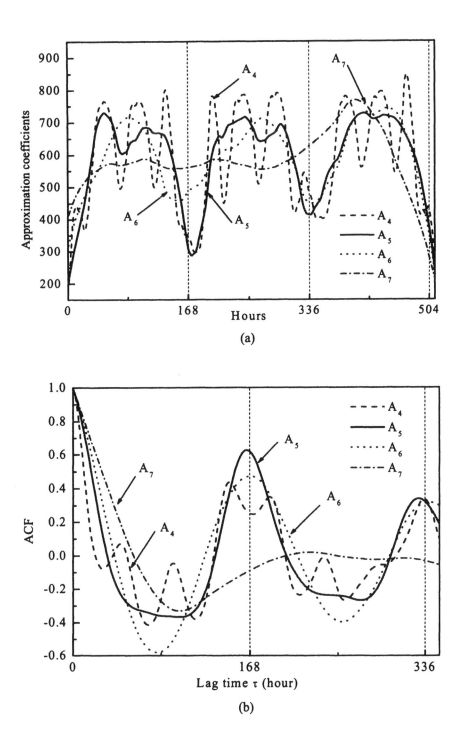

Figure 11.7 Approximation coefficients with various decomposition levels and their ACFs: (a) Approximation coefficients, and (b) ACFs of approximation coefficients with maximum lag $\tau_{max} = 350$

Daubechies wavelet of order 4 is adopted to identify the singularity in the last 512 traffic flow data points shown in Fig. 11.2a in the range of 1,312 to 1,824 shown in Fig. 11.9a. The data include an atypical pattern, as shown in Fig. 11.2b. The three-level DWPT decomposition for this traffic flow data series is shown in Fig. 11.10. The data are resolved into eight series shown in Fig. 11.9b and identified at the third level of the decomposition tree of Fig. 11.10 as AAA_3, which represents the third-level approximation coefficients (A_3) resulting from the second-level approximation (AA_2) to DDD_3, which represents the third-level details coefficients (D_3) resulting from the second-level details (DD_2). Thus, the denoised traffic flow can be represented mathematically by the third-level DWPT decomposition coefficients as follows:

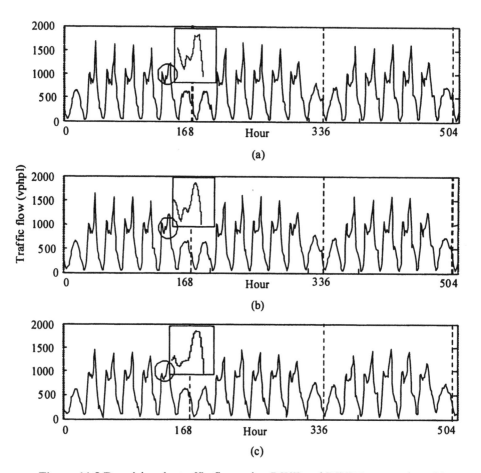

(a)

(b)

(c)

Figure 11.8 Denoising the traffic flow using DWT and DWPT approaches: (a) Original traffic flow; (b) denoised traffic flow using DWT, and (c) denoised traffic flow using DWPT. See color insert following Chapter 15.

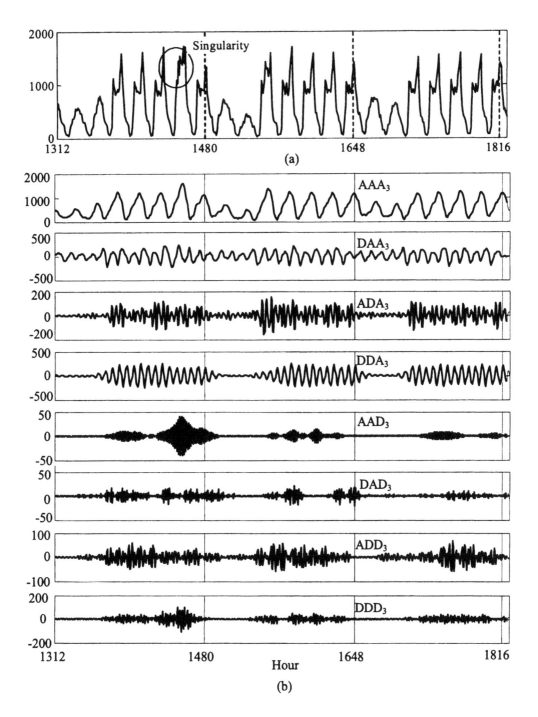

Figure 11.9 Singularity identification in the last 512 traffic flow data points shown in Figure 11.2a using DWPT and the Daubechies wavelet of order 4: (a) Denoised traffic flow with singularity, and (b) 3-level decomposition of DWPT

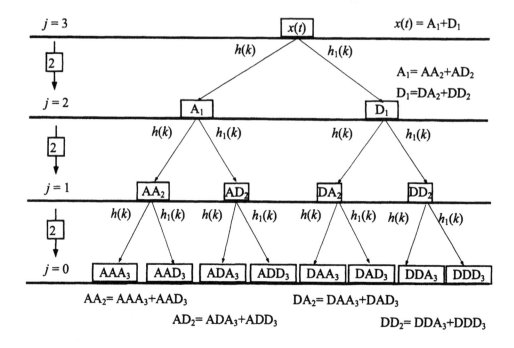

Figure 11.10 Three-level DWPT decomposition tree of the last 512 traffic flow data points shown in Figure 11.2(a)

$$x(t) = AAA_3 + AAD_3 + ADA_3 + ADD_3 + DAA_3 + DAD_3 + DDA_3 + DDD_3 \qquad (11.5)$$

Again, in Fig. 11.9b for the sake of visibility of the illustration the scale of the vertical axis is chosen differently for different levels of details.

In Fig. 11.9b, the singular oscillatory behavior is clearly identified in the coefficients AAD_3 (as well as in DDD_3 but not as conspicuously as in AAD_3) in the range of 1,350 to 1,357, where there is an abnormal enlargement of the localized oscillatory amplitude compared with other typical traffic patterns. The singularity in the decomposition coefficients results from an atypical increase of the traffic flow (Fig. 11.2b). A similar analysis was performed using the DWT with five levels of decomposition. The results are shown in Fig. 11.11. It is found that the DWT with five levels of decomposition cannot identify the singularity. Further, increasing the number of decomposition levels does not help identify singularities.

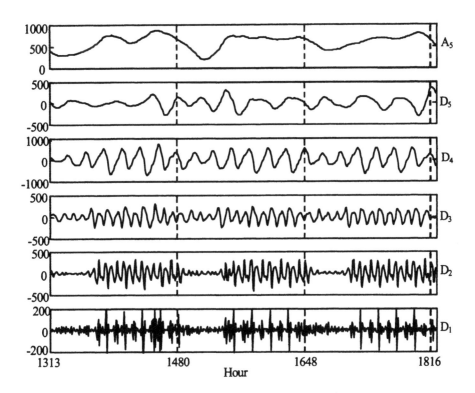

Figure 11.11 DWT multiresolution analysis coefficients for the last 512 data points of the traffic flow series shown in Figure 11.2 using the Daubechies wavelet of order 4 with five levels of decomposition

11.5 Concluding Remarks

A hybrid wavelet packets-ACF method was presented for analysis of traffic flow time series and determining its self-similar, singular, and fractal properties. The proposed method provides a powerful tool in removing the noise and identifying the singularity in the traffic flow. The DWPT analysis provides more coefficients representing additional subtle details of the signal and therefore can be used to denoise the signal even more effectively than the conventional wavelet transform. The DWPT can not only provide greater flexibility for detecting the oscillatory or periodic behavior and the fractal properties of time series, but it can also be used to denoise the signal even more effectively than the DWT. The 3-level DWPT analysis can capture the subtle properties of the traffic flow including singularities without any loss of details. The results presented in this chapter are used in the next chapter to develop a powerful traffic flow forecasting model.

Chapter 12

WAVELET NEURAL NETWORK FOR TRAFFIC FLOW FORECASTING

12.1 Introduction

Accurate and timely forecasting of traffic flow is of paramount importance for effective management of traffic congestion in intelligent transportation system (ITS). Ross (1982) and Okutani and Stephanedes (1984) apply a filtering technique for forecasting short-term traffic volume based on the past traffic flow. Davis and Nihan (1991) present a nonparametric forecasting model based on the *k*-nearest-neighbor approach for forecasting short-term freeway traffic flow. However, both the filtering technique and the nonparametric model can forecast the short-time traffic flow with reasonable accuracy only when the traffic flows are relatively constant. More recently, the linear statistical time series models, such as the autoregressive integrated moving average (ARIMA) model (Williams *et al.* 1998) and seasonal ARIMA model (Lee and Fambro 1999) have been used to improve the accuracy of short-time freeway traffic flow forecasting. However, the linear characteristics of these time series models cannot capture the dynamics and nonlinearity existing in the traffic flow, thus limiting their ability to forecast long-term traffic flow (more than a few hours) accurately.

Over the past decade, a number of papers have been published on the application of neural network models for forecasting traffic flow, taking advantage of their ability to capture the indeterministic and complex nonlinearity of time series (Smith and Demetsky 1994 & 1997; Florio and Mussone 1996; Park *et al.* 1998; Yun *et al.* 1998; Yasdi 1999; Smith *et al.* 2002; Park 2002; Dharia and Adeli 2003). Smith and Demetsky (1994 & 1997) compare the backpropagation neural network (BPNN) model with the ARIMA model for predicting short-term traffic flow. They conclude that the BPNN model is superior to the linear statistical ARIMA model because the former is more sensitive to the dynamics of traffic flow than the latter and does not experience the overprediction characteristics of the ARIMA model.

Yun *et al.* (1998) investigate the performance of a BPNN model, a *finite impulse response* (FIR) model (a linear filtering method), and a *time-delayed recurrent* model (a dynamic BPNN) for forecasting the traffic volume. They use three different traffic flow data sets collected from interstate highways, intercity highways, and urban intersections with very different characteristics in terms of volatility, period, and fluctuations. Their study shows that the time-delayed recurrent BPNN model outperforms the other models in forecasting very randomly moving traffic flow. In contrast, the FIR model demonstrates better forecasting accuracy than the time-delayed recurrent BPNN network for relatively regular periodic data. However, the BPNN model has its inherent shortcomings such as lack of an efficient constructive model (for example, requiring arbitrary selection of the number of hidden nodes), slow convergence rate resulting in excessive computation time, and entrapment in a local minimum as pointed out by Adeli and Hung (1995) and others.

In order to overcome the aforementioned shortcomings of the BPNN model, Park *et al.* (1998) use a radial basis function neural network (RBFNN) model for forecasting the short-term traffic flow. They concluded that the RBFNN model *"provided the best performance and required less computational time than BPN."* Park (2002) and Yin *et al.* (2002) propose a fuzzy-neural network approach for forecasting the short-term traffic flow. They conclude that the hybrid method requires less computation time and provides higher prediction accuracy than the BPNN model. However, these methods still lack an efficient constructive model.

The measured traffic flow using sensors (e.g., loop and sonic detectors) is usually in the form of time series. To achieve high accuracy in the traffic flow prediction model, in the opinion of the authors, one has to adopt more than a purely statistical approach. Rather, the model has to incorporate the dynamics of the traffic flow. The authors developed a hybrid discrete wavelet packet transform (DWPT)–autocorrelation function (ACF) method for analysis of traffic flow time series and determining their *self-similar* (scale-invariance property of the time series), *singular* (out of the ordinary), and *fractal* (irregular patterns) properties as described in Chapter 11. The methodology provides a powerful tool for identifying singularities in the traffic flow.

In order to develop an efficient long-term traffic flow forecasting model, the following rationale and points are considered in developing the model: (a) Traffic flow is highly complex and not amenable to accurate mathematical modeling. Therefore, nonparametric methods and adaptive algorithms are required to learn and recognize patterns in an effective manner. (b) Prior knowledge of flow behavior should be used wherever possible to simplify the algorithm and improve performance. (c) The forecasting algorithm must be capable of real-time operation. As a result, computationally intensive operations must be avoided.

It has been demonstrated that adroit integration of wavelets with neural networks can result in a powerful approach for pattern recognition with enhanced feature detection capability (Zhang and Benveniste 1992; Zhang 1997; Adeli and Karim 2000; Samant and Adeli 2000 & 2001; Karim and Adeli 2002a&b; Ghosh-Dastidar and Adeli 2003; Adeli and Ghosh-Dastidar 2004). A summary of the main concepts of wavelets was presented in Chapter 3. In this chapter, a novel dynamic time-delay recurrent wavelet neural network (WNN) model is presented for forecasting traffic flow. The model incorporates the self-similar, singular, and fractal properties discovered in the traffic flow described in Chapter 11. The goal is to achieve high accuracy for traffic flow forecasting.

In the following sections, first the input vectors are constructed from the denoised data. Then, a time-delay recurrent WNN model is created by integrating the wavelet frame and dynamic wavelet neural network. The modified Gram-Schmidt algorithm and the Akaike's final prediction error criterion are used to select the wavelets and determine the minimum number of wavelets required, respectively. Finally, the model is trained and validated using the real traffic flow data obtained from the North Carolina Department of Transportation.

12.2 Constructing Input Vectors

The forecasting problem is to find a general nonlinear mapping function based on the input-output data sets. The traffic flow forecasting models presented in the literature are based on the immediate past traffic flow only. The prediction is done independent of the time of the day and the day of the week. In this chapter, the time of the day and the day of the week are included as two additional variables in the model, called current status variables, with the goal of capturing the inherent self-similarity of the traffic flow. As such, the discrete dynamic input–output mapping is expressed as follows:

$$y(t) = f(\mathbf{y}^{t-1}) + g(\mathbf{x}^t) + e(t) \tag{12.1}$$

where $\mathbf{x}^t$ and $\mathbf{y}^{t-1}$ represent the current flow status input vector and feedback traffic flow input vector at time t (representing the discrete time series values up to time t), respectively, $f(.)$ and $g(.)$ are scalar nonlinear mapping functions, and $e(t)$ is the error between the actual and predicted values of the future traffic flow output $y(t)$. The objective is to find the relationship between the past observation vector $\mathbf{y}^{t-1}$ as described by function $f(.)$ as well as the status input vector $\mathbf{x}^t$ at forecasting time t as described by function $g(.)$ and the future output $y(t)$.

An hourly average traffic flow time series (in vehicles per hour) is used in this chapter to demonstrate the effectiveness of the proposed model. The model can also be adopted for other traffic flow time series (for example in vehicles per second) with a slight modification of the status input. The status input vector at time t in Eq. (12.1), $\mathbf{x}^t = [h, d]'$, consists of two variables: the hour of time t (time of the day), denoted by h ($h = 1$, 2, ..., 24, representing 24 hours of a day), and the day of time t (day of the week), denoted by d ($d = 1, 2, ..., 7$, representing Monday to Sunday, respectively), where the prime denotes the transpose of a vector. The status vector $\mathbf{x}^t$ is independent of traffic flow.

The past observation vector $\mathbf{y}^{t-1}$ consists of $D + 2$ variables: the value at t-168 time (one week ago), the value at time t-24 (one day ago), and the continuous D discrete inputs right before the moment t, where D is the size of the past observations. Traffic states evolve in time. Actual observations of traffic flow in fields are typically time-series measurements of a scalar quantity at a fixed spatial point which often demonstrate a strong periodicity over 24-hour intervals and a much stronger periodicity over weekly intervals (Jiang and Adeli 2004a). The two status variables in $\mathbf{x}^t$ and the two past observation variables in $\mathbf{y}^{t-1}$ are employed to quantitatively identify the self-similarity of the traffic flow time series.

An hourly average traffic flow time series is denoted by discrete quantities $\{y_i\}$, $i = 1, 2, ..., N$, where N is the number of data points. For example, if hourly average traffic flow data are available for a period of one week then N is equal to $24 \times 7 = 168$. The input vector at time t, $\mathbf{y}^{t-1}$, in the nonparametric forecasting model is represented by the state space vector, $\mathbf{y}_k$, as follows:

$$\mathbf{y}_k = [y_{k-168}, y_{k-24}, y_{k-1}, y_{k-2}, ..., y_{k-D}]', \quad k = 1, 2, ..., N_a \qquad (12.2)$$

where y_{k-168} represents the traffic flow one week ago, y_{k-24} represents the traffic flow one day ago, and the other variables represent the D continuous discrete traffic flows before the forecasting time t. The data point subscript, k, satisfies the following condition:

$$k \leq N - 168, \; n \in Z \tag{12.3}$$

where Z is the set of all integers. Based on Eq. (12.3), the maximum number of state space points, N_a, is equal to

$$N_a = N - 168 \tag{12.4}$$

The dimension of space in the state space representation of the problem and the input vector is $D + 2$. A proper choice of the input dimension plays an important role in reconstructing an appropriate multivariate input. If D is too small, the model produces inaccurate forecasting results. If D is too large, it can also lead to overprediction and inaccurate results in the nonparametric neural network model, in addition to increasing the computational cost. A trial-and-error approach is usually used to find the most suitable value for the input dimension in the nonparametric approach (Smith and Demetsky 1994 & 1997; Florio and Mussone 1996; Park et al. 1998; Yun et al. 1998; Yasdi 1999). However, the trial-and-error approach a) does not provide a rational basis for the selection of the input dimension, b) is cumbersome, c) is computationally time-consuming, and d) does not guarantee accurate prediction results.

Computational complexity increases exponentially with an increase in the input dimension of the neural network. In addition, the size of the corresponding training data set has to be increased exponentially in order to achieve the same level of forecasting accuracy, as described in Section 5.1. In this work, the statistical autocorrelation function (ACF) is used for selection of the input dimension of a given time series to overcome the dimensionality curse in the nonparametric approach to traffic flow forecasting. Statistically, autocorrelation measures the degree of association between data in a time series separated by different time lags (see Section 11.3). The ACF is evaluated for various values of the lag time and the results are plotted (for hourly traffic flows, the lag time will be in hours). Wherever the ACF curve intersects the lag time axis its value is zero, indicating that $y(t-D)$ and $y(t)$ are linearly independent. The lag time corresponding to the first point of intersection is chosen as the *optimum* input dimension.

12.3 Wavelet Neural Network (WNN) Model

12.3.1 Issues to Be Considered

The dynamic system function represented by $f(\mathbf{y})$ in Eq. (12.1) is approximated using the wavelet transform functions and wavelet coefficients expressed in a general form as follows (Daubechies 1992):

$$\bar{f}(\mathbf{y}) = \sum_m w_m \sum_{\mathbf{a,b}} \psi_{\mathbf{a,b}}(\mathbf{y}), \; m = 1,\ldots, K, \; \mathbf{a, b} \in \Re, \; \psi \in L^2(\Re) \qquad (12.5)$$

where K is the number of wavelets to be discussed in the next section, $\bar{f}(\mathbf{y})$ represents the approximation of $f(\mathbf{y})$, w_m represents the discrete wavelet transform coefficient, and $\psi_{\mathbf{a,b}}(\mathbf{y})$ is the two-dimensional wavelet expansion functions obtained from the basic wavelet function $\varphi(t)$ by simple scaling and translation

$$\psi_{\mathbf{a,b}}(\mathbf{y}) = (1/\sqrt{|\mathbf{a}|})\varphi(\frac{\mathbf{y-b}}{\mathbf{a}}), \; \mathbf{a, b} \in \Re, \varphi \in L^2(\Re) \qquad (12.6)$$

in which the parameter vectors $\mathbf{a} \neq 0$ and $\mathbf{b}$ denote the frequency (or scale) and the time (or space) location vectors corresponding to the multidimensional input vector $\mathbf{y}$, respectively, and $\Re$ is the set of real numbers. The notation $L^2(\Re)$ represents the square summable constructed state space vectors.

Several issues need to be considered in the wavelet transformation represented by Eq. (12.5). First, a choice is made between the continuous wavelet transform (CWT) and the discrete wavelet transform (DWT). The CWT represents the time series more accurately than the DWT because it makes subtle information visible. However, it has two disadvantages: 1) it requires a large number of wavelet coefficient vectors $\mathbf{a}$ and $\mathbf{b}$ which are the variables to be determined in the WNN model, thus making the size of the WNN network impractically large, and 2) it requires integrating over every possible scale and dilation, thus making it computationally expensive. Consequently, the DWT is used in the traffic flow forecasting WNN model and applied to perform the wavelet transform only at discrete scales $\mathbf{a}$ and times $\mathbf{b}$.

Second, an appropriate wavelet has to be selected. Often orthogonal wavelets such as Daubechies wavelets (Daubechies 1988 & 1992) are widely used because they are efficient for signal decomposition and reconstruction. To satisfy the orthogonality condition, the wavelet function has to satisfy restricting constraints: the integral of the wavelet must be equal to zero and the integral of the square of the wavelet function must be equal to one. Such constraints make the orthogonal wavelets non-differentiable. In the new traffic flow forecasting WNN model, errors between the approximated and actual outputs are minimized using a mathematical optimization approach which requires derivatives of the wavelet function. As such, a non-orthogonal differentiable wavelet function, the Mexican hat function, is used in the WNN model.

Compared with other wavelet functions, the Mexican hat wavelet function has several characteristics that are advantageous in this work: 1) it has an analytical expression and therefore can be used conveniently for decomposing multidimensional time series, 2) it can be differentiated analytically, 3) it is a non-compactly supported but rapidly vanishing function (Jiang and Adeli 2004), and 4) it is computationally efficient. The Mexican hat wavelet function is expressed as follows (Fig. 12.1):

$$\varphi(\varsigma) = (D - \varsigma^2)\exp(-\varsigma^2/2) \tag{12.7}$$

where $\varsigma = \left\| \dfrac{\mathbf{y} - \mathbf{b}}{\mathbf{a}} \right\|$ for a multidimensional input vector $\mathbf{y}$ in which $\|\mathbf{z}\| = \sqrt{\sum_m |z_m|^2}$ indicates the Euclidean distance.

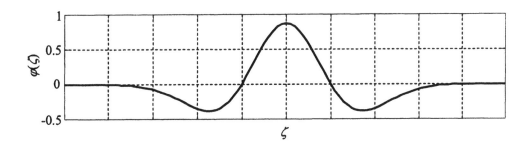

Figure 12.1 Mexican hat wavelet function

To avoid intensive computations for every possible scale a and dilation b (the two wavelet parameters used to decompose the time series into the time and frequency domains) in this chapter, dyadic values in Eq. (3.3) are used to initialize both scaling and dilation parameters in the DWT analysis and repeated as follows:

$$a_j = 2^j, \ b_{j,k} = k2^j, \ j, k \in Z \qquad\qquad \text{(3.3) (Repeated)}$$

where k and j denote the time and the frequency indices, respectively.

Third and finally, the DWT coefficients created from the nonorthogonal wavelet function decomposition contain redundant coefficients, which need to be truncated. Unlike the orthogonal wavelets where no subspace overlaps with another, the nonorthogonal wavelet coefficients are not independent of one another. Redundant wavelet bases (coefficients) always exist in the DWT decomposition of signals using nonorthogonal wavelet functions. Such DWT wavelets with redundant bases are collectively referred to as *wavelet frame*.

Frame as a *redundant basis* was first proposed by Duffin and Schaeffer (1952) in the nonharmonic Fourier analysis and introduced into the wavelet theory by Daubechies (1988 & 1992). A wavelet frame is a redundant system where a more than necessary number of wavelets are allowed to reconstruct a signal. Such redundancy provides two advantages: a) it provides flexibility in the design of wavelets, as one can choose from a subset of redundant wavelets, and b) it adds some extra features such as adaptable translation parameters desirable in multidimensional decomposition required for traffic flow forecasting. It should be noted that only nonorthogonal wavelets can be integrated with a conventional neural network to form a WNN. The redundant wavelet bases, however, add to the computational cost, making the size of the traffic flow forecasting WNN model prohibitively large and its real-time implementation impractical. The solution is to create an efficient and practically usable truncated wavelet frame by keeping only a subset of the redundant wavelets necessary to obtain accurate results and removing the rest. Those wavelet bases that do not affect the accuracy are eliminated in order to improve the efficiency of the forecasting model. The truncated wavelet frame is employed in the dynamic time-delay WNN model for traffic flow forecasting because it can effectively model approximation of the system dynamics with the multidimensional input–output time series.

12.3.2 Wavelet Frame

A wavelet frame is defined as a family of wavelet functions, ψ, if two constants A and B exist such that $A > 0$, $B < \infty$, satisfying the following conditions (Daubechies 1988 & 1992):

$$A\|\mathbf{y}\|^2 \le \sum_{m=1}^{L_w} \left(\langle \mathbf{y}, \psi_m \rangle \right)^2 \le B\|\mathbf{y}\|^2 \qquad (12.8)$$

where the input vector $\mathbf{y}$ is a multidimensional state space vector, ψ_m represents the m-th wavelet function vector with the dilation vector $\mathbf{a}_m$ and translation vector $\mathbf{b}_m$, and L_w is the number of wavelets used for approximating the dynamics of the freeway system ($L_w \ge K$), which depends on the DWT decomposition level in the initialization of the wavelet parameters (j in Eq. 3.3). The term $\langle \mathbf{y}, \psi \rangle$ represents the inner product of two scalars: $\langle \mathbf{y}, \psi \rangle = \sum_{p=1}^{D} y_p \psi_p$. The term $\|\mathbf{y}\|^2$ denotes the inner product of the state space vector series: $\|\mathbf{y}\|^2 = \langle \mathbf{y}, \mathbf{y} \rangle$. The parameters A and B are called the *frame bounds*, providing a quantitative measure on how many wavelet bases are redundant. When A = B = 1, ψ_m represents the orthogonal wavelet bases with $\|\psi_m\| = 1$ and no redundant wavelet parameters.

Two questions need to be answered when best approximating the function represented by Eq. (12.5) using the wavelet frame (Eq. 12.8). First, which wavelets (ψ_m) need to be chosen from the wavelet decompositions? Second, how many wavelets (K) are required? The two questions will be answered in Sections 12.3.3 and 12.3.5.

As noted previously, the nonorthogonal Mexican hat wavelet function is not compactly supported but is rapidly vanishing. Its support, $S_{\mathbf{a},\mathbf{b}}$, is defined by (Daubechies 1988)

$$S_{\mathbf{a},\mathbf{b}} = \{ \mathbf{y} \in \aleph : |\psi_m(\mathbf{y})| > \varepsilon \max_{\mathbf{y}} |\psi_m(\mathbf{y})| \}, \ m = 1, \ldots, L_w \qquad (12.9)$$

where $\aleph$ is the entire input state space, $\psi_m(\mathbf{y})$ is the m-th wavelet with regard to the input vector $\mathbf{y}$, and ε is a given small positive number.

The *empty* wavelets whose supports do not contain any data are removed from the wavelet decomposition based on the following two steps. First, find the index set of all wavelets, I_m, whose supports contain $\mathbf{y}_m$ as follows:

$$I_m = \{(\mathbf{a},\mathbf{b}): \mathbf{y}_m \in S_{\mathbf{a},\mathbf{b}}\},\ \mathbf{y}_m \in \Gamma,\ m = 1, ..., L_w \qquad (12.10)$$

where Γ represents the time domain of the sample input time series. Next, the union of I_m's gives the indexes of the wavelets whose supports contain at least one data point. A non-empty wavelet frame is created as follows:

$$W = \{\psi_m(\mathbf{y}):(\mathbf{a},\mathbf{b}) \in I_1 \cup I_2 \cup \cdots \cup I_{L_w}\} \qquad (12.11)$$

where $\cup$ denotes the union operator.

12.3.3 Modified Gram-Schmidt algorithm

Equation (12.11) represents the wavelet frame with a complete set of nonzero wavelet coefficients. The modified Gram-Schmidt algorithm (Zhang 1997) is applied in this chapter to select which wavelet bases to be eliminated in approximating Eq. (12.5). The algorithm first selects one wavelet from the complete nonzero wavelet coefficient sets which best approximates the measured data. Next, this wavelet is combined with the remainder of the wavelet sets one at a time and the combination which best approximates the measured data is determined. The procedure is repeated for all nonzero wavelet coefficients. In order to improve the computational efficiency, subsequently selected wavelets are orthonormalized with respect to those selected earlier. The procedure is used to find out the best combination of wavelets with the best approximation performance for all possible number of wavelets.

Let $\boldsymbol{\varphi}_{l_q} = [\psi_{l_k}(y_1), \psi_{l_k}(y_2), ... \psi_{l_k}(y_D), \psi_{l_k}(y_{24}), \psi_{l_k}(y_{168})]'$ denote the selected wavelet vector at the current iteration number q (the multidimensional input vector $\mathbf{y}$ results in a multidimensional wavelet set), in which l_q is the index of the wavelet selected from W and $\psi_{l_k}(y_p) = (1/\sqrt{|a_p|})\varphi_{l_q}[(y_p - b_p)/a_p]$, $p = 1, 2, ..., D, 24, 168$, $\mathbf{q}_{l_j}$ be the orthonormalized $\boldsymbol{\varphi}_{l_j}$ $(j = 1, ..., q-1)$, and $\boldsymbol{\psi}_{q-1} = \{\boldsymbol{\varphi}_{l_1}, \boldsymbol{\varphi}_{l_2}, ..., \boldsymbol{\varphi}_{l_{q-1}}\}$ represent the wavelets selected in the previous iteration $q-1$. The iterative modified Gram-Schmidt algorithm consists of the following steps:

1) Initialize the variables at $q = 0$: $l_0 = 0$, $\mathbf{q}_{l_0} = 0$, and let $\mathbf{p}_j^{(0)} = \mathbf{\psi}_j$ for all $j \in I^{(0)}$, where $I^{(0)} = \{1, 2, ..., J\}$, in which J is the number of the nonzero wavelet coefficients in the non-empty wavelet frame W.

2) Update variables in iteration q:

$$\mathbf{p}_j^{(q)} = \mathbf{p}_j^{(q-1)} - (\mathbf{\psi}_j^T \mathbf{q}_{l_{q-1}})\mathbf{q}_{l_{q-1}}, j \in I^{(q)} \tag{12.12}$$

$$I^{(q)} = I^{(q-1)} - \{j: \mathbf{p}_j = 0\} \tag{12.13}$$

where the superscript denotes the iteration number. It should be noted that in the original Gram-Schmidt algorithm, $\mathbf{p}_j = \mathbf{\psi}_j - \sum_{k=1}^{q-1}(\mathbf{\psi}_j^T \mathbf{q}_{l_k})\mathbf{q}_{l_k}$ is used, which causes numerical instability in selecting the wavelets to be truncated as reported by Zhang (1997). In this work, the Gram-Schmidt algorithm has been modified by using Eq. (12.12) in order to overcome numerical instability.

3) Update the following variables if $I^{(q)}$ is non-empty:

$$l_q = \text{INT}[\max_{j \in I} \frac{(\mathbf{p}_j^T \mathbf{Y})^2}{\mathbf{p}_j^T \mathbf{p}_j}] \tag{12.14}$$

$$\alpha_{jq} = \mathbf{\phi}_{l_q}^T \mathbf{q}_{l_j}, j = 1, ..., q-1 \tag{12.15}$$

$$\alpha_{qq} = \sqrt{\mathbf{p}_{l_q}^T \mathbf{p}_{l_j}} \tag{12.16}$$

$$\mathbf{q}_{l_q} = \alpha_{qq}^{-1} \mathbf{p}_{l_q} \tag{12.17}$$

$$\tilde{w}_{l_q} = \mathbf{q}_{l_q}^T \mathbf{Y} \tag{12.18}$$

$$I^{(q)} = I^{(q-1)} - \{l_q\} \tag{12.19}$$

where $\text{INT}(z)$ is an operator used to obtain the integer number of z and $\mathbf{Y} = [y_1, y_2, \cdots, y_{N_a}]'$ is the vector of actual outputs.

4) Stop the iteration if $I^{(q)}$ is empty. Otherwise, set $q = q + 1$ and return to step 2.

The procedure produces a set of wavelets resulting in the following approximation of the dynamic system function:

$$\tilde{f}(\mathbf{y}) \approx \sum_{j=1}^{K} w_{l_j} \sum_p \frac{1}{\sqrt{|a_p|}} \varphi_j \left(\frac{y_p - b_p}{a_p} \right), \quad p = 1, 2, \dots, D, 24, 168 \qquad (12.20)$$

where $\varphi(.)$ is the selected wavelet, K is the number of wavelets selected, and the coefficients w_{l_j}'s are determined by solving the following set of linear equations:

$$\mathbf{Aw} = \tilde{\mathbf{w}} \qquad (12.21)$$

in which $\mathbf{w} = [w_{l_1}, \dots, w_{l_s}]'$, $\tilde{\mathbf{w}} = [\tilde{w}_{l_1}, \dots, \tilde{w}_{l_s}]'$, and $\mathbf{A}$ is an upper triangular matrix:

$$\mathbf{A} = \begin{bmatrix} \alpha_{1,1} & \alpha_{1,2} & \alpha_{1,3} & \cdots & \alpha_{1,K-1} & \alpha_{1,K} \\ 0 & \alpha_{2,2} & \alpha_{2,3} & \cdots & \alpha_{2,K-1} & \alpha_{2,K} \\ 0 & 0 & \alpha_{3,3} & \cdots & \alpha_{3,K-1} & \alpha_{3,K} \\ \vdots & \vdots & \ddots & \ddots & \vdots & \vdots \\ 0 & 0 & \cdots & 0 & \alpha_{K-1,K-1} & \alpha_{K-1,K} \\ 0 & 0 & \cdots & 0 & 0 & \alpha_{K,K} \end{bmatrix} \qquad (12.22)$$

Equation (12.20) is an approximation of the dynamic system function represented by Eq. (12.5) using the truncated wavelet frame.

12.3.4 Time-Delay Recurrent WNN Model

Traffic flow data are a sequence of ordered measurements in the form of time series produced by individual sensors. In order to extract useful information from the time series for accurate forecasting, a dynamic time-delay WNN model with a recurrent feedback topology is created. The model is a combination of a wavelet neural network and a conventional neural network with a sigmoid activation function. The nonlinear estimator in the WNN model is expressed as

$$\bar{f}(\mathbf{x}, \mathbf{y}) = \sum_{j=1}^{K} w_j \sum_p \varphi_j \left(\frac{y_p - b_p}{a_p} \right) + \sum_{i=1}^{2} c_i g(x_i) + b_0, \quad p = 1, 2, \dots, D, 24, 168 \qquad (12.23)$$

where c_i's are the weights corresponding to the status inputs x_i and b_0 is a bias term. The sigmoid activation function is used in the conventional neural network part of the model representing the status variables as follows (Hagan $et\ al.$ 1996):

$$g(x_i) = 1/[1 + \exp(-x_i)] \qquad (12.24)$$

The parameters a_p and b_p in the wavelet basis, weights w_j and c_i, and the bias b_0 are the parameters of the model estimated by training the network to be discussed in a subsequent section. The total number of unknown parameters in the model is denoted by n_p, which is equal to

$$n_p = 2KD + K + 3 \qquad (12.25)$$

Figure 12.2 shows the architecture of the dynamic time-delay WNN model with a recurrent feedback topology for traffic flow forecasting. It consists of an input layer, a hidden layer, and an output layer. There are $D + 4$ input nodes in the input layer, including two status variables, traffic flows one week ago and one day ago, and D feedback output traffic flows. There are $K + 2$ hidden nodes in the hidden layer, including two sigmoid activation nodes and K wavelet nodes. The determination of the optimum number of wavelet neurons (K) will be discussed in the next section. The node in the output layer represents the predicted traffic flow, as expressed by Eq. (12.23).

12.3.5 Optimizing the Number of Wavelets

The optimum number of wavelets, K, is the minimum number of wavelets needed in the truncated wavelet frame to achieve accurate results. This is an important parameter in constructing an effective WNN model for accurate traffic flow forecasting. Zhang (1997) uses the generalized cross-validation method to select the number of wavelets. In this chapter, the Akaike's final prediction error (AFPE) criterion is applied to select the optimum number of wavelets to be used in the WNN model. The criterion describes how close an approximation model is to the actual one (Box *et al.* 1994; Ljung 1999; Hung *et al.* 2003). In this method an error function is defined in the following form:

$$h(K) = \frac{1 + n_p / N_a}{2(1 - n_p / N_a)} \left\{ \frac{1}{N_a} \sum_{i=1}^{N_a} [\tilde{f}(\mathbf{x}_i, \mathbf{y}_i) - y_i]^2 \right\} \qquad (12.26)$$

where the term in the brackets represents the mean of the total squared final prediction error, $\tilde{f}(\mathbf{x}_i, \mathbf{y}_i)$ represents the approximation of the actual time series y_i obtained from

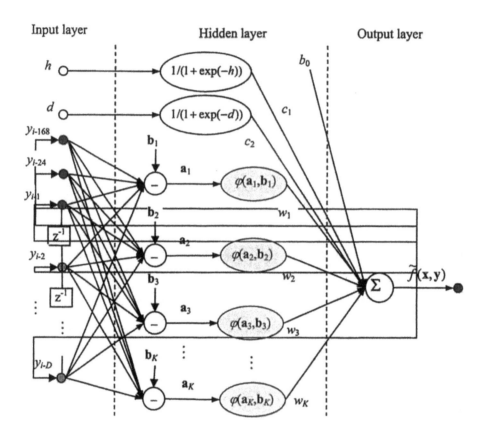

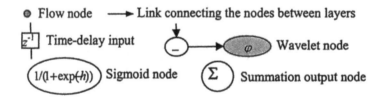

y_i = Traffic flow (i = 1, 2, …, N_a); N_a = number of data sets; b_0 = bias; D = size of the past observations; K = number of wavelet neurons; $\mathbf{a}_k, \mathbf{b}_k$ = wavelet parameter vectors (k = 1,2,…, K); $\mathbf{x}$ = status vector; $\mathbf{y}$ = input traffic flow; c_1 ,c_2 = weights of the links connecting the two status nodes in the hidden layer to the output node; w_k = weight of the link connecting node k in the hidden layer to the output node.

Figure 12.2 Architecture of dynamic time-delay recurrent WNN model

Eq. (12.23), N_a is the sample size or number of data sets defined by Eq. (12.4), K is the number of wavelets to be determined, and n_p is the number of parameters defined by Eq. (12.25).

The variation of the Akaike's final prediction error with the number of the wavelets used in the WNN model is a concave curve with a minimum. The wavelet number corresponding to the minimum point on this curve denotes the optimum number of wavelets in the WNN model.

12.4   Training the Model

To estimate the parameters of the recurrent WNN model, an error function is defined as the sum of the squares of the differences of the desired and predicted values. The parameters are updated such that this error function is minimized. This is an unconstrained mathematical optimization problem where the error function is the objective function and the parameters are the variables. The unconstrained optimization problem is solved using the delta rule [also known as the least mean square (LMS) algorithm (Eq. 5.15) or the Widrow–Hoff learning algorithm] (Hagan *et al.* 1996).

However, one has to be cognizant of the *overgeneralization* problem, also known as *overfitting* problem in the statistics literature. In this work, we employ an *optimum generalization* strategy in order to avoid overgeneralization and achieve the most accurate results as discussed previously in Section 8.4.3. This is done by dividing the available data sets for training into two groups: a training group and a checking group. The latter consists of only a fraction, in the order of 10-20%, of the total data sets available for training. The mean squared error of the checking group is used to monitor the network performance. The iteration number corresponding to the minimum checking error is where the training of the network is stopped; the values of the network parameters obtained at this iteration provide the optimum generalization results. After training of the recurrent WNN forecasting model using the optimum generalization strategy, a third group of data sets, the testing data sets, are used to evaluate the model.

12.5   Numerical Example

The average traffic flow data in vehicles per hour per lane (vphpl) along a two-lane freeway in the state of North Carolina with one lane closure, obtained from the North

Carolina Department of Transportation (NCDOT), are used in this chapter to validate the dynamic time-delay WNN model. The traffic flow data are collected on Durham freeway (NC 147) at 0.1 mile south of SR 1171 in the state of North Carolina as described previously in Section 11.4.1. The northbound urban traffic flow data are used in this chapter.

12.5.1 Finding the Dimension of Input Traffic Flow Vector

Figure 12.3a shows the time series plot of 1,824 hourly traffic flow data continuously recorded over a period of 76 days from October 1 to December 15, 2000 (that is, $N = 1,824$). It displays a strong seasonal periodical pattern of 168 hours (one week) as expected. Fig. 12.3b shows the variation of the ACF for the traffic flow data of Fig. 12.3a, with a lag time of up to 32 hours, showing a daily periodicity. The ACF curve intersects the lag time axis roughly at $D = 6$, indicating that $y(t–6)$ and $y(t)$ are linearly independent. The lag time $D = 6$ is therefore chosen as the *optimum* value to be used in the input dimension $(D + 2)$. In other words, an 8-dimensional input traffic flow vector, y_k, including y_{k-168} and y_{k-24}, is constructed for the given traffic flow time series (Eq. 12.2).

12.5.2 Training the Dynamic WNN Model

For accurate forecasting the training data should be representative of actual traffic patterns. As mentioned earlier, any prior knowledge of flow behavior should be used to improve the performance. Outliers or atypical traffic patterns should be discarded. The wavelet transform of the measured traffic flow shown in Fig. 12.3a indicates a singular oscillatory behavior in the range of 1,450 to 1,457 (Jiang and Adeli 2004), where there is an atypical increase of the traffic flow compared with other typical traffic patterns. Furthermore, there is a singular weekly traffic pattern in the range of 1,249 (November 22) to 1,344 (November 25), as seen in Fig. 12.3a, where there is an atypical decrease of the traffic flow. Consequently, only the first 1,200 traffic flow data points from October 1 to November 21, 2000 (50 days) are used to train the dynamic time-delay WNN model. Since the model uses the traffic flow one week ago (168 hours ago), $1,200 – 168 = 1,032$ input vectors are created using the 1,200 data points (Eq. 12.4). The 1,032 input vectors are divided into three sets: 672 are used for training, 168 are used for checking, and finally 192 are used for testing the forecasting model.

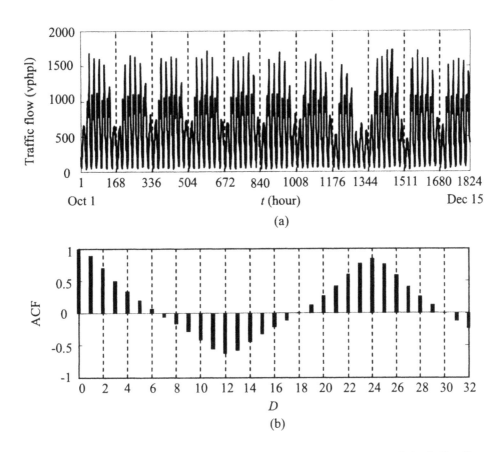

Figure 12.3 Traffic flow measured in a two-lane freeway in the state of North Carolina with one lane closure and its ACF: (a) Time series plot, and (b) histogram of the ACF for the traffic flow data

A four-level wavelet decomposition is performed on the 672 training data sets available using the Mexican hat wavelet (Fig. 12.1). The empty wavelets whose supports do not contain any data are eliminated. The result is a non-empty wavelet frame with 29 wavelets. Figure 12.4 shows the variation of the Akaike's final prediction error versus the number of wavelets used. The wavelet number corresponding to the minimum point on this curve, $K = 18$, is the optimum number of wavelets in the WNN model. Based on the AFPE criterion and using the modified Gram-Schmidt algorithm, 18 out of the 29 wavelets are chosen for use in the WNN model. As such, there are 18 wavelet nodes in the hidden layer of the network (Fig. 12.2).

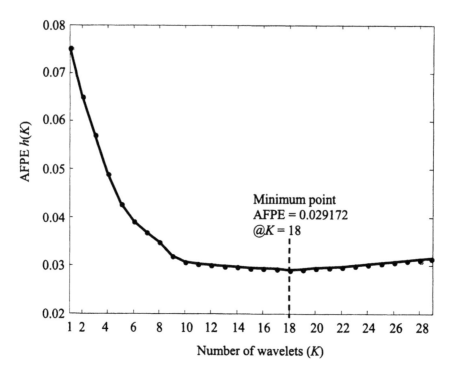

Figure 12.4 Akaike's final prediction error versus the number of wavelets

A value of 0.1 is selected by trial-and-error for the learning ratio in the delta training rule. Convergence results for training the network based on 672 training sets and 168 checking sets are displayed in Fig. 12.5. The vertical axis in this figure is the normalized square root of the mean sum of squares of errors (NSRMSE) calculated as follows:

$$\text{NSRMSE} = \sqrt{\frac{1}{N_a}\sum_{i=1}^{N_a}\left(\tilde{f}(\mathbf{x}_i,\mathbf{y}_i)-y_i\right)^2} \Bigg/ \sqrt{\frac{1}{N_a}\sum_{i=1}^{N_a}(\mathbf{y}_i-\bar{y})^2} \qquad (12.27)$$

where $\bar{y}=\dfrac{1}{N_a}\sum_{i=1}^{N_a}y_i$ is the mean of the actual time series y_i.

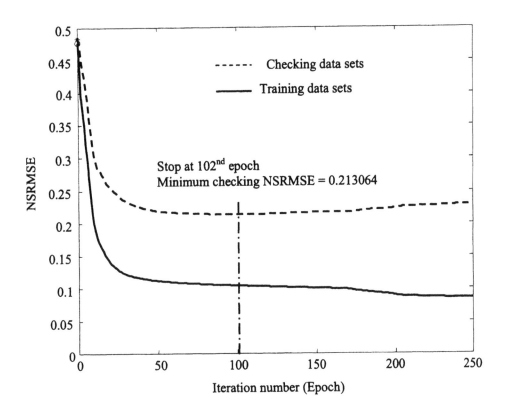

Figure 12.5 Convergence curves for training the dynamic time-delay WNN model

The epoch at which the network converges during training varies from situation to situation. In Fig. 12.5, the training of the network is stopped at the 102nd epoch, at the minimum point of the curve corresponding to the checking data. The model is implemented in MATLAB 6.1 on a Windows XP Professional platform and a 1.5-GHz Intel Pentium 4 processor. The CPU time for training the model is 4 minutes and 52 seconds using the 672 sets of training data.

Figure 12.6 shows a comparison of the measured traffic flows and the estimated flows using the dynamic time-delay WNN model. Figure 12.6a shows the traffic flow and Fig. 12.6b shows the absolute error in vphpl using the 672 training data sets. Note that the differences in the two curves in Fig. 12.6a are not noticeable. The absolute error is generally less than 100 vphpl (less than 5%) except at three points where atypical patterns may exist.

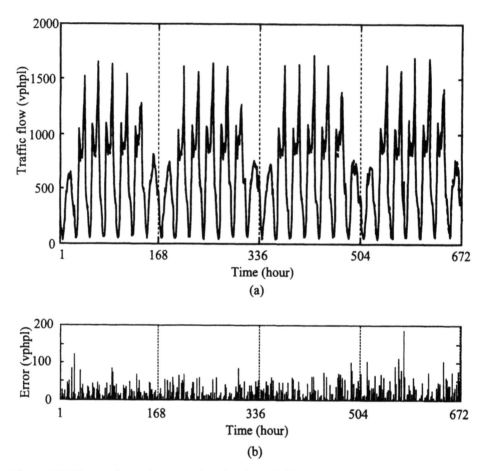

Figure 12.6 Comparison of measured and estimated flows using the 672 training data sets: (a) Traffic flow, and (b) absolute errors between actual and predicted traffic flows

12.5.3 Testing the Dynamic WNN Model

The input to the model is the hourly traffic flows for the past week. The model uses these data to forecast the traffic flow in the future for any period from one hour to any number of days. The predicted outputs during previous time intervals are fed into the input layer of the dynamic WNN model to forecast the future traffic flow (Fig. 12.2).

The trained dynamic WNN forecasting model is tested using the 192 sets of traffic flow mentioned earlier. Figure 12.7 shows a comparison of the measured traffic flows and the estimated flows using the WNN model for a period of up to 8 days. Figure 12.7a shows the traffic flow and Fig. 12.7b shows the absolute error in vphpl using the 192 testing data sets. The absolute error is less than 160 vphpl (less than 9%). Considering the complexity of the problem, the predicted results show that the proposed model can capture effectively continuous and subtle information about the traffic flow signals (the estimated curve in Fig. 12.7a follows the measured curves pretty accurately).

12.6 Concluding Remarks

Our research advances the conventional neural network significantly by presenting a *dynamic wavelet neural network* model as a powerful approach for capturing the dynamics of the traffic flow and for pattern recognition with enhanced feature detection capability. Many articles have been published on neural network applications in traffic and transportation engineering, e.g, see the review article by Adeli (2001). Adeli and Samant (Adeli and Samant 2000; Samant and Adeli 2000) introduced the concept of wavelets in transportation engineering for the first time in 2000.

In this chapter, a dynamic time-delay wavelet neural network model was presented for forecasting of traffic flows in freeways through adroit integration of a pattern recognition paradigm, neurocomputing, a signal processing methodology, wavelets, and a statistical tool, autocorrelation function. A nonorthogonal wavelet, the Mexican hat wavelet is used to create the WNN. The concept of the wavelet frame was introduced and exploited in the model to provide flexibility in the design of wavelets and to add extra features such as adaptable translation parameters desirable in traffic flow forecasting. The statistical autocorrelation function is used for selection of the optimum input dimension of traffic flow time series in the WNN model.

The model incorporates both the time of the day and the day of the week of the prediction time. As such, it can be used for long-term traffic flow forecasting in addition to short-term forecasting. Short-term traffic flow forecasting would be of great interest in on-line ITS applications and long-term forecasting would be of great interest in planning applications.

An hourly average traffic flow time series was used in this chapter to demonstrate the effectiveness of the proposed model, only as an example and because hourly traffic data were available to the authors. The model can also be adopted for other traffic flow time series, for example in vehicles per minute or second, with a slight modification of the status input. The model was validated for long-term forecasting using actual freeway traffic data. It will be as effective for short-term traffic flow forecasting if trained using traffic data in vehicles per minute or data aggregated in 2, 5, or 10 minute intervals. Even with limited data used for training, the model's accuracy is within 10%, which is very good considering the highly complicated nature of the forecasting problem. Further, the model is efficient with an excellent convergence rate and can be implemented in real time.

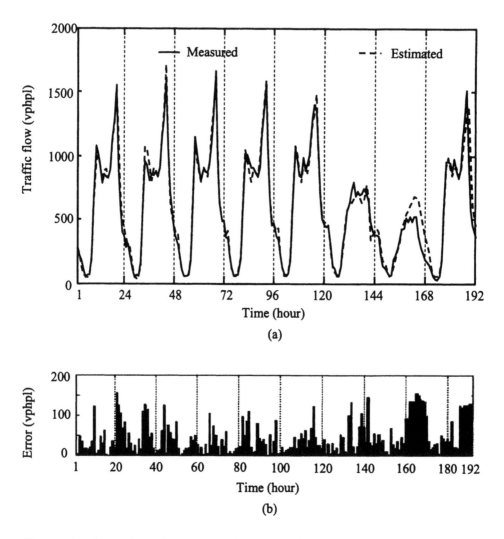

Figure 12.7 Comparison of measured and estimated flows using the 192 testing data sets: (a) Traffic flows, and (b) absolute errors between actual and predicted traffic flows

Chapter 13

DYNAMIC FUZZY WAVELET NEURAL NETWORK FOR STRUCTURAL SYSTEM IDENTIFICATION

13.1 Introduction

The goal of structural system identification research is to develop a mathematical model for a structural system based on a set of inputs and corresponding output measurements. When structures are damaged during a strong ground motion changes occur to their dynamic characteristics. Structural system identification finds applications in 1) determination of the structural properties such as the stiffness and natural periods and frequencies, 2) nondestructive damage evaluations, where input-output measurements are used to assess nondestructively the damage severity and location in an existing structure, 3) health monitoring of the global or local conditions of structures, and 4) structural control of smart structures which requires evaluation of dynamic response of structures with various structural rigidities, masses, and damping (Saleh and Adeli 1998a; Adeli and Saleh 1999).

The complicated mechanics of earthquake and the significant expense associated with detailed measurements of structural response make accurate system identification of structures a challenging problem. This problem has attracted the attention of a good number of researchers in recent years. A review of the state of the art up to 1995 is presented by Ghanem and Shinozuka (1995). Examples of recent work are Chassiakos and Masri (1996), Gurley and Kareem (1999), Loh *et al.* (2000), Masri *et al.* (2000), Hung *et al.* (2003), Kijewski and Kareem (2003), Jiang and Adeli (2005a), Adeli and Jiang (2006) , and Jiang *et al.* (2007).

There are two fundamentally different approaches for the solution of the system identification problem: parametric method and nonparametric method. The parametric method is a straightforward approach in which the parameters of an actual system model are used directly to represent physical quantities such as the structural stiffness and damping ratio. Estimation algorithms such as the Kalman filter method, maximum likelihood method, and recursive least square method are used to determine the parameters of the system model. Parametric methods have been applied successfully to identify the dynamic properties of linearized and time-invariant equivalent structural systems (Juang 1994; Ljung and Glad 1994; Ghanem and Shinozuka 1995).

Wavelet-based approaches have been developed for parametric identification of simple multi-degree-of-freedom (MDOF) systems (Staszewski 1997 & 1998; Ghanem and Romeo 2000; Lamarque et al. 2000; Hans et al. 2000; Kijewski and Kareem 2003; Chang et al. 2003). Staszewski (1997 & 1998) uses a continuous Grossman-Morlet wavelet transform to decompose the impulse response of MDOF systems into the time-scale domain. Then, the structural parameters are estimated based on the *ridges* (where the energy of the signal is mainly concentrated in the time-scale plane) and *skeletons* of the wavelet transform (the values of the wavelet transform at its ridges). This method is used for parametric estimation of the nonlinear and time variant simple MDOF systems. Ghanem and Romeo (2000) describe a wavelet-based approach for parametric identification of single and simple MDOF linear time-varying dynamic systems. They use the wavelet transform to decompose the governing differential equations into discrete forms. Then, the structural parameters are obtained from the solution of the transformed differential equations using the least square minimization technique. The authors report accurate estimation of the damping and stiffness parameters of a simple linear time-invariant system. They also demonstrate that the presence of noise affects the accuracy of the estimated damping significantly but has little effect on stiffness. Kijewski and Kareem (2003) use Morlet wavelet decomposition for identifying the frequency and damping parameters of simple MDOF systems.

These parametric identification algorithms, however, depend strongly on the accuracy of the measured data. They cannot provide the required accuracy and reliability needed for complex system identifications of real-life structures due to the complicated nonlinear nature of the behavior of civil structures, and incomplete, incoherent, and noise-contaminated measurements of structural response to strong ground motions.

Nonparametric system identification approaches have been widely used in health monitoring (e.g., Nakamura et al. 1998; Masri et al. 1996 & 2000) and structural control of building structures (e.g., Bani-Hani and Ghaboussi 1998a&b; Brown and Yang 2001). In the nonparametric approach, the input-output map is characterized and determined by a system model that may not have any explicit physical meaning. In general, the system model does not represent any physical quantity directly, but it is trained to approximate a physical structure and predict the structural responses. As such, the approach does not require complete and coherent measurements of the structural response to strong ground motions. It has better adaptability than the parametric methods (Billings and Tsang 1989; Billings et al. 1990; Sjöberg et al. 1995; Thomson et al. 1996; Aguirre et al. 2000).

In nonparametric system identifications the nonlinear autoregressive moving average with exogenous inputs (NARMAX) approach is widely used for mapping the input-output relationship (Ljung and Glad 1994; Ljung 1999) due to the following advantages: 1) it does not impose any restriction on the nature of system input excitation leading to its general applicability, 2) it uses constant parameters independent of the time variable, which makes it relatively easy to incorporate prior knowledge about the system into the model, and 3) it can model nonlinearity effectively in contrast to linear approaches such as the autoregressive moving average method (ARMA).

The key issue in nonparametric identification methods is an effective approach for estimation of the coefficients of the NARMAX. Sjöberg et al. (1995) describe a number of soft computing approaches for finding the NARMAX coefficients such as the multilayer neural networks and fuzzy logic. Loh et al. (2000) use the Levenberg-Marquardt backpropagation (BP) neural network for nonparametric identification of a five story test frame. Masri et al. (1993, 1996 & 2000) investigate the use of the BP neural network method for structural system identification and health monitoring. Their study shows that the neural network method is a practical tool in detecting changes in nonlinear structures with unknown constitutive properties and topologies. However, those neural network methods suffer from some common drawbacks such as lack of an efficient constructive model (for example, requiring arbitrary selection of the number of hidden nodes), slow convergence rate, and entrapment in a local minimum.

Hung et al. (2003) used the wavelet neural network (WNN) model, which was proposed by Zhang and Benveniste (1992) for signal processing, for system identification of a five-story test frame subjected to simulated earthquake loadings on a shaking table. However, the WNN method for structural system identification also suffers from 1) lack of an efficient constructive model, 2) the need to find the model parameters such as the input vector dimension by trial and error, 3) slow convergence rate when there exists noise in the measured data, and 4) low identification accuracy when there exists imprecision in the measured data.

In this chapter, an innovative multi-paradigm dynamic time-delay fuzzy wavelet neural network (WNN) model is presented for nonparametric identification of structures using the NARMAX approach. The model is based on the integration of four different computing concepts: dynamic time-delay neural network, wavelet, fuzzy logic, and the reconstructed state space concept from the chaos theory. The goal is to improve the accuracy and adaptability of nonparametric system identification of structures under earthquake loadings. The results of this chapter will help structural engineers better understand the behavior of structures during earthquakes and design more effective earthquake-resistant structures.

13.2   Creating State Space Vectors

The NARMAX approach has proven to be a powerful tool for mapping the nonlinear input-output relationship in system identification (Juang 1994; Ljung and Glad 1994; Thomson *et al.* 1996; Aguirre *et al.* 2000). In this approach, the general discrete dynamic input-output mapping is expressed as follows:

$$y(t) = f(\mathbf{x}^{t-1}, \mathbf{y}^{t-1}, \mathbf{e}^{t-1}) + e(t) \tag{13.1}$$

where $\mathbf{x}^t$, $\mathbf{y}^t$, and $\mathbf{e}^t$ represent the input, output, and zero-mean noise vectors at time t (representing the discrete time series values up to time t), respectively, $f(.)$ is a scalar nonlinear mapping or approximation function, and $e(t)$ is the error between the actual and estimated values of the future output $y(t)$. The objective is to find the relationship between the past observations $\mathbf{X}(t-1) = [\mathbf{x}^{t-1}, \mathbf{y}^{t-1}]'$ and the future output $y(t)$. We need to model the mapping function for effective structural system identification.

In this chapter, first the noise in the signals is removed using the discrete wavelet packet transform (DWPT) method as described in Section 3.3. The DWPT provides more coefficients than the conventional discrete wavelet transform (DWT) representing additional subtle details of a signal. Next, the dynamic system is approximated as a functional representation of the lagged past denoised inputs and outputs. The reconstructed state space concept from the chaos theory (see Chapter 4) is used to simulate the multivariate properties of the nonlinear system effectively. The input dimension or the size of the state space vector in the NARMAX approach, D, is the summation of dimensions of the structural inputs, D_x (number of discrete time steps used for any input time series), and the feedback outputs, D_y (number of discrete time steps used for the corresponding past output time series).

A proper choice of the dimension for the NARMAX approach plays an important role in reconstructing an appropriate multivariate input. If D is too small, the model produces inaccurate identification results. If D is too large, it can also lead to overestimation and inaccurate results in addition to increasing the computational cost. System identification researchers usually use a trial-and-error approach to find the most suitable value for the input dimension in the NARMAX approach (Zhang 1997; Hung *et al.* 2003). However, the trial-and-error approach a) does not provide a rational basis for the selection, b) is cumbersome, c) is computationally time-consuming, and d) does not guarantee accurate system identification results.

An input time series is denoted by discrete quantities $\{x_i\}$, $i = 1, 2, ..., N$, where N is the number of data points. The input vector at time t, $\mathbf{x}'$, in the NARMAX approach (Eq. 13.1), is represented by the state space vector, $\mathbf{x}_k$, as follows:

$$\mathbf{x}_k = [x_k, x_{k+\tau}, x_{k+2\tau}, ..., x_{k+(D_x-1)\tau}]', k = 1, 2, ..., N_a \tag{13.2}$$

where the prime denotes the transpose of a vector and the parameters D_x and τ are the dimension (usually referred to as embedding dimension, d_E, in the chaos theory, see Chapter 4) and the lag time index used to reconstruct the state space, respectively. The value of τ is set equal to 1 in this chapter. The data point subscript, k, satisfies the following condition:

$$k \leq N - D_x, \; n \in Z \tag{13.3}$$

where Z is the set of all integers. Based on Eq. (13.3), the maximum number of state space points, N_a, is equal to

$$N_a = N - D_x \tag{13.4}$$

Equation (13.2) represents the state space vector with one lag time reconstructed from the sensor data. The feedback input vector $\mathbf{y}'$ in Eq. (13.1) is defined similar to Eq. (13.2) with D_y dimension. Thus, the state space $\mathbf{X}_k = [\mathbf{x}_k, \mathbf{y}_k]'$ is obtained. In the theory of chaos, the set of points used to simulate the evolution trajectory in the original state space is referred to as an *attractor*, as explained in Section 4.1. The purpose of the attractor is to unfold the time series back to a multivariate state space representing the original physical system. An attractor is the geometric invariance in the state space representation of a time series. The invariants of the dynamic system producing the time series are preserved if the time series is transformed into a sufficiently large reconstructed state space (defined in terms of its dimension). A proper choice of the parameters D_x and D_y plays an important role in accurately reconstructing the multivariate state space.

For the structural system identification problem, the false nearest neighbor (FNN) method described in Section 4.3.3 was found to be more accurate than the fill-factor and average integral local deformation methods and therefore is used in this chapter to find the optimum embedding dimensions D_x and D_y. The FNN approach is based on the assumption that a small embedding dimension results in state space points that are far apart in the original state space to be considered neighboring points in the reconstructed state space. See Section 4.3.3 for details of the method.

13.3   Constructing the Dynamic Fuzzy WNN Model

13.3.1   Creating the Fuzzy WNN Model

The dynamic system identification function represented symbolically by $f(.)$ in Eq. (13.1) is approximated using the wavelet transform functions and wavelet coefficients expressed in a general form as follows (Daubechies 1998 & 1992):

$$\bar{f}(\mathbf{X}_k) = \sum_{i=1}^{K} w_i \sum_{j=1}^{D} \varphi\left(\frac{X_{kj} - b_{ij}}{a_{ij}}\right), k = 1, \ldots, N_a, \ a, b \in \Re, \ \varphi(.) \in L^2(\Re) \quad (13.5)$$

where $D = D_x + D_y$ is the dimension of the input vector $\mathbf{X}_k$, X_{kj} represents the *j*-th item in the *k*-th input vector $\mathbf{X}_k = [x_k, x_{k+\tau}, x_{k+2\tau}, \ldots, x_{k+(D_x-1)\tau} x_k, y_{k+\tau}, y_{k+2\tau}, \ldots, y_{k+(D_y-1)\tau}]'$, K is the number of wavelets obtained using the modified Gram-Schmidt algorithm described in Section 12.3.3, and the Akaike's final prediction error method (AFPE) described in Section 12.3.5. In the modified Gram-Schmidt algorithm, first the wavelet which best approximates the measured data is selected from the complete nonzero wavelet coefficient sets. Next, this wavelet is combined with the remainder of the wavelet sets one at a time and the combination which best approximates the measured data is determined. The procedure is repeated for all nonzero wavelet coefficients. The symbol $\bar{f}(\mathbf{X}_k)$ represents the approximation of $f(\mathbf{X}_k)$, w_i represents the discrete wavelet transform coefficient, and $\varphi(.)$ is the two-dimensional wavelet expansion functions with scaling and translation (the multidimensional input $\mathbf{X}_k$ results in a multidimensional wavelet sets). The Mexican hat wavelet (Fig. 12.1) is used due to its analytical expression convenient for decomposing multidimensional time series, its differentiability, and its *non-compactly supported but rapidly vanishing* feature (the function is nonzero on an infinite interval but approaches to zero rapidly) (Zhou and Adeli 2003) which is advantageous in this work (see Section 12.3.1). The parameters $a_{ij} \neq 0$ and b_{ij} denote the frequency (or scale) and the time (or space) location parameters corresponding to the multi-dimensional input vector $\mathbf{X}_k$, respectively, and $\Re$ is the set of real numbers.

The power of wavelets for time series analysis stems from their capability of decomposing the signal into spatial (or time) and frequency (or scale) domains simultaneously as described in Section 3.1. Equation (13.5) is used to construct a wavelet neural network for effective identification of general nonlinear systems. Unlike the conventional multilayer neural network whose activation function, such as a sigmoid function, has infinite energy (the function is nonzero on an infinite interval and area under the function is infinity), the proposed WNN uses a wavelet function with a spatial-spectral zooming property, which influences the output of the model only in the finite range of input time series. This property a) reduces the undesirable interaction effects among the nodes of the neural network thus in general improving the accuracy of the system identification and b) accelerates the neural network training process thus improving its computational efficiency. The flip side of the WNN model is that local imprecision in the training data may result in a large local output error because of the same spatial-spectral zooming property.

The fuzzy logic theory is a powerful tool for effective representation of imprecision existing in the time series data measured by sensors (Zadeh 1978). To overcome the aforementioned shortcoming of the WNN model, a fuzzy WNN model is created based on the following two steps. First, the fuzzy c-means clustering (FCM) algorithm is used as a data mining tool to divide the sensor data into clusters with common features. Next, a fuzzy-wavelet model is created by combining the fuzzy clusters of the reconstructed state space vectors with the nonorthogonal wavelets defined by Eq. (13.5).

The FCM algorithm performs a fuzzy partitioning of the data set into classes, where the concept of fuzzy membership is used to represent the degree to which a given data set belongs to some cluster. The state space vectors reconstructed earlier are clustered based on their similarities. The fuzzy clustering problem is defined as a constrained optimization problem as follows:

Minimize

$$f_\beta(\mathbf{c}) = \sum_{k=1}^{N_a}\sum_{i=1}^{K} A_{ki}^\beta \|\mathbf{X}_k - \mathbf{c}_i\|^2 \tag{13.6}$$

Subject to the constraints:

$$\sum_{i=1}^{K} A_{ki} = 1, \ 1 \le k \le N_a \tag{13.7}$$

$$A_{ki} \ge 0, \ 1 \le k \le N_a, 1 \le i \le K \tag{13.8}$$

where $f_\beta(\mathbf{c})$ is the objective function, $\mathbf{X}_k$ is the k-th input state space vector, A_{ki} is the membership function or the degree of membership of the k-th variable in the i-th fuzzy implication rule, $\mathbf{c}_i = [c_{1i}, c_{2i}, ..., c_{Di}]'$ represents the D-dimensional center of the i-th cluster, K is the total number of fuzzy implication rules, and N_a is the number of the training data sets defined by Eq. (13.4). The parameter β represents the degree of fuzziness in the data. A value in the range $1 < \beta \le 2$ is often chosen (larger values are selected for fuzzier data situations). A value of $\beta = 1.5$ is chosen in this chapter. The parameter K is also equal to the number of the clusters as well as the number of the wavelets used in the WNN model.

The variables of the optimization problem represented by Eqs. (13.6) to (13.8) are the membership grades, A_{ki}. The total number of the variables is equal to K times N_a. This formulation leads to a nonconvex optimization problem that does not always yield a global optimal solution using the conventional optimization method (Al-Sultan and Fediki 1997). In order to overcome this shortcoming, an iterative procedure is employed for the solution of the optimization problem and the values of the membership grades A_{ki} of a vector based on its Euclidean distance from the class center c can be obtained as follows (Bezdek 1981):

$$A_{ki}^{m+1} = \left[\sum_{j=1}^{K} \left(\left\| \mathbf{X}_k - \mathbf{c}_i^m \right\|^2 \middle/ \left\| \mathbf{X}_k - \mathbf{c}_j^m \right\|^2 \right)^{1/(\beta-1)} \right]^{-1}, \quad 1 \le k \le N_a, \ 1 \le i \le K \qquad (13.9)$$

where the superscript m denotes the iteration number. The iterative procedure used in the FCM algorithm is stated as follows:

1)   Initialize the c vectors randomly.

2)   Find the values of the membership grades A_{ki} of a vector by Eq. (13.9). Then, calculate the value of the objective function using Eq. (13.6).

3)   Update the class centers for the data pattern matrix $\mathbf{X}$ using the following equation:

$$\mathbf{c}_i^m = \sum_{j=1}^{n} A_{ji}^m \mathbf{x}_j \middle/ \sum_{j=1}^{n} A_{ji}^m, \quad 1 \le i \le K \qquad (13.10)$$

4)   Update the objective function by substituting Eq. (13.10) into Eq. (13.6).

5)   Check for convergence criteria using a given tolerance of ε ($\varepsilon = 0.00001$ is used in this chapter):

$$\left| f_\beta(\mathbf{c}_i^m) - f_\beta(\mathbf{c}_i^{m-1}) \right| < \varepsilon, \quad 1 \le i \le K \qquad (13.11)$$

and a given maximum number of iterations, M_{max} (a value of 300 found to be appropriate in this chapter). If convergence criteria are not met update the membership grades using Eq. (13.9) and return to step 2).

Next, a fuzzy logic inference mechanism is developed in the FCM algorithm using a set of IF-THEN fuzzy implication rules in the following form (Sugeno and Kang 1988):

$$\text{IF } A_{ki}^m : \mathbf{X}_k, \text{ THEN } \mathbf{c}_i^m = \sum_{k=1}^{N_a} A_{ki}^m \mathbf{X}_k \middle/ \sum_{k=1}^{N_a} A_{ki}^m, \quad i = 1, \dots, K, \ k = 1, \dots, N_a \qquad (13.12)$$

where $A_{ki}^m : \mathbf{X}_k$ indicates the degree of membership of $\mathbf{X}_k$ in the i-th fuzzy implication rule at iteration m is A_{ki}^m. Figure 13.1 shows a conceptual example of two clusters obtained from the data points in a two-dimensional state space using the FCM clustering algorithm. The partitioning of the data is mathematically expressed in the fuzzy partition matrix $A = [A_{ki}]$ whose elements are the membership degrees of the data vectors $X_k = [x_1, x_2]$ in the fuzzy clusters c_i.

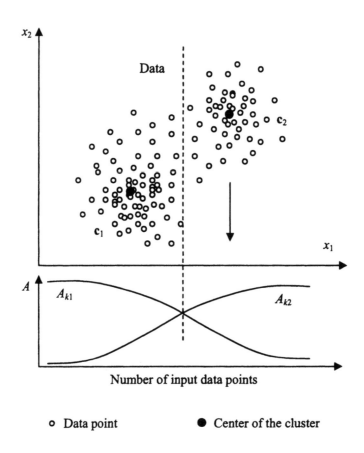

Figure 13.1 Two clusters obtained using the FCM clustering algorithm

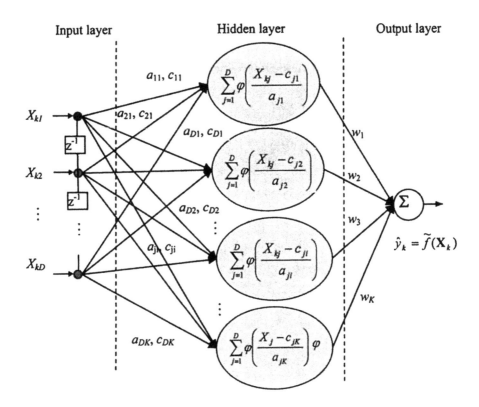

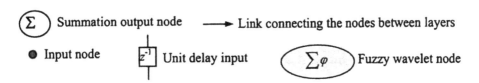

$\mathbf{X}_k$ = Input vector = $[X_{k1}, X_{k2}, ..., X_{kD}]'$, $(k = 1, 2, ..., N_a)$; N_a = number of data sets; D = dimension of the model (= $D_x + D_y$); K = number of fuzzy wavelets; a_{ji} = wavelet parameter connecting the j-th input node to the i-th fuzzy-wavelet node in the hidden layer $(j = 1, 2, ..., D; i = 1, 2, ..., K)$; $\hat{y}_k$ = output of fuzzy wavelet; $\mathbf{c}_i$ = i-th center of the clusters created by FCM algorithm = $[c_{i1}, c_{i2}, ..., c_{iD}]^T$; w_j = weight of the link connecting the j-th node in the wavelet layer to the output node.

Figure 13.2 Architecture of fuzzy-WNN

The fuzzy WNN model is formed using the fuzzy wavelets in the following three steps: 1) construct the WNN model, 2) find fuzzy clusters using the FCM algorithm, and 3) replace the conventional translation parameters, b_{ij} in Eq. (13.5), by the fuzzy clusters c_{ij} to form the fuzzy-wavelets. Figure 13.2 shows the feedforward architecture of the fuzzy WNN model. It consists of an input layer, a hidden fuzzy-wavelet layer, and an output layer. There are $D = D_x + D_y$ input nodes in the input layer. The optimum number of fuzzy-wavelet nodes is found using the AFPE criterion (Ljung 1999) described in Section 12.3.5. This number is set equal to the number of fuzzy clusters required in the FCM algorithm. In Fig. 13.2, the ellipses in the hidden layer represent the fuzzy wavelet functions. The output in the output layer represents the approximated structural response (Eq. 13.5).

13.3.2 Creating the Dynamic Fuzzy WNN Model

Both earthquake records and structural responses are a sequence of ordered measurements in the form of time series produced by individual sensors. In order to extract useful information from the time series for accurate system identification of a structure using the fuzzy WNN model, the input data from a single time series cannot be treated as independent data points. The data order in the time series must be preserved in temporal processing to avoid distorting the time signal and changing its frequency content. Consequently, a dynamic time-delay fuzzy WNN model with a recurrent feedback topology is created for structural system identification.

In order to capture the linear characteristics of the nonlinear system identification problems, a linear term is added to Eq. (13.5) to form the fuzzy WNN model expressed as

$$\bar{f}(\mathbf{X}_k) = \sum_{i=1}^{K} w_i \sum_{j=1}^{D} \varphi\left(\frac{X_{kj} - c_{ij}}{a_{ij}}\right) + \sum_{j}^{D} b_j X_{kj} + d \qquad (13.13)$$

where b_j is the weight of the link of the j-th input to the output, and d is the weight of a bias term ($p_0 = 1$, see Fig. 13.3). The double summation in Eq. (13.13) represents a linear combination of the weighted sum of the wavelets. Equation (13.13) approximates the structural dynamics using a few wavelets in combination with the linear characteristics.

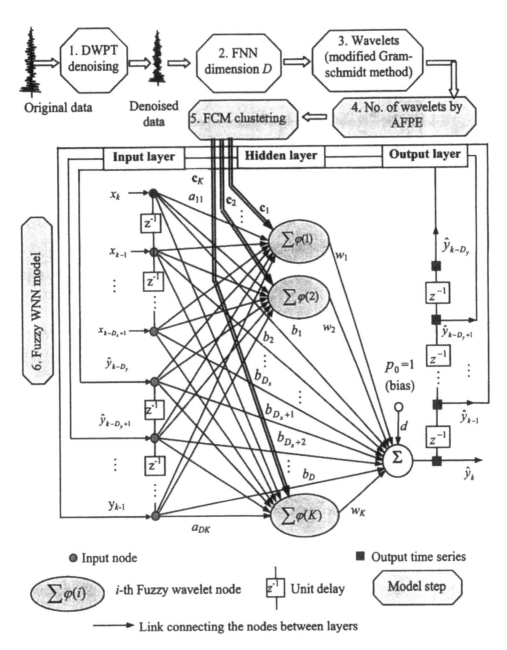

x_k = Input time series (k = 1, 2, ..., N_a); $\hat{y}_k$ = output time series; d = weight of the bias term (p_0 = 1); D_x = dimension of structural inputs x_k; D_y = dimension of feedback inputs y_k; D = input dimension of the model (= $D_x + D_y$); K = number of fuzzy wavelets; a_{ji} = wavelet parameter connecting the j-th input node to the i-th fuzzy-wavelet node in the wavelet layer (j = 1, 2, ..., D; i = 1, 2,..., K); b_j = weight of the link connecting the j-th node in the input layer to the output node; w_i = weight of the link connecting the i-th node in the wavelet layer to the output node

Figure 13.3 Architecture of dynamic time-delay fuzzy WNN model

Figure 13.3 shows the architecture of the multi-paradigm dynamic time-delay fuzzy WNN model for structural system identification. It consists of six steps: 1) denoise the original time series sensor data using the DWPT-based approach (see Section 11.2), 2) construct the state space vectors using the FNN approach (see Section 4.3.3), 3) select wavelets using the modified Gram-Schmidt algorithm (see Section 12.3.3), 4) minimize the number of the required wavelets using the Akaike's final prediction error criterion (see Section 12.3.5), 5) find fuzzy clusters using the FCM algorithm (see Section 13.3.1), and 6) construct the dynamic time-delay fuzzy WNN model.

The topology of the dynamic fuzzy WNN model consists of an input layer, a hidden (fuzzy wavelet) layer, and an output layer. The input layer has $D = D_x + D_y$ nodes representing the constructed state space vector of structural inputs $\mathbf{x}$ and feedback outputs $\mathbf{y}$. In addition to D_x structural inputs, D_y outputs, $y_{k-1}, y_{k-2},..., y_{k-D_y}$, are fed back to the input layer at a time. The fuzzy wavelet layer consists of K fuzzy-wavelet nodes. The output layer has only one node which is the predicted structural response (Eq. 13.13).

Unlike the conventional static neural networks that are based on single-valued input nodes, the proposed model is a dynamic neural network that preserves the time sequence of the input vectors and memorizes the past of the time series sensor data. Each data set in the input layer is a state space vector, a delay input with lag time of 1. The delay inputs are aggregated nonlinearly through the fuzzy wavelet processing nodes in the fuzzy wavelet layer to memorize the past.

13.4   Training the Dynamic Fuzzy WNN Model

A hybrid learning algorithm, Levenberg-Marquardt-Least-Squares (LM-LS) algorithm, is developed for estimating the parameters of the fuzzy WNN model: parameters a_{ij} in the wavelet basis, weights w_i and b_j, and d (Eq. 13.13). The LM algorithm, an approximate combination of the Gauss-Newton and steepest descent algorithms, is used to estimate the parameters a_{ij} of nonlinear wavelet functions. This algorithm is presented in detail in the next chapter.

13.5   Numerical Example

13.5.1   Example Structure

Experimental results on a ½-scaled five-story steel frame reported in the recent literature were used to validate the dynamic time-delay fuzzy WNN model for system identification of structures (Loh *et al.* 2000; Hung *et al.* 2003). The structure is a 3m long, 2m wide, and 6.5m high steel frame (Fig. 13.4). Lead blocks were nearly uniformly distributed on each floor such that the mass of each floor was approximately equal to 3,664kg. This test structure was subjected to the basic excitation of the original Kobe earthquake with five different scales (20%, 32%, 40%, 52%, and 60%) on a shaking table at the National Center for Research on Earthquake Engineering in Taiwan (Hung *et al.*, 2003). These excitations are denoted as Kobe1, Kobe2, Kobe3, Kobe4, and Kobe5, respectively. Acceleration responses were measured at the four corners of each floor in two horizontal directions over a period of 25 seconds at increments of 0.001 seconds. Thus, around 25,000 output data points are recorded in a single test. Figure 13.5 shows the Kobe1 earthquake ground motion record and the corresponding acceleration responses of the five floors of the test frame. It should be noted that in Fig. 13.5 the vertical scales for floor accelerations are different from those for the ground acceleration.

In order to improve computational efficiency, all acceleration responses were normalized to the gravity acceleration (g). For the sake of comparison with the results of WNN presented by Hung *et al.* (2003), only the response data at the first, second, and third floors are used for training and testing data sets in this chapter, which is the same scheme used by Hung *et al.* (2003), as depicted in Fig. 13.4. The model is used to identify the response of the second floor and compare the results with those of Hung *et al.* (2003).

Figure 13.6a shows the measured data points from 7 seconds to 8 seconds of the structural response at the 1st floor under the five earthquake excitations mentioned earlier. It is found that at each floor all structural acceleration response curves have the same general shape at various earthquake excitations without outliers. The response patterns of the structure subjected to different levels of excitations are not changed significantly, but the peak ground values of acceleration increase as the excitation level increases, as expected.

13.5.2 Data Denoising

The measured data contain noise. The DWPT-based denoising approach and Daubechies wavelets of order 4 are employed for denoising all measured data. As an example, Fig. 13.6b shows the denoised results of the measured data points shown in Fig. 13.6a. It is

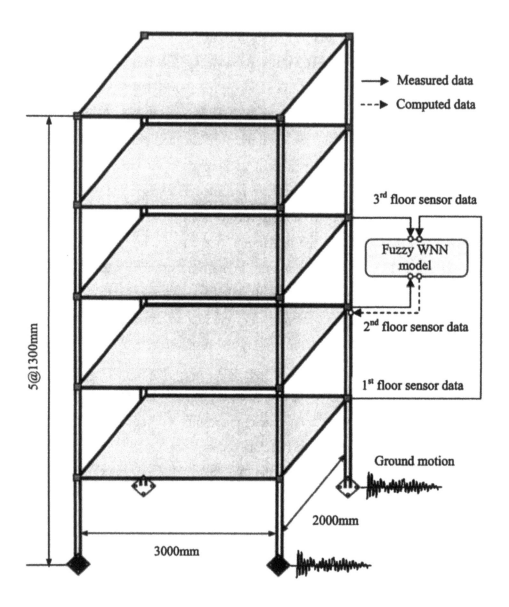

Figure 13.4 Five-story test steel frame model

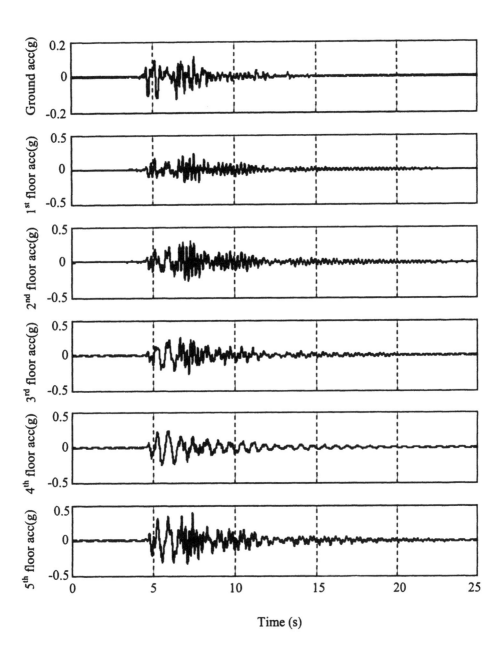

Figure 13.5 Kobe1 earthquake ground motion record and acceleration responses of the five-story test frame

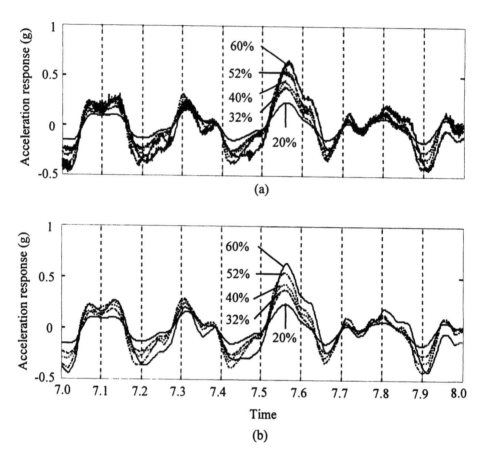

Figure 13.6 Data points from 7 seconds to 8 seconds of the structural response at the 1st floor under five earthquake excitations: (a) Measured, and (b) denoised. See color insert following Chapter 15.

observed that the DWPT-based de-noising approach smoothes the contaminated data effectively while at the same time preserves the subtle features in the signal.

13.5.3 Constructing the Model

The 14,000 denoised data points from 4 seconds to 18 seconds of the structural response at the first, second, and third floors are extracted from the 25,000 data points for constructing the fuzzy WNN model. For reconstructing the state space vectors, the same optimum embedding dimension of 6 is obtained using the FNN method for the three floor acceleration responses of the structure under different earthquake excitation levels. This is not unexpected because the general shape of the structural response under various excitation levels does not change significantly. The same dimension is therefore used for both measured inputs and feedback inputs of the fuzzy WNN model ($D_x = 6$ and $D_y = 6$). The number of input vectors created is thus equal to $N_a = 14,000 - 6 = 13,994$ (Eq. 13.4).

Next, a four-level wavelet decomposition is performed on the 13,994 training vectors using the Mexican hat wavelet. The empty wavelets whose supports do not contain any data are eliminated, resulting in 157 non-empty wavelets. Based on the AFPE criterion and using the modified Gram-Schmidt algorithm, it was concluded that 2 out of the 157 non-empty wavelets are sufficient to construct an effective fuzzy WNN model. As such, there are 2 wavelet nodes in the hidden layer of the network (Fig. 13.3).

13.5.4 Training the Model

The acceleration responses of the first, second, and third floors during the previous $D_x=6$ time intervals are used as inputs, and the current acceleration response of the second floor is used as the output of the dynamic time-delay fuzzy WNN model (Fig. 13.4). The five sets of experimental data obtained at different excitation levels (Kobe1 to Kobe5) are used to train the model, one set at a time, resulting in five trained models. The dynamic time-delay fuzzy WNN model converges very fast after only two training iterations using the hybrid adaptive LM-LS learning algorithm. All absolute errors between the computed and measured responses are less than 0.002g (less than 0.6%).

13.5.5 Testing the Model

The measured acceleration responses of the first and third floors from the previous $D_x = 6$ time intervals and the computed acceleration response of the second floor from the previous $D_y = 6$ time intervals are fed into the input layer to calculate the current acceleration response of the second floor. Each trained model is tested using all five different levels of excitations (Kobe1 to Kobe5). As such, 25 different sets of system identification results are produced. The root mean square errors (RMSE) between the measured and estimated output is computed for all 25 test cases. The results are shown as a bar diagram in Fig. 13.7. The horizontal axis represents the five different excitation levels. As an example, Fig. 13.8 shows the identified structural responses of the second floor subjected to Kobe1 earthquake using the model trained by the structural responses from Kobe2.

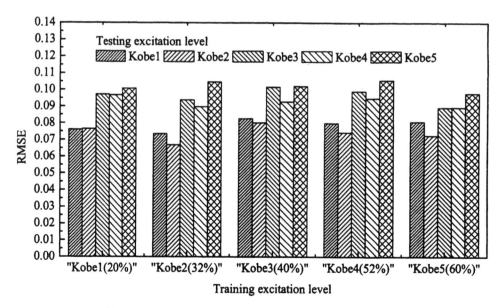

Figure 13.7 RMSE bar diagram of system identification results

Several observations are made. First, all RMSE values for identifying the structural responses under smaller Kobe1 and Kobe2 excitations are less than 8.2%, which is slightly larger than 7% reported by Hung *et al.* (2003). However, the maximum absolute errors between the computed and measured responses are around 0.04g, the same as that reported by Hung *et al.* In all likelihood, the responses of the structure under earthquakes Kobe1 and Kobe2 represent the response of the undamaged structure.

Second, the RMSE values for identifying the structural responses under the stronger Kobe3, Kobe4, and Kobe5 excitations are less than 11%, which is significantly lower than 28% reported by Hung *et al.* (2003). As an example, Fig. 13.9 shows the identified structural responses of the second floor subjected to Kobe5 earthquake using the model trained by the structural responses from Kobe3. The maximum absolute errors

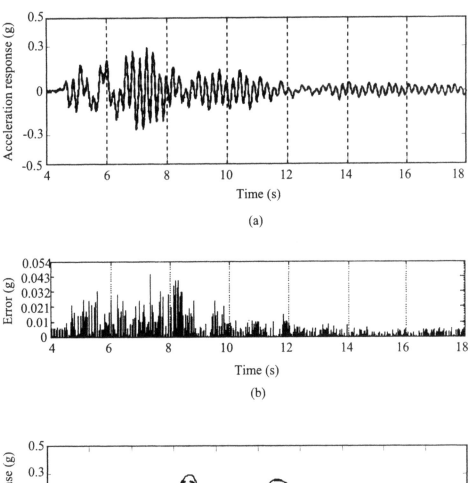

(a)

(b)

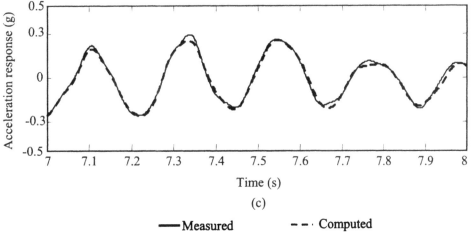

(c)

——Measured   – – · Computed

Figure 13.8 Identified structural responses of the second floor subjected to Kobe1
earthquake using the model trained by the structural responses from Kobe2 earthquake:
(a) Comparison of acceleration responses from 4s to 18s; (b) absolute error, and (c)
comparison of acceleration response from 7s to 8s

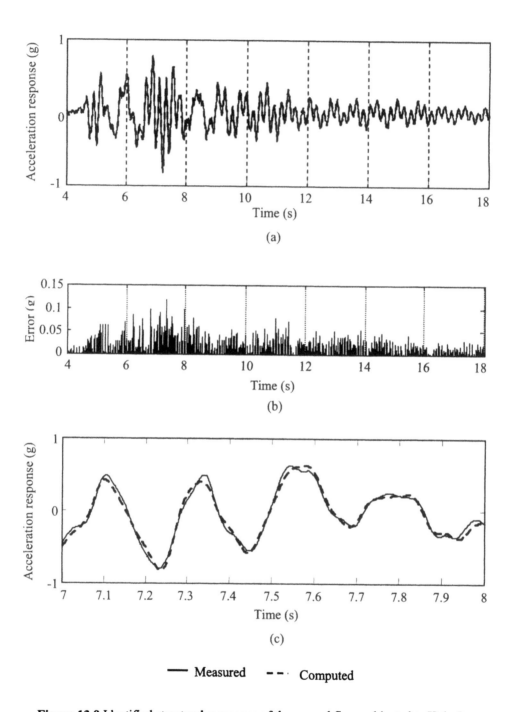

Figure 13.9 Identified structural responses of the second floor subjected to Kobe5 earthquake using the model trained by the structural responses from Kobe3: (a) Comparison of acceleration responses from 4s to 18s; (b) absolute error, and (c) comparison of acceleration response from 7s to 8s

between the computed and measured responses are less than 0.14g compared with 0.2g reported by Hung *et al.* (2003). It is concluded that the model trained using the structural response data under a lower excitation can be used to identify the structural responses under a higher excitation accurately.

Third, the maximum RMSE value (11%) for identifying the structural responses under the stronger Kobe3, Kobe4, and Kobe5 excitations is larger than the corresponding value (8.2%) for identifying the structural responses under lower excitations (Kobe1 and Kobe2). These results imply changes in the structural properties during larger excitations Kobe3, Kobe4, and Kobe5 because of damage or inelastic behavior. However, the changes do not appear to be significant.

Figure 13.10 shows the identified structural responses of the second floor subjected to a lower Kobe2 excitation using the model trained by the structural responses from the stronger Kobe5 excitation. In this figure the maximum absolute error is around 0.05g, which is significantly lower than the value of 0.2g reported in Hung *et al.* (2003). It is concluded that the model trained using the structural response data under a larger excitation can still be used to identify the structural responses under a lower excitation accurately.

13.6 Concluding Remarks

Accurate system identification requires capturing the dynamic characteristics of time series response data. In this chapter, a dynamic time-delay fuzzy wavelet neural network model was presented for nonparametric identification of structures using the NARMAX approach. The following conclusions are made.

1) The wavelet packet denoising technique employed in the model speeds up the training convergence and improves the system identification accuracy. Figure 13.11 shows comparisons of the identification results using the denoised and original structural response data under Kobe4 excitation using the model trained by structural responses from Kobe1. The model is implemented in a combination of C++ programming language and MATLAB 6.1 (Mathworks 2001) on a Windows XP Professional platform and a 1.5GHz Intel Pentium 4 processor. The CPU time for training the model is 54 seconds using the denoised data and 129 seconds using the original data. More significantly, the maximum identification error increases to 15%

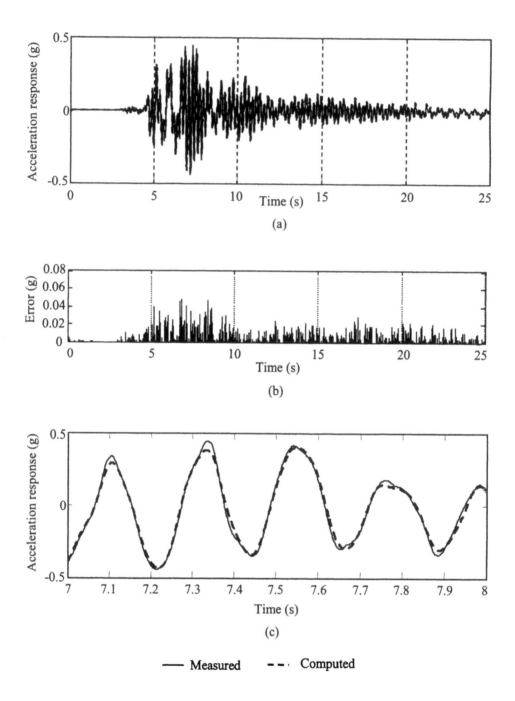

Figure 13.10 Identified structural responses of the second floor subjected to lower Kobe2 excitation using the model trained by the structural responses from the stronger Kobe5 excitation: (a) Comparison of acceleration responses from 0s to 25s; (b) absolute error, and (c) comparison of acceleration response from 7s to 8s

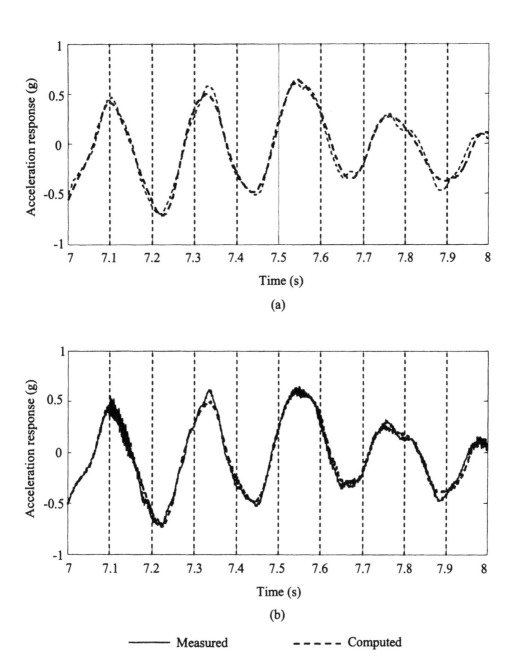

Figure 13.11 Comparisons of the identification results using the denoised and original structural response data under Kobe4 excitation using the model trained by structural responses from Kobe1: (a) Comparison of acceleration responses using denoised sensor data, and (b) comparison of acceleration responses using original sensor data

when the original data are used to train the model (the maximum identification error is 7% when data are denoised).

2) In order to preserve the dynamics of time series, the reconstructed state space concept from the chaos theory is employed to construct the input vector. Rather than choosing the dimension of the input space arbitrarily or by trial and error, which is the norm in the structural system identification field, an optimum value for the input space dimension is found using the FNN method.

3) Wavelets are employed in two different contexts. First, for denoising the data as explained under 1). Second, it is used in combination with two soft computing techniques, neural networks and fuzzy logic, to create a new pattern recognition model to capture the characteristics of the time series sensor data accurately and efficiently. The result is a fuzzy WNN model. The fuzzy WNN model a) balances the global and local influences of the training data due to the complementary properties of the soft computing technique, neural network, and the signal processing tool, wavelet transform, b) incorporates the imprecision existing in the sensor data effectively due to the use of the fuzzy clustering technique, c) provides more accurate system identifications, and d) results in fast training convergence thus significantly reducing the computational requirements.

4) The number of fuzzy WNN nodes in the hidden layer is selected by the Akaike's Final Prediction Error criterion, thus avoiding the arbitrary selection of this important neural network parameter.

5) The model provides more accurate nonlinear approximation because it is based on the integration of chaos theory (based on nonlinear dynamics theory), wavelets (a signal processing method), and two complementary soft computing methods, i.e., fuzzy logic and neural network.

Chapter 14

NONLINEAR SYSTEM IDENTIFICATION OF HIGH-RISING BUILDING STRUCTURES

14.1 Introduction

Chapter 13 presents a multi-paradigm dynamic time-delay fuzzy wavelet neural network (WNN) model for nonparametric identification of structures using the nonlinear autoregressive moving average with exogenous inputs (NARMAX) approach. The model adroitly integrates four different computing concepts: dynamic neural networks, wavelets, fuzzy logic, and chaos theory with the goal of improving the accuracy and adaptability of nonparametric system identification. The model balances the global and local influences of the training data and effectively incorporates the imprecision existing in the sensor data.

The focus of this chapter is on application of the dynamic time-delay fuzzy WNN model to three-dimensional high-rising building structures taking into account their geometric nonlinearity. It should be noted that the training of a dynamic neural network is substantially more complicated and time-consuming than the training of conventional neural networks because in the former both input and output are not single-valued but in the form of time-series. An effective training algorithm is essential for identification accuracy and real-time implementation of the dynamic fuzzy WNN model for health monitoring or control of large-scale structures. In this chapter, a hybrid adaptive learning algorithm, called Levenberg-Marquardt-Least-Squares (LM-LS) algorithm, is presented for training and adjusting the parameters of the dynamic fuzzy WNN model. The LM algorithm, an approximate combination of the Gauss-Newton and steepest descent algorithms, is employed to train the parameters of the nonlinear wavelet functions. The approximation avoids the second-order differentiation required in the Gauss-Newton algorithm and overcomes the numerical instabilities encountered in the steepest descent algorithm. The LS algorithm is employed to determine the parameters of the linear part of the dynamic fuzzy WNN approximator. Next, a backtracking inexact linear search algorithm is developed to automatically update the iteration step length with the goal of accelerating the learning convergence rate of the model and achieving high computational efficiency.

14.2 Dynamic Time-Delay Fuzzy WNN Model

Figure 14.1 shows the three stages and the flow diagram of the dynamic fuzzy WNN model presented in Chapter 13 for system identification of structures.

Stage 1 Denoising Time Series Data

Seismic records and measured data using sensors are usually in the form of time-series and almost always contain noise from the data acquisition process (e.g., instrument noise). The noise affects the accuracy and efficiency of structural system identification, health monitoring, and structural control models. Therefore, the first step is to denoise the sensed data.

The discrete wavelet packet transform (DWPT) technique for denoising a contaminated signal was presented in Sections 3.3 and 11.2. The DWPT-based approach provides more effective denoising results than the conventional DWT because the DWPT method provides more coefficients representing additional subtle details of a signal. In this work, the DWPT-based data denoising technique is used for denoising the structural response. The denoised signals are used for structural system identification to improve its accuracy.

Stage 2 Constructing Dynamic Fuzzy WNN model

The details of this stage are presented in Chapter 13. The procedure to construct the dynamic fuzzy WNN model is briefly sumarized in five steps as follows:

1) A state space vector is constructed from the denoised time series sensor data. The resulting state space vector is used as the input vector of the NARMAX approach to capture the invariants of the dynamic system represented by the time series.

2) Wavelets are selected based on an orthogonalization procedure using the modified Gram-Schmidt algorithm (see Section 12.3.3).

3) The Akaike's final prediction error (AFPE) criterion is employed to find the optimum (minimum) number of wavelets in the wavelet neural networks to produce accurate system identification results (see Section 12.3.5).

4) The denoised time series are clustered using a fuzzy c-means clustering algorithm to create *fuzzy-wavelets*, an integration of the wavelets with fuzzy clusters, where the translation parameters in the conventional wavelet are replaced by fuzzy clusters. Fuzzy-wavelets are aimed to enhance the ability of wavelets to capture any existing imprecision in the time series and extract useful information from them.

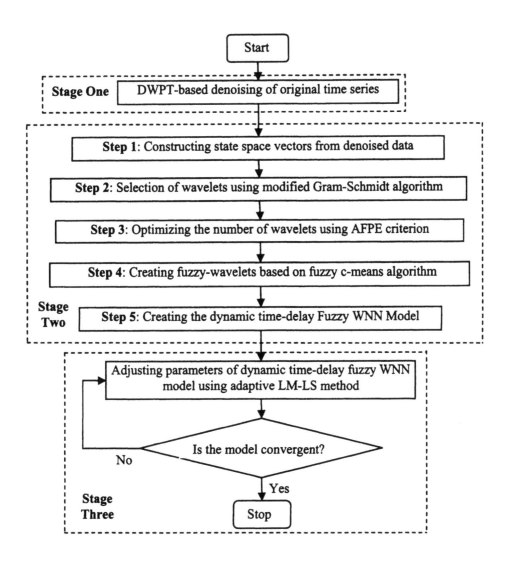

Figure 14.1 Stages and flow diagram for dynamic fuzzy WNN model

5) A dynamic time-delay fuzzy WNN model with a recurrent feedback topology is created using the fuzzy-wavelets. This model can represent the time-dependent mapping relationship in the input-output measurements effectively.

Stage 3 Training Dynamic Fuzzy WNN model

A hybrid-learning algorithm, called the adaptive LM-LS algorithm, is developed for adjusting the parameters of the dynamic fuzzy WNN model. The algorithm is described in the next section.

14.3   Adaptive LM-LS Learning Algorithm

14.3.1   Error Objective Function

The wavelet coefficients (w_i), wavelet dilation parameters (a_{ij}), input weights (b_j), and the bias weight (d) in Eq. (13.5) are found through an optimization process by minimizing an error objective function of the fuzzy WNN training error, $E(a_{ij}, w_i, b_j, d)$. This objective function is defined as the sum of the square of the training errors between the actual and approximated values of the input-output time series as follows:

$$E(a_{ij}, w_i, b_j, d) = \left\| \hat{\mathbf{Y}} - \mathbf{Y} \right\|^2 = \sum_{k=1}^{N_a} \left[\sum_{i=1}^{K} w_i \sum_{j=1}^{D} \varphi\left(\frac{X_{kj} - c_{ij}}{a_{ij}} \right) + \sum_{j}^{D} b_j X_{kj} + d - y_k \right]^2 \quad (14.1)$$

where $\left\| \hat{\mathbf{Y}} - \mathbf{Y} \right\| = \sqrt{\sum_{N_a} \left| \bar{f}(\mathbf{X}_k) - y_k \right|^2}$ is the Euclidean distance; $\hat{\mathbf{Y}} = [\ \bar{f}(\mathbf{X}_1)\ \ \bar{f}(\mathbf{X}_2)$ $\ldots \bar{f}(\mathbf{X}_{N_a})\ ]'$ and $\mathbf{Y} = [y_1\ \ y_1\ \ \cdots\ \ y_{N_a}]'$ represent the column matrices of the computed and actual structural response time series output vectors, respectively.

Minimizing the error objective function is an unconstrained optimization problem. This problem is solved in two steps iteratively. In the first step, the parameters w_i, b_j, and d are found by solving a linear programming problem using the least squares approach. Next, the parameters of the nonlinear part of the fuzzy WNN model, a_{ij}'s, are found by solving a nonlinear programming problem using the Levenberg-Marquardt approach. The result is an efficient and robust training algorithm.

14.3.2 Hybrid LM-LS Method

Equation (14.1) is expressed in matrix form as follows:

$$E(\mathbf{a},\omega) = \sum_{k=1}^{N_a} [\mathbf{v}_k'(\mathbf{a})\omega - y_k]^2 \tag{14.2}$$

where $\mathbf{v}_k(\mathbf{a}) = [\varphi_k(\mathbf{a}_1), \varphi_k(\mathbf{a}_2), \dots, \varphi_k(\mathbf{a}_K), X_{k1}, X_{k2}, \dots, X_{kD}, 1]'$ represents the input vector with $K + D + 1$ terms, in which the vector $\mathbf{a}_i = [a_{i1}, a_{i2}, \dots, a_{iD}]'$ represents the variables in the nonlinear function

$$\varphi_k(\mathbf{a}_i) = \varphi(z_{ik}) = \varphi\left(\left\|\frac{\mathbf{X}_k - \mathbf{c}_i}{\mathbf{a}_i}\right\|\right), \quad i = 1, 2, \dots, K; \; k = 1, 2, \dots, N_a \tag{14.3}$$

where $z_{ik} = \left\|\dfrac{\mathbf{X}_k - \mathbf{c}_i}{\mathbf{a}_i}\right\| = \sqrt{\sum_{j=1}^{D}[(X_{kj} - c_{ij})/a_{ij}]^2}$ represents the Euclidean distance,

$\mathbf{c}_i = \{c_{i1}, c_{i2}, \dots, c_{iD}\}'$ is the i-th cluster, and $\varphi(z_{ik}) = (D - z_{ik}^2)\exp(-z_{ik}^2/2)$ is a Mexican hat wavelet function (Fig. 12.1).

In Eq. (14.2), the vector $\omega = [w_1, w_2, \dots, w_K, b_1, b_2, \dots, b_D, d]'$ represents the parameters of the linear part of the fuzzy WNN model as well as the weights w_i's representing the linear combination of the nonlinear functions $\varphi_k(\mathbf{a}_i)$ (Eq. 13.5). For any given $\mathbf{a}$ representing the parameters of the nonlinear part, the optimal values of the linear parameters, $\omega^*(\mathbf{a})$, are obtained using the LS algorithm as follows:

$$\omega^*(\mathbf{a}) = [\mathbf{V}(\mathbf{a})\mathbf{V}'(\mathbf{a})]^{-1}[\mathbf{V}(\mathbf{a})\mathbf{Y}] \tag{14.4}$$

where $\mathbf{V} = [\mathbf{v}_1, \mathbf{v}_2, \dots, \mathbf{v}_{N_a}]'$ is an $N_a \times (K+D+1)$ matrix ($\mathbf{v}_k$ is defined previously). Substituting Eq. (14.4) into Eq. (14.2) the objective function becomes

$$E(\mathbf{a}) = E(\mathbf{a}, \omega^*(\mathbf{a})) = \sum_{k=1}^{N_a} [\mathbf{v}_k'(\mathbf{a})\omega^*(\mathbf{a}) - y_k]^2 \tag{14.5}$$

We now have a nonlinear programming problem. Since the wavelet basis of the Mexican hat wavelet is continuously differentiable, the necessary conditions for minimizing Eq. (14.4) at some point $\mathbf{a}^* \in \mathfrak{R}^2$ are

$$\nabla E(\mathbf{a}) = \partial E(\mathbf{a}, \omega) / \partial \mathbf{a}\big|_{\omega=\omega^*} = 0 \qquad (14.6)$$

where the symbol ∇ denotes the differential in terms of the parameters $\mathbf{a}$. Since $\mathbf{a}$ is a $D \times K$ matrix, Eq. (14.6) produces $D \times K$ sets of equations. The solution of Eq. (14.6), $\mathbf{a}^*$, represents the optimum dilation parameters required in the fuzzy WNN model. Substituting Eq. (14.5) into Eq. (14.6) and after differentiation we find

$$\nabla E(\mathbf{a}) = 2 \sum_{k=1}^{N_a} [\omega^* e_k \, \partial v_k'(\mathbf{a}) / \partial \mathbf{a}] \qquad (14.7)$$

$$\partial v_k'(\mathbf{a}_i) / \partial \mathbf{a}_i = (2 + D - z_{ik}^2) \psi_k(\mathbf{a}_i) / \mathbf{a}_i \,, \, i = 1, 2, ..., K, k = 1, 2, ..., N_a \quad (14.8)$$

where $e_k = \bar{f}(\mathbf{X}_k) - y_k$ and $\psi_k(\mathbf{a}_i) = [\varphi_k(\mathbf{a}_1), \varphi_k(\mathbf{a}_2), ..., \varphi_k(\mathbf{a}_K)]$, in which $\varphi_k(\mathbf{a}_i)$ is defined by Eq. (14.3).

The Newton algorithm has superior convergence efficiency compared with the steepest descent method in that both size and direction of the vector are modified simultaneously. It is widely used for the solution of nonlinear programming problems. The Newton formula is expressed as

$$\mathbf{a}^{m+1} = \mathbf{a}^m - \mathbf{H}_m^{-1} \nabla E(\mathbf{a}^m) \qquad (14.9)$$

where m is the iteration number and $\mathbf{H}_m = \nabla^2 E(\mathbf{a}^m)$ is the Hessian matrix of size $D \times K$ to be calculated for any given $\mathbf{a}^m \in \mathfrak{R}^2$.

Newton's method, which is based on quadratic approximation, however, suffers from three shortcomings (Hagan et al. 1996; Hooshdar and Adeli 2004). First, the gradient of E and the inverse of the Hessian matrix $\mathbf{H}_m$ must be evaluated in each iteration, which for our dynamic time series problem with thousands of time steps adds to the computational cost significantly. Second, the Newton algorithm does not converge when 1) $\mathbf{H}_m$ is not positive definite for any $\mathbf{a}^m$ (that is, when $|\mathbf{H}_m| < 0$) or 2) the solution $\mathbf{a}^{m+1}$ obtained from Eq. (14.9) at iteration $m + 1$ is not close to $\mathbf{a}^m$ obtained at iteration m indicating the inaccuracy of the quadratic approximation of E. Finally, Newton's algorithm requires the computation of the inverse of the Hessian $\mathbf{H}_m$ in each iteration, which may result in numerical instability when $\mathbf{H}_m$ is a near-singular matrix.

In order to overcome the aforementioned shortcomings, in this chapter, the LM

method is employed to solve the unconstrained nonlinear optimization problem defined
by Eq. (14.5) based on the following equation (Dennis and Schnable 1983):

$$\mathbf{a}^{m+1} = \mathbf{a}^m - [\mathbf{G}^T\mathbf{G} + \lambda_m\mathbf{I}]^{-1}\mathbf{G}'e \qquad (14.10)$$

where $e = \sum_{k=1}^{N_a} e_k$ is the accumulated error for all training data, $\mathbf{G} = \nabla E(\mathbf{a}^m)$ is a
gradient matrix with dimension $D \times K$, and $\mathbf{I}$ = identity matrix with the same dimension as
$\mathbf{G}'\mathbf{G}$. The parameter λ_m represents an adaptive step chosen so that $E(\mathbf{a}^{m+1}) < E(\mathbf{a}^m)$
leading to a minimum $E(\mathbf{a}^*)$ along the steepest decent direction, which will be discussed
in the next section.

The LM algorithm is an approximate combination of the steepest descent and
Gauss-Newton algorithm. It avoids the second-order differentiation and the computation
of the Hessian matrix by approximating the Hessian matrix with the term $\mathbf{G}'\mathbf{G}$ and
improves numerical stability by adding the additional term $\lambda_m\mathbf{I}$. When the term $\mathbf{G}'\mathbf{G}$ is
neglected, this method becomes the steepest descent method. When λ_m is equal to zero, it
becomes the Gauss-Newton method. The value of the steepest descent step length, λ_m,
affects the behavior of the optimization algorithm significantly and plays a critical role in
the algorithm convergence. The trial-and-error approach is the standard method used for
determining the step length (Hooshdar and Adeli 2004). In the following section, a
method is presented for determining λ_m with the goal of achieving a quickly convergent
algorithm.

14.3.3 Backtracking Inexact Linear Search Algorithm

A backtracking inexact linear search algorithm (Dennis and schnable 1983; Adeli and
Hung 1995) is employed to update the iteration step length λ_m in this work. In this search
algorithm, the sum of square of errors (SSE) for all training data is used to determine
whether or not the step length needs to be backtracked as follows:

$$SSE = \sum_{k=1}^{N_a} \left(\bar{f}(\mathbf{X}_k) - y_k\right)^2 \qquad (14.11)$$

In the m-th iteration, the parameter λ_m is determined in two stages. In the first stage, it is assumed $\lambda_m = 1$, the vector $\mathbf{a}^m$ is updated using Eq. (14.10), and the resulting gradient matrix, $\mathbf{G}$, and the SSE value are computed. When the current SSE value is less than its value in the previous iteration, both the updated $\mathbf{a}^m$ and current SSE are recorded for use in the second stage.

In the second stage, a small value is initially selected for λ_m (a value of 0.005 is chosen in this work), the step parameter λ_m is backtracked using the gradient matrix obtained in the first stage, and the vector $\mathbf{a}^m$ and the step length λ_m are updated iteratively using the following three steps:

1) Check the SSE value.
   a. If the current SSE is larger than the previous value, set $\lambda_m = 2\lambda_m$, go to step 2;
   b. If the training convergence criteria to be discussed in the next section are reached, stop;
   c. Otherwise, set $\lambda_m = \lambda_m/2$ and recompute the SSE value, go to step 2;
2) Update variables $\mathbf{a}$ using Eq. (14.10) and recompute the SSE value.
3) Check the backtracking stopping criterion using a given minimum ($\lambda_{min} = 10^{-5}$ is used in this work) and a given maximum λ_m ($\lambda_{max} = 10$ is used in this work). If the stopping criterion is not reached, return to step 1).

14.3.4 Convergence Criteria

Batch training is used in the LM-LS algorithm where parameters are updated only after all training data are presented to the network. Three convergence/stopping criteria are used to terminate the iterations of the LM-LS algorithm. First, a minimum reciprocal condition index, κ_{min}, is used to check whether or not the matrix $\mathbf{J} = \mathbf{G}'\mathbf{G} + \lambda_m\mathbf{I}$ with dimension $D \times K$ used to approximate the Hessian matrix is near-singular (a value of 10^{-7} is used in the examples presented in a subsequent section). This index is defined as follows (Mathworks 2001):

$$\kappa = 1/([\mathbf{J}^{-1}]_1[\mathbf{J}]_1)$$ (14.12)

where $[\mathbf{J}]_1 = \max_{1 \le j \le DK} \sum_{i=1}^{DK} |J_{ij}|$ is the maximum of the sums of the terms of the columns in the square matrix **J**.

The second criterion is a minimum value for the normalized root of the sum of the square of errors (RSSE) for all training data points as follows:

$$\text{RSSE} = \sqrt{\sum_{k=1}^{N_a}(\hat{y}_k - y_k)^2} \Bigg/ \sqrt{\sum_{k=1}^{N_a}(\hat{y}_k - \bar{y})^2} \qquad (14.13)$$

where y_k and $\hat{y}_k = \bar{f}(\mathbf{X}_k)$ represent the measured and predicted outputs, and $\bar{y}$ is the mean of the measured outputs. RSSE is also an indicator of the performance of the fuzzy WNN model. In the examples presented in a subsequent section a minimum value of $\varepsilon = 0.005$ is specified. The last criterion is the maximum number of iterations or epochs ($N_{\max} = 50$ is used in the examples presented in Section 14.5). The adaptive LM-LS learning algorithm for training the dynamic fuzzy WNN model is summarized in the form of a flowchart in Fig. 14.2.

14.4 Application of LM-LS Algorithm for System Identification

Figure 14.3 shows how the LM-LS algorithm is used in training the dynamic time-delay fuzzy WNN model for structural system identification of high-rising building structures. This is done in four steps (identified by numbers 1 to 4 in Fig. 14.3):

1) Measured data are denoised using the DWPT-based approach.

2) Denoised data of the three-dimensional (3D) structural responses are used as input-output data sets, $\mathbf{X}_k$ and y_k ($k = 1, 2, \ldots, N_a$), where the k-th input state space vector, $\mathbf{X}_k$, is constructed from the denoised time series sensor data using the concepts from chaos theory (see Chapter 4) and y_k is the corresponding k-th denoised measured output.

3) The dynamic fuzzy WNN model with a recurrent feedback topology is created as described in stage two of Fig. 14.1.

4) The dynamic fuzzy WNN model is trained using the adaptive LM-LS algorithm and the denoised input-output data sets. In every iteration, all the training data sets are applied to the model and the three convergence criteria are checked.

An advantage of the nonparametric approach to system identification and the dynamic fuzzy WNN model is that it can be trained using the data for any given number

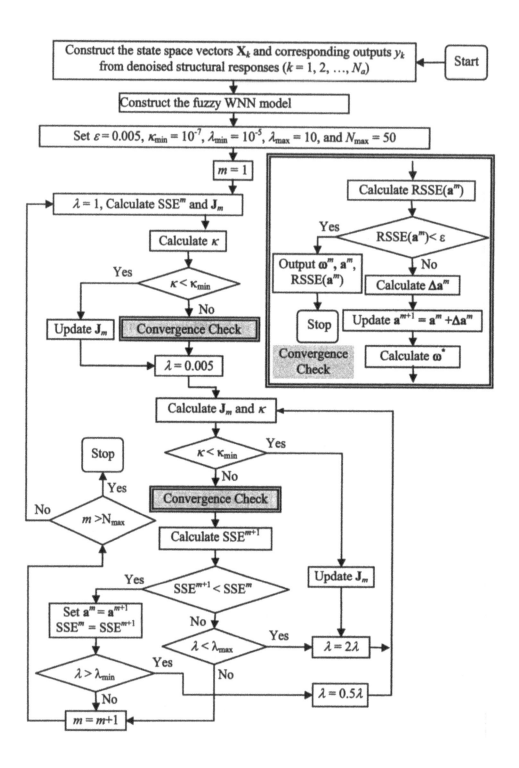

Figure 14.2 Adaptive LM-LS algorithm for training the dynamic fuzzy WNN model

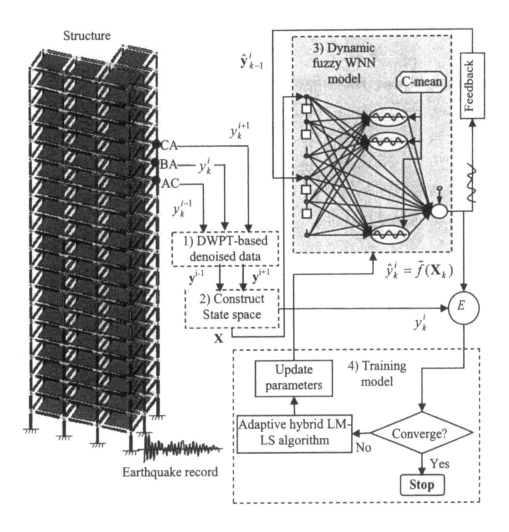

y_k^i = k-th measured structural response (k represents time step) at i-th story;

$\mathbf{y}^i = [y_1^i \quad y_2^i \quad \cdots \quad y_k^i]'$ = all measured structural responses at i-th story;

$\mathbf{X} = [\mathbf{X}_1 \quad \mathbf{X}_2 \quad \cdots \quad \mathbf{X}_{N_a}]'$ = input state space vector for all structural response data;

$\bar{f}(\mathbf{X}_k)$ = fuzzy WNN mapping function represented by Eq. (13.5); $\hat{y}_k^i = \bar{f}(\mathbf{X}_k)$ = k-th computed output at i-th story; $\hat{y}_{k-1}^i$ = feedback structural response; E = error function represented by Eq. (14.1)

Figure 14.3 Training dynamic time-delay fuzzy WNN model for structural system identification using the adaptive LM-LS algorithm

of floors. For training the model, the structural responses at several floors below and/or above the floor to be identified during the previous time intervals are used as inputs, and the current response of the floor to be identified is used as the output (the floors do not have to be contiguous). In the examples presented in the next section, the structural responses are selected at three different floors arbitrarily.

The model can be trained using responses from a given set of dynamic excitations such as earthquake records. But, it can be used to identify the structural response under a new set of dynamic excitations.

After training, the computed output data at i^{th} story, $\hat{y}_{k-1}^i$, and denoised input sensor data vectors at i-1st and i+1st stories, $\mathbf{y}^{i-1}$ and $\mathbf{y}^{i+1}$, are used as the past time input to determine the subsequent output at i^{th} story by Eq. (13.5). The dynamic fuzzy WNN model is applied for system identification of two irregular steel high-rising building structures using data from the horizontal component of the Chichi earthquake of Richter magnitude 7.3 which occurred at Foothills, Western Taiwan, on September 21, 1999 (Fig. 14.4c).

14.5   Numerical Examples

The dynamic responses of each example structure subjected to the Chichi earthquake with six different scales (20%, 40%, 60%, 80%, 90%, and 100%) are obtained through a nonlinear finite element analysis. These excitations are denoted as Chi1, Chi2, Chi3, Chi4, Chi5, and Chi6, respectively. The geometry of a high-rising building structure changes appreciably due to large deformation of the structure and P-Δ effects of columns. Therefore, in order to accurately model the dynamical response of the 3D high-rising building subjected to the ground excitations, geometric nonlinearity should be considered in the structural analysis. Both large displacement and P-Δ effects are taken into account in the two example structures presented in this chapter. A damping ratio of 2% is used in all examples and cases. Acceleration and displacement responses are obtained at each floor in the x horizontal direction at increments of 0.01 seconds over a period of 40 seconds. Thus, the number of sample data, N in Eq. (13.5), is 4,000. In order to demonstrate the model can identify both displacements and accelerations accurately, accelerations are used in Example 1 and displacements are used in Example 2.

14.5.1 Example 1: Twelve-Story Steel Building

<u>Example Structure and its Response</u>

This example is an irregular 12-story three-dimensional moment-resisting steel building structure with vertical setbacks as shown in Fig. 14.4. This structure was created by Adeli and Saleh (Saleh and Adeli 1998a; Adeli and Saleh 1999) for the study of structural control. It consists of 148 members, 77 nodes, and 462 degrees of freedom. The yield stress of steel is 50 ksi (344.8 N/mm^2). The members of the structure are W shapes as indicated in Fig. 14.4. The frame is subjected to the combination of uniformly distributed static gravity loads of 4.8 kN/m^2 and the Chichi earthquake record shown in Fig. 14.4c. The earthquake excitation in the form of a time history function is applied on the structure in the x-direction at all support joints. For dynamic analysis, floors are assumed to be rigid, each with two translational and one rotational degree of freedom.

Figure 14.5 shows the acceleration responses of joints A (on the 10th floor), B (on the 9th floor), and C (on the 8th floor) identified in Fig. 14.4a subjected to the original Chichi earthquake (Fig. 14.4c). All acceleration responses have been normalized to the gravity acceleration (g).

<u>Denoising Data</u>

The computed structural responses contain noise because the measured earthquake records contain noise. The DWPT-based denoising approach described in Sections 3.3 and 11.2 and Daubechies wavelets of order 4 are used for denoising all computed structural response data. As an example, Fig. 14.6a shows the acceleration response from 10 seconds to 20 seconds shown in Fig. 14.5c for joint C of the structure subjected to the original Chichi excitation. Figure 14.6b shows the denoised signal of the same acceleration response. Figure 14.6c shows the noise in the data (note that the scale of the vertical axis in this figure has been enlarged for visualization purposes). It is observed that the denoised data (low-frequency signal) shown in Fig. 14.6b approximates the original signal shown in Fig. 14.6a effectively without loss of any significant feature. The noise (high-frequency signal) shown in Fig. 14.6c is therefore ignored in the subsequent analysis for system identification.

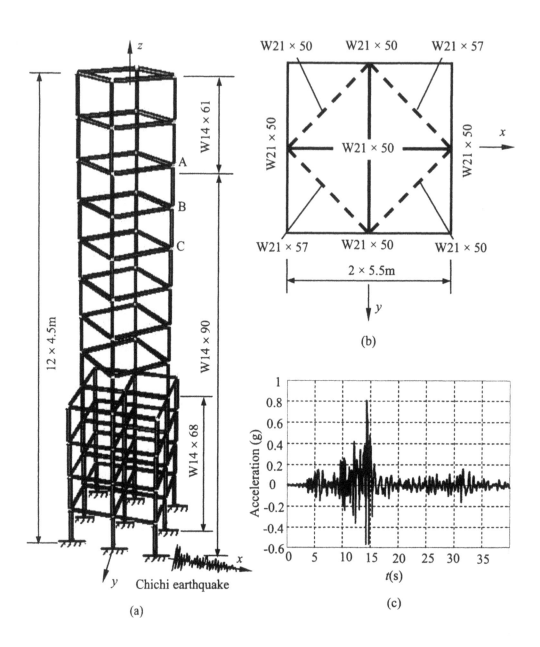

Figure 14.4 Twelve-story moment-resisting steel frame with a vertical setback subjected to Taiwan Chichi earthquake: (a) Perspective view; (b) plan view, and (c) Chichi earthquake, Taiwan, Sep 21, 1999

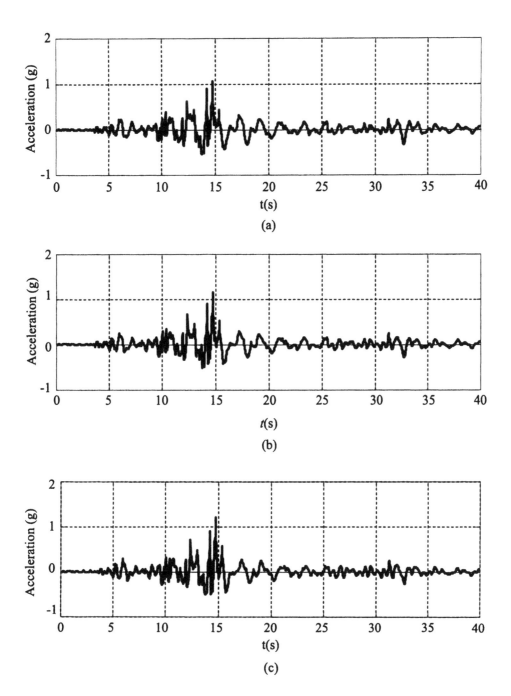

Figure 14.5 Acceleration responses of the twelve-story steel frame at joints A, B, and C subjected to the original Chichi earthquake excitation: (a) Joint A (10th floor); (b) Joint B (9th floor), and (c) Joint C (8th floor)

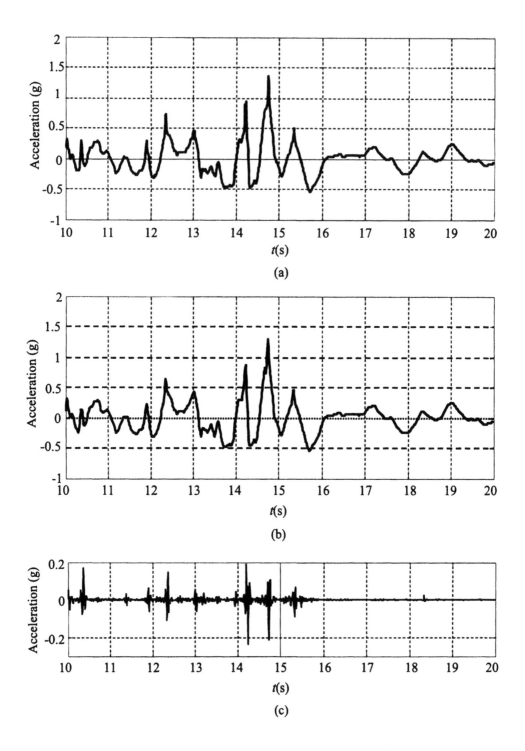

Figure 14.6 Denoising acceleration responses at joint C of the 12-story frame subjected to the original Chichi earthquake excitation using the DWPT approach: (a) Original signal; (b) denoised signal, and (c) noise

Constructing Fuzzy WNN Model

The 4,000 denoised data points of the structural acceleration response at joints A, B, and C are used for constructing the fuzzy WNN model. For the structural system identification problem, the false nearest neighbor (FNN) method described in Section 4.3.3 was found to be more accurate than other approaches for finding the optimum embedding dimension and is therefore used in this work. The optimum embedding dimensions of 2, 3, and 3 are obtained using the FNN method for the acceleration responses at joints A, B, and C of the structure under different scales of excitations, respectively. The number of input vectors N_a is obtained by

$$N_a = N - \max\{D_x, D_y\} \tag{14.14}$$

where N is the number of sample data and $\max\{D_x, D_y\}$ denotes the maximum of D_x and D_y. In this example, $N_a = 4{,}000 - \max\{2, 3\} = 3{,}997$.

Next, a four-level wavelet decomposition is performed on the 3,997 training vectors using the Mexican hat wavelet function (Fig. 12.1). The empty wavelets whose supports do not contain any data are eliminated, resulting in 39 non-empty wavelets. Based on the Akaike's final prediction error (AFPE) criterion presented in Section 12.3.3 and using the modified Gram-Schmidt algorithm presented in Section 12.3.5, it is concluded that 6 out of the 39 non-empty wavelets are sufficient to construct an effective fuzzy WNN model based on the results presented in Fig. 14.7, which shows the variation of AFPE with the number of wavelets (the minimum point of the curve yields the optimum value). As such, there are six fuzzy wavelet nodes in the hidden layer of the network.

Training the Fuzzy WNN Model

The acceleration responses of joints A, B and C (Fig. 14.4) during the previous $D_A = 2$, $D_B = D_C = 3$ time intervals are used as inputs, and the current acceleration response of joint B is used as the output of the dynamic time-delay fuzzy WNN model. The six sets of computed structural response data obtained at different excitation levels of Chi1 to Chi6 are used to train the model, one set at a time, resulting in six trained models. The training of the model using the adaptive LM-LS learning algorithm converges very fast after only seven training iterations. The model is implemented in a combination of C++

programming language and MATLAB 6.1 (Mathworks 2001) on a Windows XP Professional platform and a 1.5GHz Intel Pentium 4 processor. The CPU time for training the model using the denoised time series data is only 27 seconds. All absolute errors between the computed and measured responses are less than 0.03g (less than 0.5%).

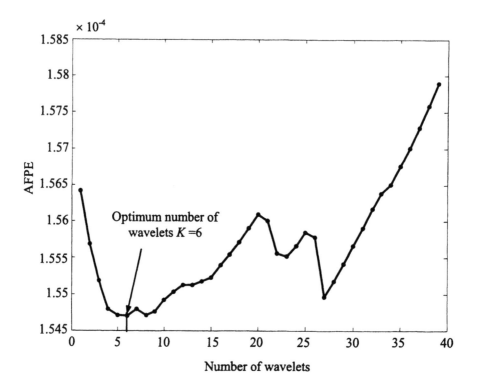

Figure 14.7 Optimum number of wavelets using Akaike's Final Prediction Error Criterion

System Identification Results

The acceleration responses of joints A and C (Fig. 14.4a) from the previous $D_A = 2$ and $D_C = 3$ time intervals and the computed acceleration response of joint B from the previous $D_B = 3$ time intervals are fed into the input layer to calculate the current acceleration response of joint B. Each of the six trained models is tested using all computed structural response data resulting from six different levels of excitations (Chi1 to Chi6). As such, 36 different sets of system identification results are produced. As a sample, the identified results using the model trained by the response data of Chi3 (60%)

excitation is presented in Table 14.1. The errors for the maximum acceleration response for all six sets of test data are less than 3.1%. As an example, Fig. 14.8 shows the identified structural responses of joint B subjected to Chi1 earthquake using the model trained by the structural responses from Chi3. The maximum absolute error is less than 0.05g.

Table 14.1 Comparison of maximum identified acceleration of joint B of twelve-story steel frame using the model trained by structural responses from Chi3 excitation

Excitation scale	Chi1	Chi2	Chi3	Chi4	Chi5	Chi6
Original max. (g)	0.2537	0.5074	0.7612	1.0149	1.1418	1.2687
Predicted max. (g)	0.2459	0.4926	0.7387	0.9849	1.1079	1.2311
Percent error (%)	3.1	2.9	2.9	3.0	3.0	3.0

The wavelet packet denoising technique employed in the model speeds up the training convergence and improves the system identification accuracy. Figure 14.9 shows comparisons of the identification results using the denoised and original structural response data from 10 seconds to 20 seconds under Chi6 excitation using the model trained by structural responses from Chi3 (note that the scale of the vertical axis in Figs. 14.9b and 14.9d has been enlarged for visualization purposes). The CPU time for training the model is 30 seconds using the denoised data and 41 seconds using the original data. More significantly, the structural responses from 12.4 seconds to 13.5 seconds can not be identified when the original data are used to train the model.

14.5.2 Example 2: Twenty-Story Steel Building

Example Structure and its Response

This example is a 20-story three-dimensional moment-resisting steel frame with plan irregularity as shown in Figs. 14.10 and 14.11. This structure was used by Liew and Jiang (Liew *et al*. 2001; Jiang *et al*. 2002) for study of the nonlinear three-dimensional behavior of steel structures. The structure has 460 beams and columns, 210 nodes, and 1,260 degrees of freedom. For dynamic analysis, floors are assumed to be rigid, each with two translational and one rotational degree of freedom.

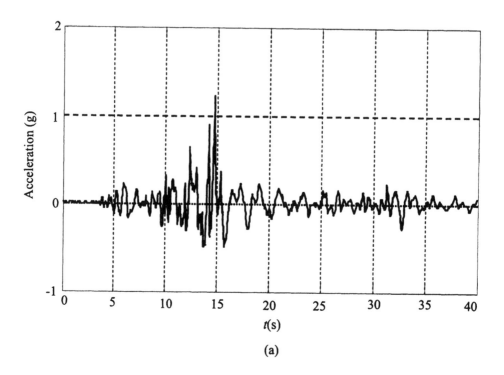

(a)

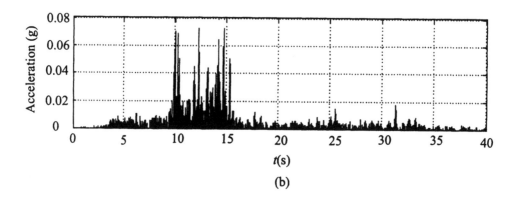

(b)

Figure 14.8 Identified structural responses of Joint B subjected to Chi1 earthquake
using the model trained by structural responses from Chi3 earthquake: (a) Acceleration
responses, and (b) absolute errors

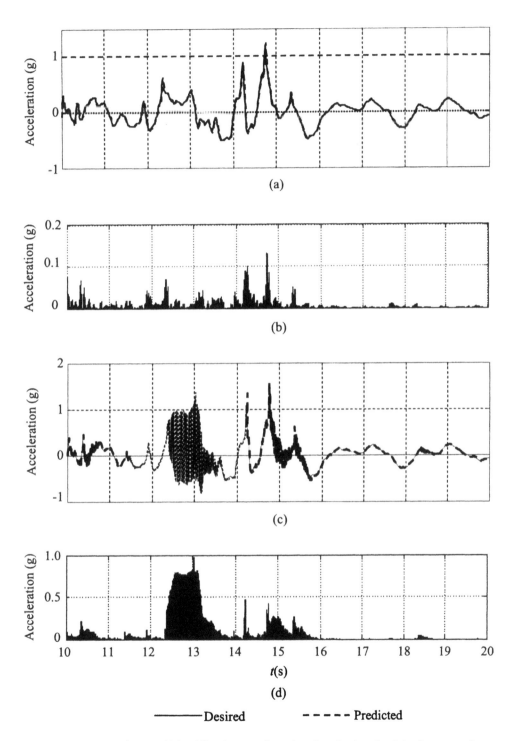

Figure 14.9 Comparisons of identification results using denoised and original structural response data from 10s to 20s under Chi6 excitation using the model trained by structural responses from Chi3: (a) Comparison of acceleration responses using denoised data; (b) absolute errors of (a); (c) comparison of acceleration responses using original data, and (d) absolute errors of (c)

The yield stress of steel is 50 ksi (344.8 N/mm^2). The members of the structure are W shapes as indicated in Figs. 14.10 and 14.11. The frame is subjected to the combination of uniformly distributed static gravity loads of 4.8 kN/m^2 and different levels of Chichi earthquake. The earthquake excitation in the form of a time history function is applied in the x-direction (noted in Fig. 14.10) at all support joints. Figures 14.12b and 14.12c show the displacement and acceleration responses of joint A on the top of the structure subjected to the original Chichi earthquake shown in Fig. 14.4c. Figure 14.12a shows the deformation of the structure at the point of maximum horizontal displacement with a magnification of 5,000.

The displacement responses are used for system identification in this example. Figure 14.13 shows the displacement responses of joints A (on the 20th floor), B (on the 14th floor), and C (on the 11th floor) identified in Fig. 14.12a subjected to the original Chichi earthquake (Fig. 14.4c). In this example, the three selected floors are not contiguous.

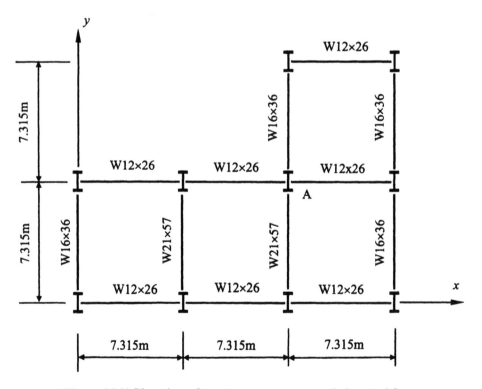

Figure 14.10 Plan view of twenty-story moment-resisting steel frame

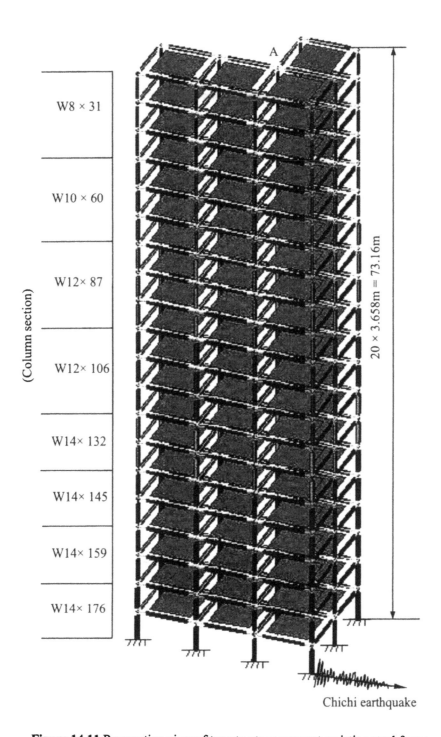

Figure 14.11 Perspective view of twenty-story moment-resisting steel frame

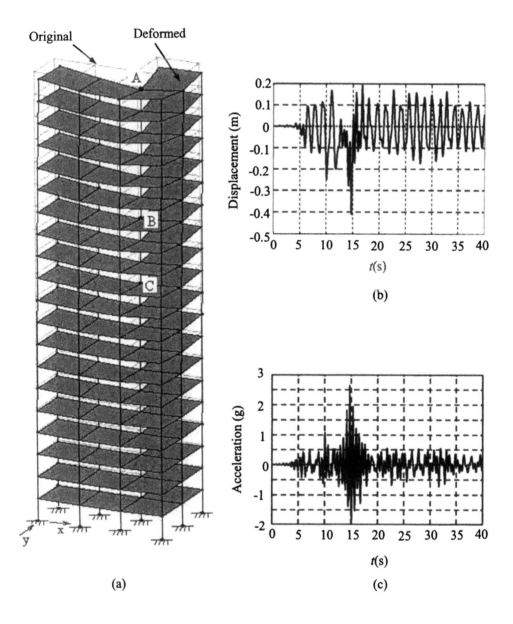

Figure 14.12 Structural response of twenty-story moment-resisting frame subjected to Taiwan Chichi earthquake: (a) Structural deformation; (b) displacement of joint A in the x-direction, and (c) acceleration of joint A in the x-direction

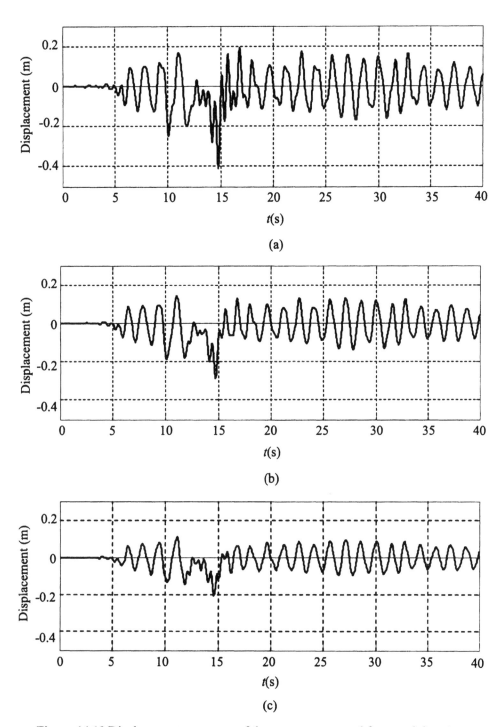

Figure 14.13 Displacement responses of the twenty-story steel frame at joints A, B, and C subjected to original Chichi earthquake: (a) Joint A (20th floor); (b) Joint B (14th floor), and (c) Joint C (11th floor)

<u>Denoising Data</u>

The 4,000 data points of the structural response at joints A, B, and C are denoised using the DWPT-based denoising approach and Daubechies wavelets of order 4.

<u>Constructing Fuzzy WNN Model</u>

The same optimum embedding dimension of 2 is obtained using the FNN method (see Section 4.3.3) for the displacement responses at joints A, B, and C of the structure subjected to different scales of excitations. The number of input vectors created is thus equal to $N_a = 4,000 - \max\{2, 2\} = 3,998$ (Eq. 14.14). A four-level wavelet decomposition is performed on the 3,998 training vectors using the Mexican hat wavelet function (Fig. 12.1). The empty wavelets whose supports do not contain any data are eliminated, resulting in 97 non-empty wavelets. Similar to Example 1, based on the AFPE criterion and using the modified Gram-Schmidt algorithm, it is concluded that 20 out of the 97 non-empty wavelets are sufficient to construct an effective fuzzy WNN model. As such, there are twenty fuzzy wavelet nodes in the hidden layer of the network.

<u>Training Fuzzy WNN Model</u>

The displacement responses of joints A, B and C (Fig. 14.12) during the previous $D_x = 2$ time intervals are used as inputs, and the current displacement response of joint B is used as the output of the dynamic time-delay fuzzy WNN model. The six sets of computed structural response data obtained at different excitation levels of Chi1 to Chi6 are used to train the model, one set at a time, resulting in six trained models. The training of the model using the adaptive LM-LS algorithm converges after only five iterations. The CPU time for training the model using the denoised time series data is 57 seconds. All absolute errors between the computed and measured responses are less than 0.03 m (less than 0.1%).

<u>System Identification Results</u>

The displacement responses of joints A and C (Fig. 14.12a) from the previous $D_x = 2$ time intervals and the computed displacement response of joint B from the previous $D_y = 2$ time intervals are fed into the input layer to calculate the current displacement response of joint B. Each of the six trained models is tested using all computed structural response data resulting from six different levels of excitations (Chi1 to Chi6). As a sample, the identified results using the model trained by the response data of Chi2 (40%) excitation is presented in Table 14.2. The errors in the maximum displacement responses for all six sets of test data are less than 2%. As an example, Figure 14.14 shows the identified structural responses of joint B subjected to Chi5 (90%) excitation using the model trained by the structural responses from Chi2. Note that the differences in the two curves in Fig. 14.14a are not noticeable.

Table 14.2 Comparison of maximum identified x-direction displacement of joint B of twenty-story steel frame using the model trained by structural responses from Chi2 excitation

Excitation scale	Chi1	Chi2	Chi3	Chi4	Chi5	Chi6
Original max. (mm)	56.77	113.54	170.32	227.09	255.47	283.86
Predicted max. (mm)	56.33	111.33	169.71	224.63	253.02	281.75
Percent error (%)	0.78	1.95	0.34	1.08	0.96	0.74

14.6   Concluding Remarks

In the dynamic time-delay WNN model for system identification of large-scale structures, wavelets are employed adroitly for two purposes. First, wavelets are used to preprocess the training data and to improve the accuracy and efficiency of the structural system identification algorithm. It was shown that the accuracy of structural system identification is improved significantly using the DWPT-based denoising approach.

Second, wavelet functions are used as activation functions in conjunction with the fuzzy clustering technique to form a fuzzy wavelet neural network. The purpose of this second application is to create a powerful pattern recognition model which can accurately capture the characteristics of the time series sensor data.

An adaptive LM-LS algorithm was presented for training the dynamic neural network model. The system identification model and the training algorithm were evaluated using two high-rising building structures taking into account geometric nonlinearities. The results of numerical validations indicate a powerful model for system identification of large building structures with nonlinear behavior. In all cases, the identification error is less than about 3%. The LM-LS training algorithm converged in less than 8 iterations in all cases, resulting in a very fast training algorithm. The results of this chapter provide a powerful tool for real-time health monitoring, nondestructive damage, and control of large structures such as high-rising building structures.

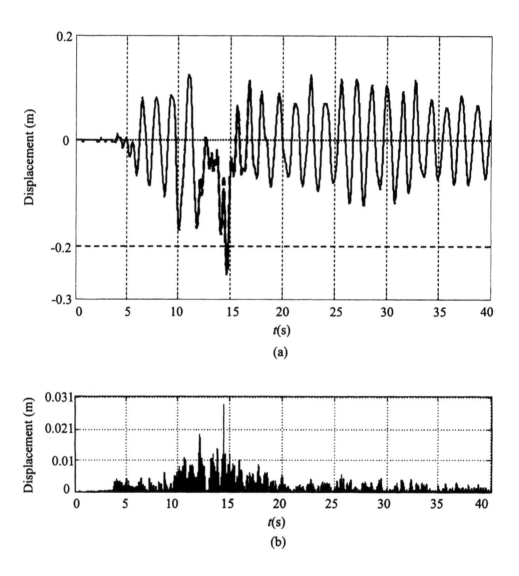

Figure 14.14 Identified structural responses of Joint B subjected to Chi5 earthquake using the model trained by the structural responses from Chi2 earthquake: (a) Displacement responses, and (b) absolute error between desired and predicted responses

Chapter 15

DAMAGE DETECTION AND HEALTH MONITORING OF STRUCTURES

15.1 Introduction

The goals of intelligent health monitoring of structures are mainly to 1) provide an early warning for structural failure in order to reduce the loss of life and properties to a minimum, 2) help structural engineers better understand the behavior of structures during earthquakes and design more effective earthquake-resistant structures, and 3) create a new generation of smart structures whose health is continuously monitored in real time by the embedded sensors (Beck *et al.* 2006; Bolton *et al.* 2006; Ching *et al.* 2006; Furukawa *et al.* 2006; Hou *et al.* 2006; Huang and Loh 2006; Ko and Ni 2005; Lam *et al.* 2006; Li *et al.* 2006; Park *et al.* 2006; Saeki and Hori 2006; Sanayei *et al.* 2006; Wong and Yao 2006; Yuen and Katafygiotis 2006; Jiang and Adeli 2005a & 2007; Adeli and Jiang 2006; Jiang and Mahadevan 2008).

Deterioration and damage generally result in a reduction of structural stiffness and mass, which changes the dynamic properties of structures such as their natural frequencies and mode shape. Since structural dynamic properties are directly related to structural mass and stiffness, changes of their dynamic properties rather than structural mass and stiffness are often used in global or local damage detection and effective intelligent health monitoring of large-scale structures under extreme loadings. This is a challenging problem due to the complicated nonlinear nature of the behavior of large-scale structures and the incomplete, incoherent, and noise-contaminated measurements of structural response under extreme loadings. In addition to the innovative hardware technologies such as smart materials, sensors, and communication and measurement tools, advanced software technologies such as data mining and automatic feature extraction techniques play an important role in developing an effective health monitoring system.

Recently civil engineering researchers have embraced the sensor technology for structural system identification and health monitoring due to advances in data processing and signal analysis capabilities and the increasing computational power of microprocessors. A good number of research papers have been published on the use of various types of sensor technologies for health monitoring of structures (e.g., Aktan *et al* 1997; Blanas *et al.* 1997; Housner *et al.* 1997; Chung 2001; Catbas and Aktan 2002; Chong *et al.* 1994 & 2003). The sensors are embedded in or attached to a structure at selected locations and are used to measure the static and dynamic responses of the structure for the purpose of monitoring the structural performance. The recently developed sensor technology provides a powerful tool for collecting structural response data. Most of the recently published sensor-based health monitoring methods (e.g., Chong *et al.* 1994 & 2003; Aktan *et al.* 1997; Housner *et al.* 1997; Chung 2001) are effective when applied to detect the health status of small structures with a few degree-of-freedom (DOF) but impractical for damage detection of large-scale structures with a large number of DOF due to 1) incomplete measurements of the dynamical responses of structures, 2) excessive costs of using a large number of sensors, and 3) prohibitive computational processing cost required for real-time processing of a huge amount of data.

In the past few years, a few researchers have presented wavelet-based damage detection methods for health monitoring of simple beams utilizing time and frequency localization capability of the wavelet transform (Zhou and Adeli 2003, Jiang and Adeli 2004a). Wavelets have been used in three different ways for detecting damage to a simple beam. In the first approach, variation of wavelet coefficients is used to detect the existence and severity of damage (Yoon *et al.* 2000; Melhem and Kim 2003). In the second approach, local perturbation of wavelet coefficients in a space domain is used to detect the existence and location of the damage (Liew and Wang 1998; Hong *et al.* 2002). In the third approach, Gabor wavelet transform is used to localize the severity of the damage (Quek *et al.* 2001). These approaches are feasible only for monitoring the local damage of a single structural member or a simple structure, but not feasible for monitoring the global conditions of large-scale or even moderate-scale structures.

Salawu (1997) reviewed the natural frequencies-based methods for structural damage detection. The frequencies obtained from periodic vibration tests are used to monitor structural behavior and assess structural conditions (for example, changes in the structural stiffness). However, these damage detection approaches are also unfeasible for detecting the global damage of large-scale structures for two reasons. First, these methods require taking into consideration all possible damage scenarios at various locations on the structure. They are therefore computationally prohibitive for damage detection of large-scale civil structures. Second, the structural modal information is needed for structural system identification and subsequent damage detection. Such modal frequency-based damage indicators are known to be very sensitive to environmental variability. In addition, the linearity or nonlinearity of the structural behavior needs to be assumed in the damage detection. The assumption has significant effects on the modal selection and the accompanying identification scheme. Different modal selection and identification schemes will result in completely different damage detection results.

In order to overcome the shortcomings of the conventional natural frequencies-based damage detection approaches, neural network models have been developed for structural damage detection due to their ability to capture the indeterministic nature and complex nonlinearity of time series (e.g., Elkordy et al. 1993; Szewezyk and Hajela 1994; Masri et al. 1996 & 2000; Nakamura et al. 1998; Xu et al. 2000; Hung and Kao 2002; Wu et al. 2002; Kao and Hung 2003; Pei et al. 2004). Masri et al. (1996 & 2000) present a backpropagation neural network (BPNN)-based nonlinear system identification approach for damage detection of unknown structural systems. First, vibration measurements from a healthy structure behaving elastically are used to train the BPNN model. Then, vibration measurements from the same structure under different levels of excitation are used as inputs to the trained network model in order to monitor the health of the structure. The errors between the prediction results and the actual measurements from the structure are used as a damage index to assess the extent of the changes in the condition of structures.

Hung et al. (2000 & 2002) present a two-step system identification-based neural network approach for damage detection of two-dimensional frame structures. An adaptive limited memory quasi-Newton second-order learning algorithm is used to train the neural network. In the first step, a neural system identification network (NSIN) is developed to identify the undamaged and damaged states of a structural system. Changes of the structural condition are detected through comparing the partial derivatives of the NSIN that represent a certain damaged state with those that represent the undamaged state. In the second step, a neural damage detection network (NDDN) is developed to detect the location and extent of the structural damage. This latter network is related to the NSIN through its partial derivatives. The weights of the NSIN are used to represent the structural properties. The output of the NDDN identifies the damage level (for example, reduced percentage of structural damping and/or stiffness) for each member in the structure.

The existing neural network-based damage detection approaches are mostly parametric identification methods. They are advantageous over the conventional frequency-based methods for damage detection of small-scale structures due to the following two reasons. First, these methods do not require any prior knowledge of the structural system to be investigated. Both linear and nonlinear structural systems can be treated with the same formulation, which is required for civil structures subjected to strong earthquake ground motions. Second, these methods are fault tolerant for representing a given structural system. However, these approaches require the inputs including the relative displacements, velocities and accelerations of all floors of a building structure subjected to the external excitations. They are impracticable for damage detection and real-time health monitoring of large-scale structures because they require the use of a large number of sensors.

Chapter 13 presented a multi-paradigm dynamic time-delay fuzzy wavelet neural network (WNN) model for nonparametric identification of structures using the nonlinear autoregressive moving average with exogenous (NARMAX) inputs. In addition, Chapter 14 presented an adaptive Levenberg-Marquardt-Least-Squares (LM-LS) algorithm with a backtracking inexact linear search scheme for training of the dynamic fuzzy WNN model. The model was applied to two high-rising moment-resisting building structures taking into account their geometric nonlinearities. In this chapter, a nonparametric system identification-based model is presented for damage detection of large-scale high-rising building structures subjected to ground motions using the dynamic fuzzy WNN model with an adaptive LM-LS learning algorithm. The proposed method does not require complete measurements of the dynamic responses of the whole structure. This is achieved by dividing a large structure into a series of sub-structures around selected floors where measurements are made. A power density spectrum method is proposed for damage detection of a building structure. The methodology for damage detection of high-rising building structures is validated using the sensed data obtained for a 1:20 scaled 38-story concrete test model.

15.2 Substructuring a High-rising Structure

Placing sensors on every floor of a high-rising building structure is expensive and may not be practical. Prior to instrumentation of an entire structure, the size or number of substructures and the number of sensors per substructure usually play a role in the cost-effectiveness of the health monitoring systems. These issues are closely related to the subject of sensor placement optimization (e.g., Guo *et al.* 2004; Papadimitriou 2004). In the proposed model sensors are placed at selected floors along the height of a high-rising building structure, called measurement floors (identified by black dots in Fig. 15.1), such as the top floor and locations of mass or stiffness changes. The structure is then divided into a series of substructures around these measurement floors (the blocks identified by dashed lines in Fig. 15.1). The measurement or sensed floor and its neighboring floors are aggregated into a substructure for the purpose of damage detection. The measurements collected by a sensor in a substructure represent the local response of the high-rising structure within that substructure.

15.3 Dynamic Fuzzy WNN Model for Damage Detection

The general discrete dynamic input-output mapping in the dynamic fuzzy WNN approach is expressed by Eq. (13.5). The dynamic fuzzy WNN model for damage detection of high-rising building structures consists of two stages: 1) training the model using dynamic response of substructures under small-level excitations and 2) detecting damage to substructures under large-level excitations using the trained model.

In the first stage, the dynamic fuzzy WNN model is trained using the LM-LS learning algorithm presented in Chapter 14. The acceleration responses of three adjacent measurement floors corresponding to substructures i-1, i, and i+1 during the previous time interval are used as input to the dynamic fuzzy WNN model. The output of the model is the current acceleration response of substructure i. Figure 15.2 shows the training diagram of the dynamic fuzzy WNN model using the adaptive LM-LS learning algorithm and substructures response data. The training is done in three steps (identified by numbers 1 to 3 in Fig. 15.2):

1) Sensed data of the sub-structural responses are used as input-output data sets, $\mathbf{X}_k$ and y_k (k = 1, 2, ..., N_a), where the k-th input state space vector, $\mathbf{X}_k$, is constructed from the time series sensor data using the reconstructed state space concepts from chaos theory

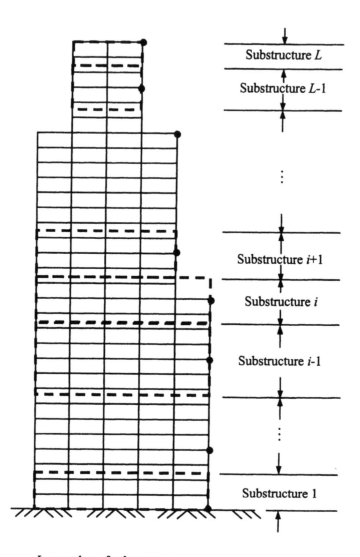

L = number of substructures
● Sensor location (measurement floor)

Figure 15.1 Substructuring scheme for a high-rising frame structure

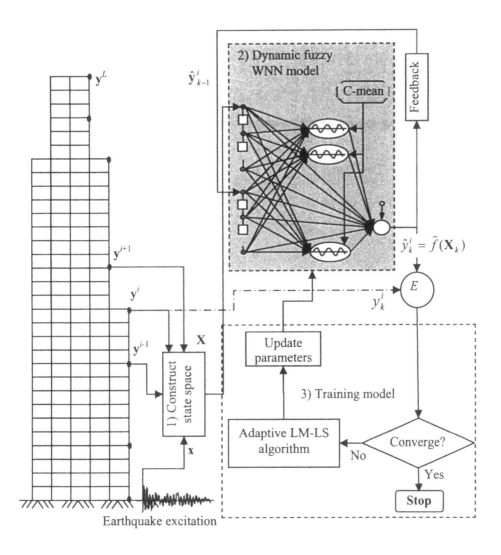

$\mathbf{x} = [x_1, x_2, ..., x_N]$ = all earthquake excitation points;

y_k^i = k-th measured structural response (k represents time step) at i-th floor;

$\mathbf{y}^i = [y_1^i \quad y_2^i \quad \cdots \quad y_N^i]'$ = all measured structural responses at i-th floor;

$\mathbf{X} = [\mathbf{X}_1 \quad \mathbf{X}_2 \quad \cdots \quad \mathbf{X}_{N_a}]'$ = input state space vector for all structural response data;

$\bar{f}(\mathbf{X}_k)$ = output of the fuzzy WNN;

$\hat{y}_k^i = \bar{f}(\mathbf{X}_k)$ = k-th computed output at i-th floor;

$\hat{\mathbf{y}}_{k-1}^i$ = feedback structural response; E = error function

Figure 15.2 Training diagram of the dynamic fuzzy WNN model using adaptive LM-LS learning algorithm and sensed substructures response data

(see Chapter 4) as follows:

$$\mathbf{X}_k = [\mathbf{x}_k \quad \mathbf{y}_k^{i-1} \quad \mathbf{y}_k^{i} \quad \mathbf{y}_k^{i+1}]' \tag{15.1}$$

where

$$\mathbf{x}_k = [x_k, x_{k+\tau_x}, x_{k+2\tau_x}, \ldots, x_{k+(D_x-1)\tau_x}]' \tag{15.2}$$

$$\mathbf{y}_k^{i-1} = [y_k^{i-1}, y_{k+\tau_y^{i-1}}^{i-1}, y_{k+2\tau_y^{i-1}}^{i-1}, \ldots, y_{k+(D_y^{i-1}-1)\tau_y^{i-1}}^{i-1}]' \tag{15.3}$$

$$\mathbf{y}_k^{i} = [y_k^{i}, y_{k+\tau_y^{i}}^{i}, y_{k+2\tau_y^{i}}^{i}, \ldots, y_{k+(D_y^{i}-1)\tau_y^{i}}^{i}]' \tag{15.4}$$

$$\mathbf{y}_k^{i+1} = [y_k^{i+1}, y_{k+\tau_y^{i+1}}^{i+1}, y_{k+2\tau_y^{i+1}}^{i+1}, \ldots, y_{k+(D_y^{i+1}-1)\tau_y^{i+1}}^{i+1}]' \tag{15.5}$$

in which x_k and y_k^i are the k-th earthquake excitation input and the corresponding structural response (k refers to the time step) of substructure i, respectively; the parameters D_x and D_y^i, referred to as embedding dimensions in the chaos theory (see Section 4.3), represent dimensions of the earthquake excitation input and structural response inputs of substructure i, respectively. The input dimension, D, is the summation of dimensions of the structural response inputs and the earthquake excitation, defined as follows:

$$D = D_x + D_y^{i-1} + D_y^{i} + D_y^{i+1} \tag{15.6}$$

The parameters τ_x and τ_y^i are the lag time indexes of the earthquake excitation input and structural response inputs of substructure i, respectively. The number of input vectors, N_a, is calculated by

$$N_a = N - \max\{(D_x-1)\tau_x, (D_y^{i-1}-1)\tau_y^{i-1}, (D_y^{i}-1)\tau_y^{i}, (D_y^{i+1}-1)\tau_y^{i+1}\} - 1 \tag{15.7}$$

where N is the number of data points in the time series and $\max\{.\}$ denotes the maximum operator. For the structural damage detection problem, the false nearest neighbor (FNN) method described in Section 4.3.3 is used to find the optimum embedding dimensions D_x and D_y^i. The average mutual information (AMI) method, a type of nonlinear autocorrelation function (ACF) as described in Section 4.2, is used in this chapter for the selection of the lag time τ_x and τ_y^i.

2) The dynamic fuzzy WNN model with a recurrent feedback topology is constructed as described in Section 13.3.

3) The dynamic fuzzy WNN model is trained using the adaptive LM-LS algorithm and the sensed input-output data sets as described in Section 14.3.

In the second stage, the measured responses of substructures i-1 and i+1, the input earthquake excitation, and the computed responses of substructure i in the previous time increment are fed into the input layer and the current acceleration response of substructure i is computed. If a structure is damaged under strong ground motions, the structural characteristics will change such that the measured response of the substructure will not correspond to the response predicted by the trained model. The difference between the actual measured response of a substructure and the response predicted by the trained model provides a quantitative measure of the changes of the condition of the substructure or damage.

Figure 15.3 shows the damage detection diagram using the trained fuzzy WNN model. The damage detection is also done in three steps (identified by numbers 1 to 3 in Fig. 15.3):

1) Construct the input state space vector, $\mathbf{X}_k$ (defined by Eq. 15.1), based on the measured responses of substructures i-1 and i+1, the input earthquake excitation, and the computed responses of substructure i in the previous time increment.

2) Predict the current structural response of substructure i using the trained model.

3) Evaluate the damage to the substructure based on the error between the predicted and measured responses of substructure i, using a damage evaluation method to be discussed in the next section.

15.4 Power Density Damage Evaluation Method

The relative root mean square (RRMS) error between the measured responses of substructures and the responses predicted by a neural network model has been used widely to measure structural damage (e.g., Masri *et al.* 1996; Wu *et al.* 2002; Pei *et al.* 2004). The RRMS error is defined as follows:

$$RRMS = \sqrt{\frac{1}{N_a - 1} \sum_{k=1}^{N_a-1} \left(\bar{f}(\mathbf{X}_k) - y_k\right)^2} \Bigg/ \sqrt{\frac{1}{N_a} \sum_{k=1}^{N_a-1} y_k^2} \qquad (15.8)$$

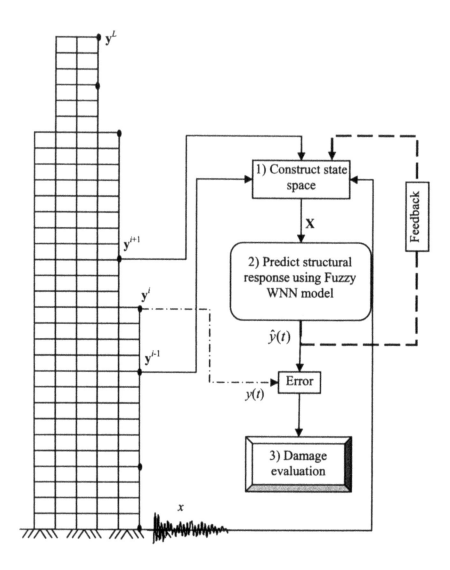

Figure 15.3 Damage detection diagram of a high-rising building structure using the trained fuzzy WNN model

where $\bar{f}(\mathbf{X}_k)$ is the output of the dynamic fuzzy WNN model representing the predicted structural response, defined by Eq. (13.5).

It is customary to assume that structural damage has occurred when the RRMS error exceeds a predefined threshold level obtained by trial and error (for example, 0.6). However, the RRMS error method is not always accurate because predictive errors can result from three different sources: 1) training the neural network model to approximate the structural properties, 2) sensing data which always contain measurement noises, and 3) possible structural damage.

In this chapter, a new damage evaluation method is proposed based on a power density spectrum method, called *pseudospectrum* (Trefethen 1992). The pseudospectrum provides a reliable solution for eigenvalues of a *non-normal* matrix (whose eigenvectors are not orthogonal), which cannot be solved easily by a standard eigenvalue solution method. The non-normal matrices exist in the chaotic motion of transient nonlinear systems represented by measured data (Stoica and Moses 1997; Trefethen 1999). The multiple signal classification (MUSIC) method is employed in this chapter to compute the pseudospectrum from the structural response time series. The structural damage is identified explicitly through comparison of the pseudospectra of the predicted output of the trained model and the measured output of the structure. It should be noted while the pseudospectrum method and the MUSIC approach are known in the mathematics literature; neither had ever been used in any structural engineering applications before the authors' paper (Jiang and Adeli 2007).

The MUSIC approach produces a higher frequency resolution even for data with a low signal-to-noise ratio (high noise) than the conventional Fast Fourier transform (FFT) method (Stoica and Moses 1997). It provides an estimate of a signal's frequency content from a set of eigenvectors of an autocorrelation matrix generated by the input signals. Given the structural response time series for a substructure i with N_a data points, $y_1, y_2, \cdots, y_{N_a}$, the pseudospectrum is estimated using the following five steps:

1) Construct an $N \times D_y^i$ state space matrix by placing the vectors of the structural response time series data based on the reconstructed state space concept column-wise as follows:

$$\mathbf{Y} = [\mathbf{y}_1 \quad \mathbf{y}_2 \quad \cdots \quad \mathbf{y}_{D_y^i}] = \begin{bmatrix} y_1 & \cdots & 0 & 0 \\ \vdots & \ddots & \vdots & \vdots \\ \hline y_{\tau_y^i(D_y^i-1)} & \cdots & y_{\tau_y^i} & y_1 \\ \vdots & \ddots & \vdots & \vdots \\ y_{N-\tau_y^i(D_y^i-1)} & \cdots & y_{\tau_y^i(D_y^i-2)} & y_{\tau_y^i(D_y^i-1)} \\ \vdots & \ddots & \vdots & \vdots \\ y_{N_a} & \cdots & y_{N-\tau_y^i(D_y^i-2)} & y_{N-\tau_y^i(D_y^i-1)} \\ \hline \vdots & \ddots & \vdots & \vdots \\ 0 & \cdots & 0 & y_{N_a} \end{bmatrix} \quad (15.9)$$

where $\mathbf{y}_k$ represents the k-th column vector in the brackets.

2) Compute the $N \times D_y^i$ correlation matrix, $\mathbf{C(Y)}$, for all the state space vectors

$$\mathbf{C(Y)}_{jk} = Cov(\mathbf{Y})_{jk} \Big/ \sqrt{Cov(\mathbf{Y})_{jj} Cov(\mathbf{Y})_{kk}} \ , j = 1, 2, \ldots, N_a; k = 1, 2, \ldots D_y^i \quad (15.10)$$

where $Cov(\mathbf{Y}) = (\mathbf{Y'Y})^{-1}$ represents the covariance matrix.

3) Calculate the eigenvectors of the correlation matrix (Eq. 15.10) and place them column-wise in a matrix $\mathbf{v}$.

4) Transform the state space matrix defined by Eq. (15.9) into the frequency domain using the FFT method with the length of $2(K_e-1)$ as follows:

$$\mathbf{e(Y)} = [1 \quad \exp(j2\pi \mathbf{Y}) \quad \exp(j4\pi \mathbf{Y}) \quad \cdots \quad \exp(j2(K_e - 1)\pi \mathbf{Y})]^H \quad (15.11)$$

where exp(.) represents a complex sinusoid of the FFT, the parameter K_e is the dimension of the eigenvectors (number of rows in $\mathbf{v}$) determined by the effective length of the FFT, and the superscript H is the conjugate transpose operator.

5) Compute the pseudospectrum defined as follows:

$$P_M(\mathbf{Y}) = 1 \Big/ \sum_{k=D_y+1}^{K_e} \left| \mathbf{v}_k^H \mathbf{e(Y)} \right|^2 \quad (15.12)$$

where $\mathbf{v}_k$ is the k-th eigenvector (the k-th column in $\mathbf{v}$).

Note that the value of an effective FFT length, K_e, is not crucial in this analysis as long as it is an integer number with a power of 2 but not larger than the total number of data points, N. Usually a value of 256 or larger provides sufficient FFT accuracy (Stoica and Moses 1997). In this chapter, a value of 1024 is chosen for the parameter K_e. Thus, $K_e = 513$ is obtained from $2(K_e - 1) = 1024$.

The values of pseudospectrum are converted into the decibel units for the convenience of comparison among a large range of values as follows:

$$\mathbf{P}_Y = 10\log_{10}[P_M(\mathbf{Y})] \tag{15.13}$$

where $\log_{10}(.)$ is a logarithm operator with a basis of 10. The length of FFT is also converted proportionally into the range 0 and 2π. Two pseudospectrums are plotted in the frequency range of 0 and π, one for the measured response and the other for the computed response. The larger the difference between the two, the greater the damage to the structure.

15.5 Numerical Example

15.5.1 Example Structure

Ni *et al.* (2006) measured the acceleration responses of a 1:20 scale test model of a typical high-rising residential building in Hong Kong under seismic excitations. Their response data are employed in this chapter to evaluate the performance of the proposed approach for detecting damage to high-rising building structures. The prototype structure is a 38-story reinforced concrete building (excluding the machine floors) with 34 stories of typical floors supported by a 3-level podium via a transfer floor. The scaled test model, shown in Figs. 15.4a and 15.4b, is 2.37m long, 2.16m wide, and 6.515m high with a symmetric floor plan. It consists of three bottom floors (F_1 to F_3), one transfer floor (F_4), 34 typical floors (F_5 to F_{38}), and three top machine floors (F_{39} to F_{41}). The locations of the sensors in the floor plan are shown in Fig. 15.4b.

The test structure is divided into 9 substructures along the elevation based on 9 selected measurement floors identified by black dots in Fig. 15.4. The resulting substructures are identified as dashed-line boxes. The sensors (accelerometers) are installed at the middle of two front elevations in two principal (x and y) directions (Fig. 15.4b).

The model structure was tested dynamically on a 5m × 5m 6DOF shaking able at the Hong Kong Polytechnic University (Ni *et al.* 2006). Two types of excitations were

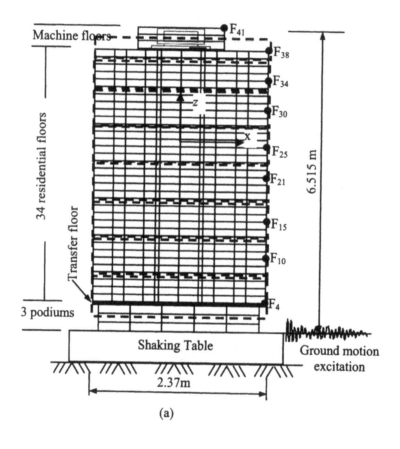

(a)

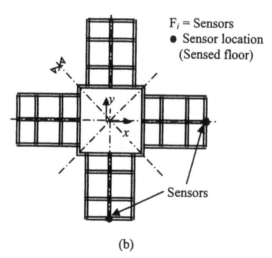

(b)

Figure 15.4 38-story concrete building model: (a) Front elevation view and (b) plan view

applied to the structure: seismic excitations (Case 1) and seismic excitations followed by white noise (Case 2). In the seismic excitation case, the model is tested with four different levels of earthquake excitations (small-, moderate-, large-, and super large-level) under three different foundation conditions of rock, medium soil and soft soil separately. These are intended to result in small, moderate, large, and very large (near collapse) amounts of damage. For the purpose of illustration, only the data in the x-direction are used in this chapter for damage detection of the structure. Refer to Ni *et al.* (2006) for the test details of the example structure.

The damage level of the test structure is defined by Ni *et al.* (2006) based on the visual inspection of the cracks in the structural members and the crack-widths. The small earthquake excitation results in fine 'hair-line' cracks in several members at the podium floors (small damage level of excitation). The number of cracks and the crack-widths along the height of the test structure increase as the excitation level increases. Under the super large-level earthquake excitation, horizontal cracks appear on the floor slabs in all the stories, with the severest at floors F_4 to F_{10}. An end wall above floor F_4 is separated from the transfer floor, resulting in the loss of the structural integrity and near collapse of the whole structure.

The input excitation and acceleration responses were measured at increments of 0.01 seconds over a period of 45 seconds for small-level excitation and 30 seconds for moderate-level and large-level excitations. Thus, around 4,500 and 3,000 data points are recorded in a single test for small-level excitation and moderate/large-level excitations, respectively. Fig. 15.5 shows the three sets of seismic excitation input data on a hard soil site (small-, moderate-, and large-level earthquake excitations) used in this chapter as Case 1 to validate the model for damage detection of structures. As an example, Fig. 15.6 shows the small-level earthquake excitation and measured acceleration responses of sample floors F_{25}, F_{30}, and F_{34}, representing the responses of their respective substructures. It is observed that the response patterns of various substructures are similar, which is an indication of no or insignificant damage.

It should be pointed out that, since experimental results indicated that most of the damage occurs in the transfer floors of the structure, the data in the transfer floors should provide higher detection accuracy. However, in order to demonstrate the effectiveness and robustness of the proposed method using partially or incompletely sensed data, the

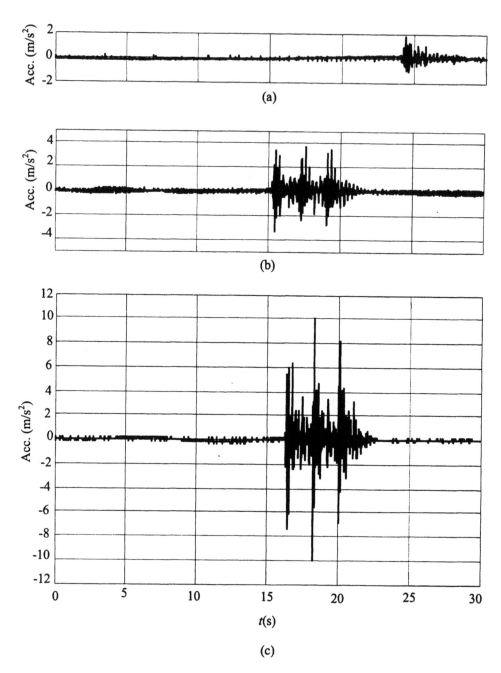

Figure 15.5 Seismic excitations for three types of damage in Case 1: (a) Small-level; (b) moderate-level, and (c) large-level

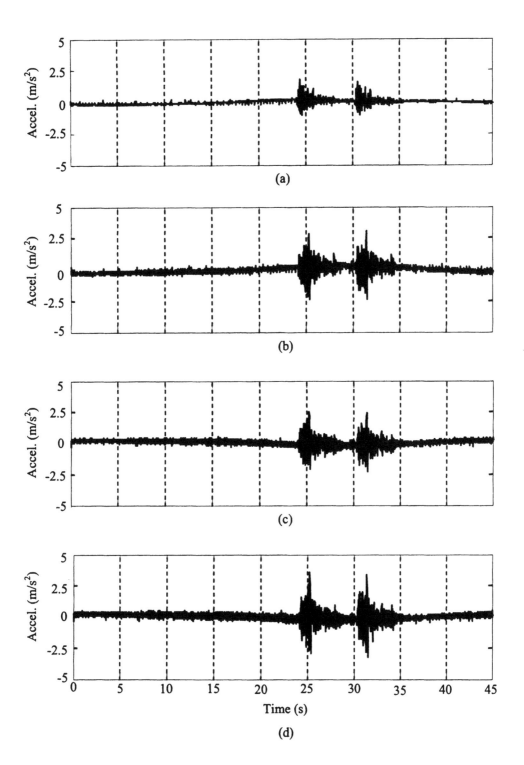

Figure 15.6 Small-level seismic excitation and the resulting floor acceleration responses (Case 1): (a) Seismic excitation; (b) F_{25} sensor data; (c) F_{30} sensor data, and (d) F_{34} sensor data

data in three sample floors, F_{25}, F_{30}, and F_{34}, are chosen in Cases 1 and 2 for the purpose of illustration.

In Case 2, five sets of 200-second white noise excitations (Ni *et al.* 2006) are used to detect five damage states: no damage, small damage, large damage, severe damage, and near collapse. The white noise random excitations are applied to the undamaged structure and the structure after applying one of the four levels of earthquake excitations described in Case 1 (small-, moderate-, large-level, and super large-level) intended to result in four damage modes (minor, moderate, severe, and near collapse). Each excitation and the resulting acceleration responses were measured at increments of 0.01 seconds, which results in about 20,000 data points. As such, five sets of acceleration measurements in the *x*-direction along with the corresponding white noise excitations are used as Case 2 to validate the damage detection model.

15.5.2 Case 1: Seismic Excitation

Constructing the Dynamic Time-Delay Fuzzy WNN Model

The 4,500 data points of the seismic excitation and the corresponding measured structural responses of floors (or substructures) F_{25}, F_{30}, and F_{34} are used for constructing the fuzzy WNN model. For reconstructing the state space vectors, the same optimum embedding dimension of 4 and lag time of 3 are obtained for both the acceleration responses of three substructures and the earthquake excitation using the FNN (see Section 4.3.3) and AMI methods (see Section 4.2), respectively ($D_x = 4$ and $D_y^{25} = D_y^{30} = D_y^{34} = 4$). The number of input vectors created is therefore equal to $N_a = 4,500 - 3(4-1) - 1 = 4,490$ (Eq. 15.7).

Next, a four-level wavelet decomposition is performed on the 4,490 training vectors using the Mexican hat wavelet (Fig. 12.1). The empty wavelets whose supports do not contain any data are eliminated, resulting in 134 non-empty wavelets. Based on the AFPE criterion (see Section 12.3.5) and using the modified Gram-Schmidt algorithm (see Section 12.3.3), it was concluded that 3 out of the 134 non-empty wavelets are sufficient to construct an effective fuzzy WNN model. As such, there are 3 wavelet nodes in the hidden layer of the network. For a discussion on the concept of empty wavelets and how the necessary number of non-empty wavelets is chosen, please refer to Section 12.3.2.

Training the Dynamic Fuzzy WNN Model

The acceleration responses at floors F_{25} and F_{34} during the previous $D_x = 4$ time steps are used as inputs, and the current acceleration response at substructure F_{30} is obtained as the output of the dynamic time-delay fuzzy WNN model (Fig. 15.2). The measured data from small-level excitation are used to train the model. Using the adaptive LM-LS learning algorithm the model converges quickly after only two training iterations. All absolute errors between the computed and measured responses are less than 0.5 m/s^2 (less than 1% in this training case). The trained dynamic WNN model will be used to predict the dynamic response of floor F_{30}.

Damage Detection

The measured acceleration responses of floors F_{25} and F_{34} (Fig. 15.4) during the previous $D_x = 4$ time steps and the computed acceleration response of the floor F_{30} during the previous $D_y = 4$ time steps are fed into the input layer. The output of the network is the estimated current acceleration response of substructure F_{30}. The trained model is used to detect the damage condition of the structure under small-, moderate-, and large-level excitations. Figure 15.7 shows the estimated structural responses of substructure F_{30} subjected to small-level seismic excitation. A small RRMS value of 0.58 is obtained for this case, indicating no or insignificant damage.

Figures 15.8 and 15.9 show the estimated structural responses of floor F_{30} subjected to moderate- and large-level seismic excitations, respectively. Their corresponding RRMS values are 0.73 and 1.1. These larger values of RRMS indicate probably damage to the structure. But, this cannot be ascertained by any level of certainty.

Figure 15.10 shows the pseudospectra of measured and estimated structural responses of floor F_{30} subjected to small-, moderate-, and large-level seismic excitations. It is observed that for small-level seismic excitation the two pseudospectra match well (Fig. 15.10a). However, as the level of the seismic excitation increases, the difference between the pseudospectra for the measured and estimated responses become more pronounced (Figs. 15.10b and c). It is noted that the three pseudospectra for the estimated structural responses (dashed lines in Fig. 15.10) show a similar pattern but the values and locations of the peaks are different. This is expected because the estimated structural responses are obtained from the neural network trained by data obtained from the

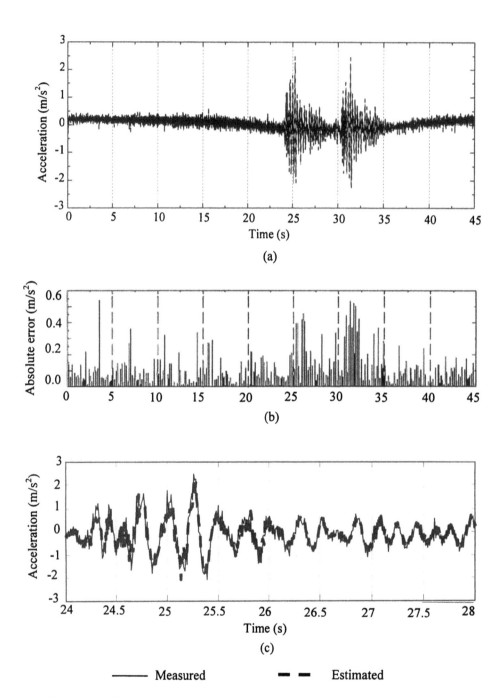

Figure 15.7 Comparison of measured and estimated floor acceleration responses
using the 4,500 data sets from the small-level seismic excitation (Case 1): (a)
Acceleration responses from 0s to 45s; (b) absolute error of acceleration responses
(RRMS = 0.58), and (c) comparison of acceleration responses from 24s to 28s. See
color insert following Chapter 15.

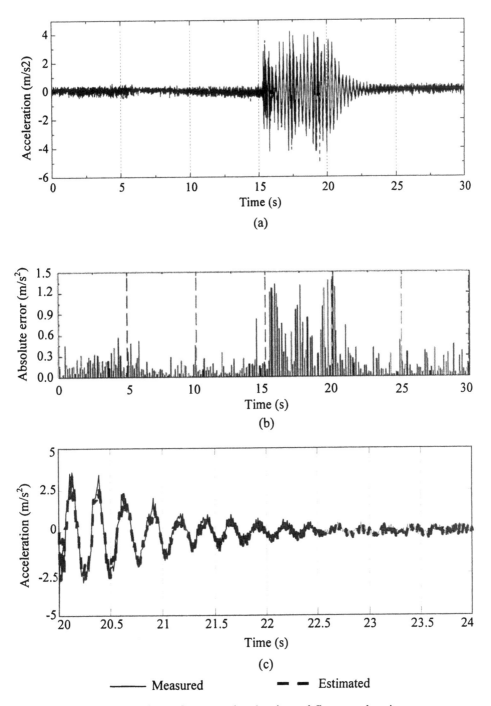

Figure 15.8 Comparison of measured and estimated floor acceleration responses using the 3,000 data sets from the moderate-level seismic excitation (Case 1): (a) Acceleration responses from 0s to 30s; (b) absolute error of acceleration responses (RRMS = 0.73), and (c) comparison of acceleration responses from 24s to 28s. See color insert following Chapter 15.

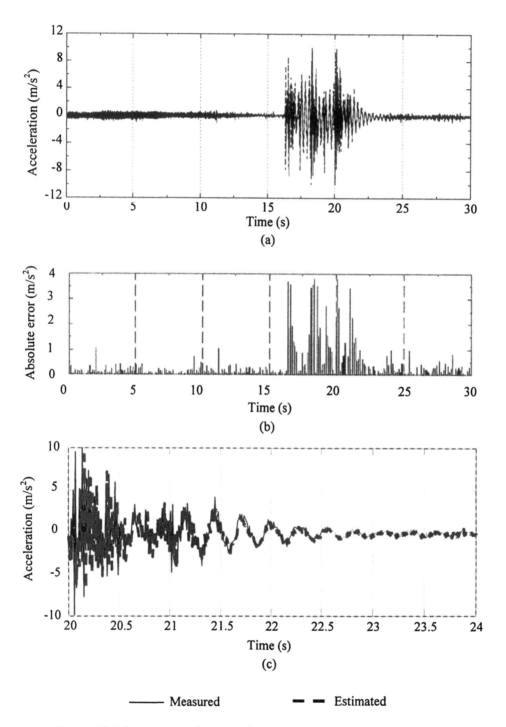

Figure 15.9 Comparison of measured and estimated floor acceleration responses using the 3,000 data sets from the large-level seismic excitation (Case 1): (a) Acceleration responses from 0s to 30s; (b) absolute error of acceleration responses (RRMS = 1.10), and (c) comparison of acceleration responses from 24s to 28s. See color insert following Chapter 15.

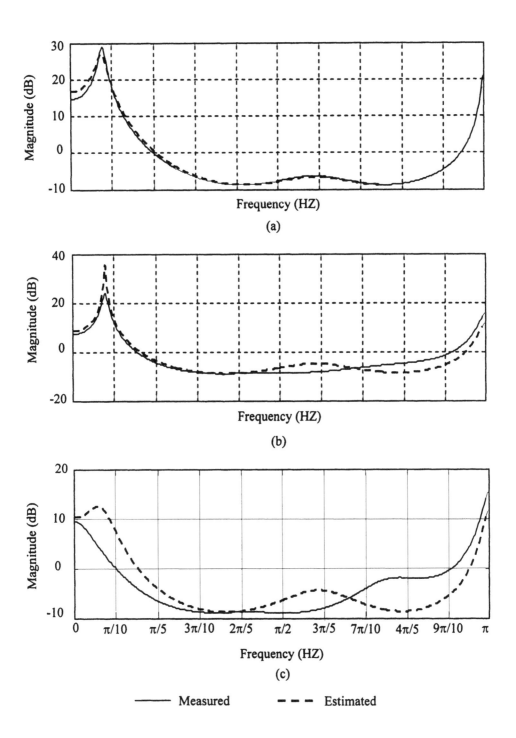

Figure 15.10 Comparison of pseudospectra of measured and estimated acceleration responses of floor F_{30} subjected to small-, moderate-, and large-level seismic excitations (Case 1): (a) Small-level seismic excitation; (b) moderate-level seismic excitation, and (c) large-level seismic excitation. See color insert following Chapter 15.

structure without damage or with insignificant damage.

15.5.3 Case 2: White Noise

<u>Constructing the Dynamic Fuzzy WNN Model</u>

This example is used to validate the proposed methodology for damage detection of high-rising structures subjected to an input excitation with a low signal-to-noise ratio. The white noise random excitations are applied to the undamaged structure and the structure after applying one of the four levels of earthquake excitations described in Case 1 intended to result in four damage modes. In this case, 12,000 x-direction data points of the white noise excitations and the corresponding structural responses at floors F_{25}, F_{30}, and F_{34} are used for constructing the fuzzy WNN model. For reconstructing the state space vectors, the same optimum embedding dimension of 6 and lag time of 7 are obtained for both acceleration responses of three floors and white noise excitation using the FNN and AMI methods, respectively ($D_x = 6$ and $D_y^{25} = D_y^{30} = D_y^{34} = 6$). The number of input vectors created is therefore equal to $N_a = 12,000 - 7(6 - 1) - 1 = 11,964$ (Eq. 15.7).

Next, a four-level wavelet decomposition is performed on the 11,964 training vectors using the Mexican hat wavelet. The empty wavelets whose supports do not contain any data are eliminated, resulting in 23 non-empty wavelets. Based on the AFPE criterion and using the modified Gram-Schmidt algorithm, it was concluded that 2 out of the 23 non-empty wavelets are sufficient to construct an effective fuzzy WNN model. As such, there are 2 wavelet nodes in the hidden layer of the network.

<u>Training the Dynamic Fuzzy WNN Model</u>

The measured acceleration responses at floors F_{25} and F_{34} during the previous $D_x = 6$ time steps are used as inputs, and the current acceleration response at floor F_{30} is obtained as the output of the dynamic time-delay fuzzy WNN model (Fig. 15.2). The measured data from small-level excitation are used to train the model. Using the adaptive LM-LS learning algorithm the model converges quickly after only four training iterations. All absolute errors between the computed and measured responses are less than 0.6m/s^2 (less than 5% in this training case). This is acceptable because of the high level of noise.

Damage Detection

The measured acceleration responses of floors F_{25} and F_{34} (Fig. 15.4a) during the previous $D_x = 6$ time steps and the computed acceleration response of the floor F_{30} during the previous $D_y = 6$ time steps are fed into the input layer. The output of the network is the estimated current acceleration response of substructure F_{30}. The trained model is used to detect the damage condition of the structure under small-, moderate-, large-, and super large-level excitations.

Figure 15.11 shows the estimated structural responses of floor F_{30} under five aforementioned damage states subjected to the white noise excitations after five seismic excitations described in Section 15.5.1. Figure 15.11a shows the estimated results for the undamaged structure. The error between the measured and estimated structural responses is due to two different sources: training the model and measured data. The RRMS increases from no damage case (RRMS = 0.61 in Fig. 15.11a) to the case of moderate damage (RRMS = 0.84 in Fig. 15.11c). The RRMS then decreases to 0.42 (Fig. 15.11d) when the structure is severely damaged and increases again to 0.85 (Fig. 15.11e) when the structure is in the near collapse condition. It is impossible to distinguish the damage levels from the error curves shown in Figs. 15.11b to 15.11e. As such, the RRMS is not an effective damage detection measure for input excitations with a low signal-to-noise ratio.

Figure 15.12 shows the pseudospectrums of measured and estimated structural responses of floor F_{30} under five aforementioned damage states subjected to the white noise excitations described in Section 15.5.1. There are two spectrum peaks in the frequency range of 0 to π. When a white noise is applied to the undamaged structure, the pseudospectra of the measured and estimated structural responses match quite well (Fig. 15.12a), especially at two spectrum peaks. However, as the level of the damage increases, the difference between the pseudospectra, especially in the vicinity of the peaks, becomes pronounced. Furthermore, the distance between the peaks in the two pseudospectra increases as the level of damage increases. The difference between the two pseudospectra obtained from the measured and estimated structural responses is an effective indicator of the changes in the structural properties and damage to the structure.

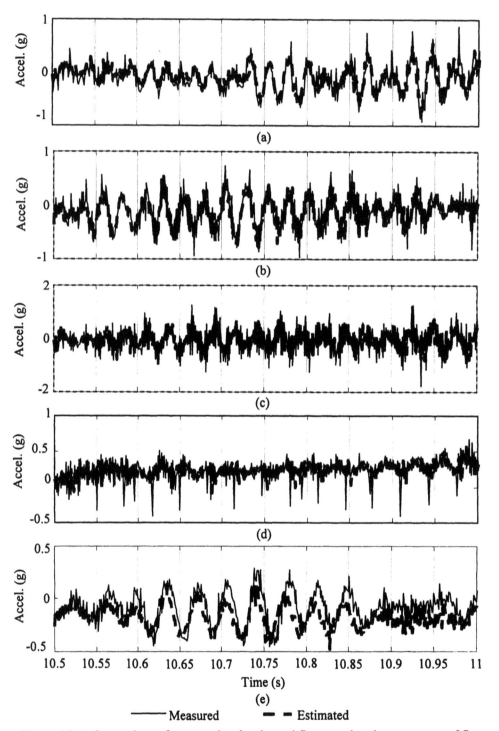

Figure 15.11 Comparison of measured and estimated floor acceleration responses of floor F_{30} under five damage states subjected to the white noise excitations (from 10.5 seconds to 11 seconds) (Case 2): (a) No damage (RRMS = 0.61); (b) minor damage (RRMS = 0.70); (c) moderate damage (RRMS = 0.84); (d) severe damage (RRMS = 0.42), and (e) near collapse (RRMS = 0.85). See color insert following Chapter 15.

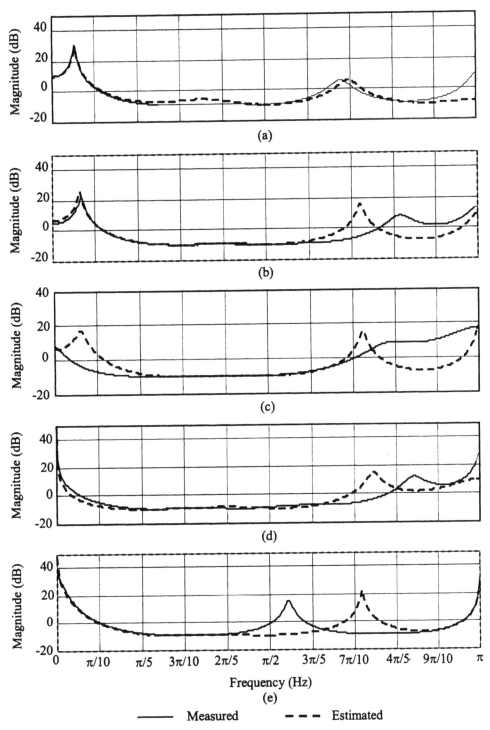

Figure 15.12 Comparison of pseudospectra of measured and estimated acceleration responses of floor F_{30} under five damage states subjected to the white noise excitations (Case 2): (a) No damage; (b) small damage; (c) moderate damage; (d) severe damage, and (e) near collapse

15.6   Concluding Remarks

The proposed dynamic fuzzy WNN model for damage detection of building structures balances the global and local influences of the training data and incorporates the imprecision existing in the sensor data effectively, thus resulting in fast training convergence and high accuracy. The LM-LS training algorithm used in the model converged in fewer than 5 iterations in all cases.

The conventional RRMS error method cannot detect the structural damage effectively especially for input excitations with a low signal-to-noise ratio. It was shown that the proposed power density pseudospectrum approach provides an effective method for damage detection in high-rising building structures. The results of this research provide a powerful tool for real-time health monitoring and nondestructive damage evaluation of high-rising building structures.

☐ Magnifier box   ⟹ Step

Figure 4.6 Mandelbrot set (http://en.wikipedia.org/wiki/Mandelbrot_set)

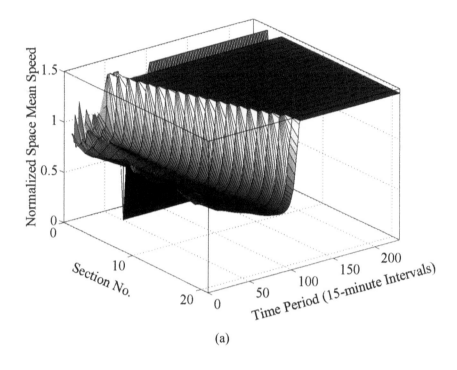

(a)

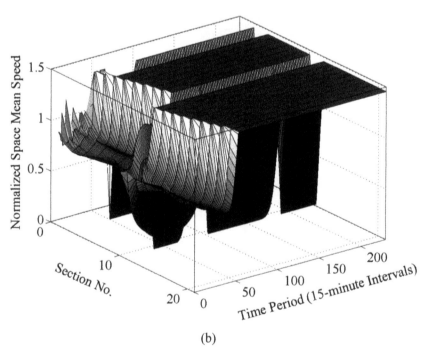

(b)

Figure 6.6 Surface plot of normalized traffic space mean speed across the freeway segment for the duration of the work zone for Example 1: (a) Without construction, and (b) with construction.

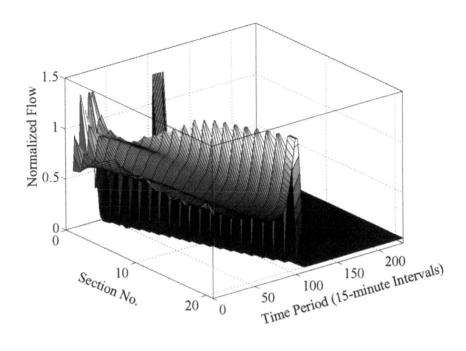

(a)

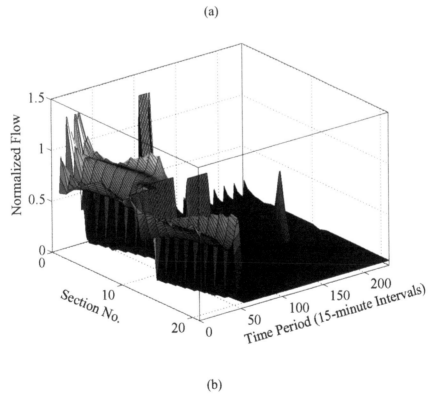

(b)

Figure 6.7 Surface plot of normalized traffic flow across the freeway segment for the duration of the work zone for Example 1: (a) Without construction, and (b) with construction.

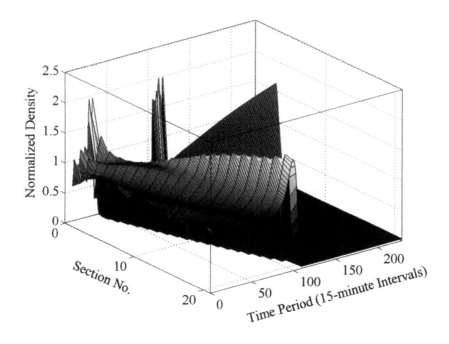

(a)

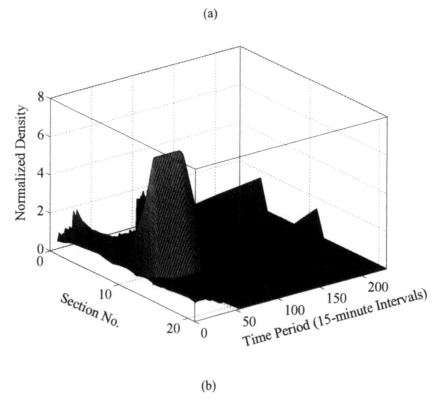

(b)

Figure 6.8 Surface plot of traffic density across the freeway segment for the duration of the work zone for Example 1: (a) Without construction, and (b) with construction.

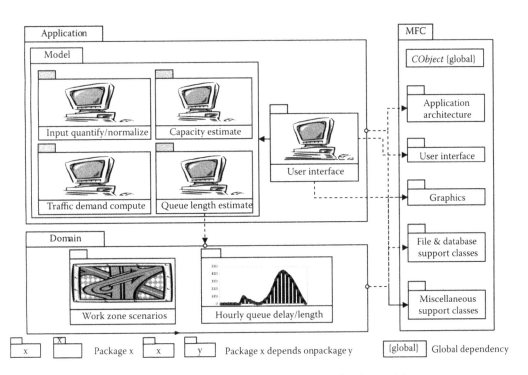

Figure 10.4 Package diagram of IntelliZone application architecture.

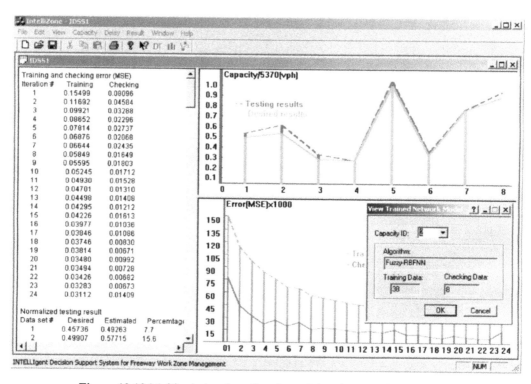

Figure 10.10 Multi-window interface for training the neural network.

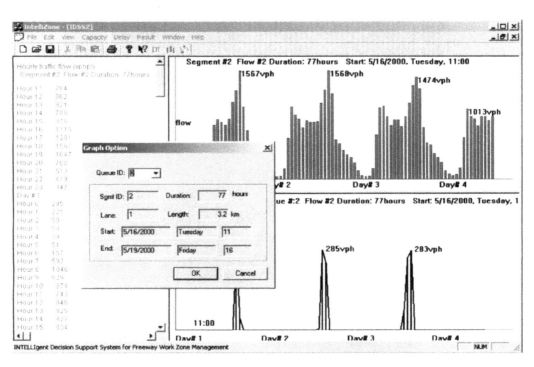

Figure 10.12 Multi-window interface for displaying the traffic flow and queue delay results.

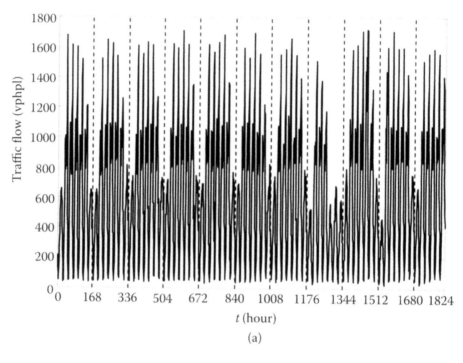

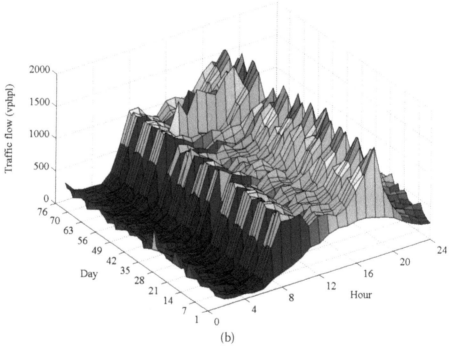

Figure 11.2 Traffic flow measured in a two-lane freeway in the state of North Carolina with one lane closure: (a) Time series, and (b) three-dimensional graph.

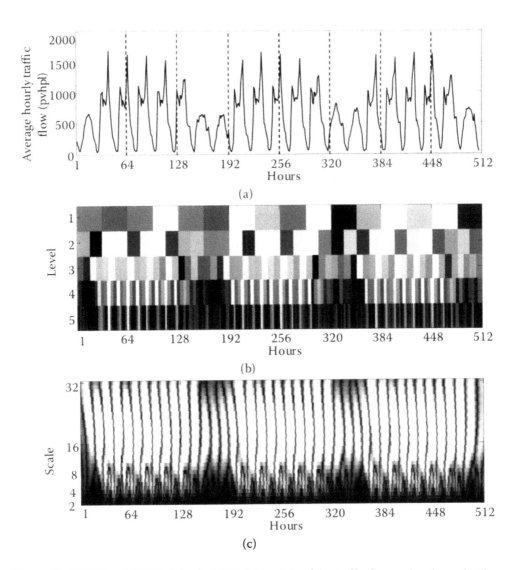

Figure 11.4 DWT and CWT of the first 512 data points of the traffic flow series shown in Figure 11.2 using the Daubechies wavelet of order 4: (a) Traffic flow; (b) absolute coefficients from DWT, and (c) absolute coefficients from CWT.

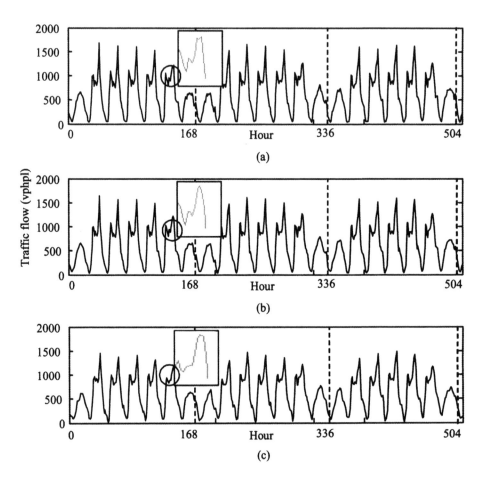

Figure 11.8 Denoising the traffic flow using DWT and DWPT approaches: (a) Original traffic flow; (b) denoised traffic flow using DWT, and (c) denoised traffic flow using DWPT.

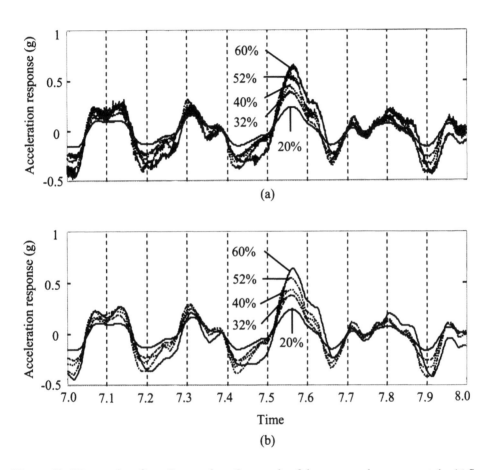

Figure 13.6 Data points from 7 seconds to 8 seconds of the structural response at the 1st floor under five earthquake excitations: (a) Measured, and (b) denoised.

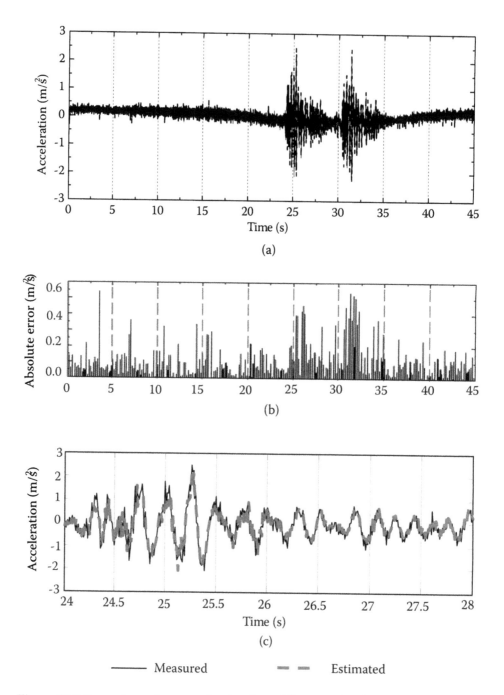

Figure 15.7 Comparison of measured and estimated floor acceleration responses using the 4,500 data sets from the small-level seismic excitation (Case 1): (a) Acceleration responses from 0s to 45s; (b) absolute error of acceleration responses (RRMS = 0.58), and (c) comparison of acceleration responses from 24s to 28s.

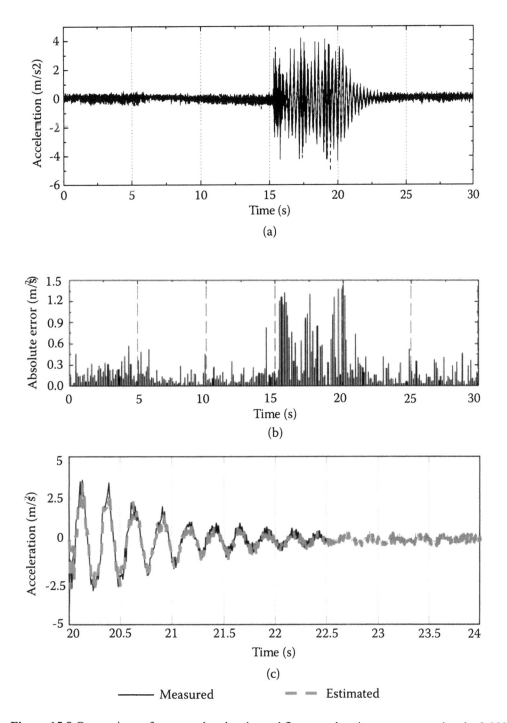

Figure 15.8 Comparison of measured and estimated floor acceleration responses using the 3,000 data sets from the moderate-level seismic excitation (Case 1): (a) Acceleration responses from 0s to 30s; (b) absolute error of acceleration responses (RRMS = 0.73), and (c) comparison of acceleration responses from 24s to 28s.

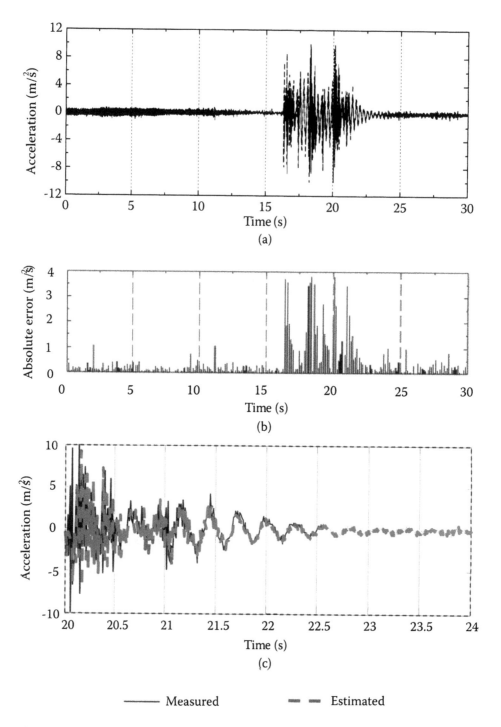

Figure 15.9 Comparison of measured and estimated floor acceleration responses using the 3,000 data sets from the large-level seismic excitation (Case 1): (a) Acceleration responses from 0s to 30s; (b) absolute error of acceleration responses (RRMS = 1.10), and (c) comparison of acceleration responses from 24s to 28s.

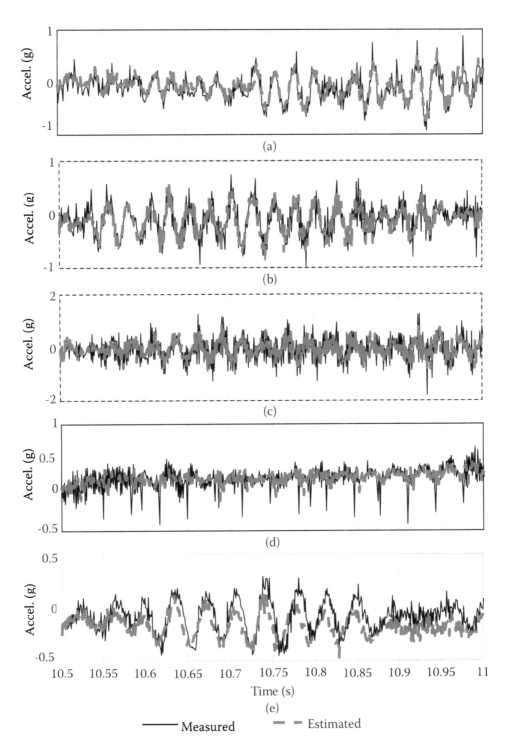

Figure 15.11 Comparison of measured and estimated floor acceleration responses of floor F_{30} under five damage states subjected to the white noise excitations (from 10.5 seconds to 11 seconds) (Case 2): (a) No damage (RRMS = 0.61); (b) minor damage (RRMS = 0.70); (c) moderate damage (RRMS = 0.84); (d) severe damage (RRMS = 0.42), and (e) near collapse (RRMS = 0.85).

Chapter 16

NONLINEAR CONTROL OF SMART STRUCTURES

16.1 Introduction

Over the past two decades, a large amount of research has been conducted on the development and implementation of active, semi-active, and hybrid control of structures (e.g., Masri *et al.* 1982; Stein and Athans 1987; Miller *et al.* 1988; Yang *et al.* 1988 & 1992; Soong 1990; Chang and Yang 1995; Agrawal and Yang 1996; Nikzad *et al.* 1996; Thomson *et al.* 1996; Tomasula *et al.* 1996; Housner *et al.* 1997; Adeli and Saleh 1997 & 1999; Edwards and Spurgeon 1998; Bani-Hani and Ghaboussi 1998a&b; Saleh and Adeli 1998a&b; Lu *et al.* 1998; Hung *et al.* 2000; Connor 2003; Adeli and Kim 2004; Kim and Adeli 2004), and various control strategies have been proposed, such as sliding mode control (Edwards and Spurgeon 1998) and optimal polynomial control (Agrawal and Yang 1996). However, most of the studies have focused on the application of classical linear control theories such as linear quadratic regulator (LQR) feedback control algorithm (Soong 1990) and linear quadratic Gaussian (LQG) control algorithm (Stein and Athans 1987). These algorithms control structural responses effectively only when the structure is small and assumed to behave linearly.

There are several motivations for considering active nonlinear control of structures. The linear control algorithms are not effective for active nonlinear control of high-rising building structures subjected to extreme loadings. High-rising building structures may experience yielding and nonlinear behavior (geometrical or material nonlinearity or both) under severe earthquakes or strong winds. Any structural damage changes the structural stiffnesses during the extreme dynamic event. In such cases, a computational model based on the assumption that the controlled structure behaves linearly would be inadequate for representing the actual behavior of the structure. Linear control algorithms cannot be used to control the response of structures in the nonlinear range effectively (Bani-Hani and Ghaboussi 1998b). On the other hand, maintaining linear behavior for a large controlled structure such as a high-rising building during an extreme dynamic event may require actuators with impractically large capacities.

Neural networks are known for their ability to model complex and nonlinear phenomena where no explicit mathematical model exists (Adeli and Hung 1995; Adeli and Park 1998; Adeli 2001) and for their adaptability for handling incomplete information. In the past decade or so, a number of articles have been published on the development and application of neural network-based or other adaptive/intelligent control algorithms for active linear/nonlinear control of mostly small structural systems (Chen *et al.* 1995; Ghaboussi and Joghataie 1995; Nikzad *et al.* 1996; Bani-Hani and Ghaboussi 1998a&b; Hung *et al.* 2000; Kim *et al.* 2000 & 2004; Brown and Yang 2001; Kim and Lee 2001). Ghaboussi and Joghataie (1995) first use a neural network-based emulator to identify the response of a three-story, two-dimensional (2D) frame structure. Then, a neural network-based controller is trained using the emulator for linear control of the structure. Structural displacement and acceleration responses during the previous two time steps and actuator electric signals during the previous three time steps are used as inputs to the neural network model. Bani-Hani and Ghaboussi (1998a&b) extended the neural network-based algorithm for nonlinear control of a benchmark three-story 2D frame structure considering material nonlinearity. The authors show the neural network-based approaches provide adequate accuracy for active control of such small 2D frame structures.

In recent years, a few journal articles have been published on active control of three dimensional (3D) high-rising building structures subjected to earthquake loads, e.g., Fur *et al.* (1996), Al-Dawod *et al.* (2004), Ohtori *et al.* (2004), Yoshida and Dyke (2005), Tan *et al.* (2005), Kim and Adeli (2005b&c), and Jiang and Adeli (2008a). Fur *et al.* (1996) use an active tuned mass damper (ATMD) system for an eight-story building structure considering the lateral-torsional coupling. The control *gain*, an important parameter required in feedback control algorithms such as LQR, is obtained through the complete feedback of position and velocity. Al-Dawod et al. (2004) present a fuzzy logic-based controller for active control of three- and twenty-story 3D symmetric building structures. Ohtori *et al.* (2004) define three-, nine-, and twenty-story symmetric steel structures for benchmark active control studies without presenting any active nonlinear control algorithm. The nine-story structure has also been investigated in a semi-active control study using magnetorheological dampers (Yoshida and Dyke 2005) and an active control study using an H_2/LQG controller (Tan *et al.* 2005).

Adeli and Kim (2004) present a novel wavelet-hybrid feedback-Linear Mean Square (LMS) algorithm for robust control of civil structures. It is shown that the wavelet transform can be used to enhance the performance of feedback control algorithms. An active mass driver benchmark structure is used to validate the effectiveness of the proposed control model. Simulation results demonstrate that the proposed model is effective for control of both steady and transient vibrations without any significant additional computational burden. Kim and Adeli (2005a) present a hybrid control system through judicious combination of a passive supplementary damping system with a semi-active tuned liquid column damper (TLCD) system. The wavelet-hybrid feedback LMS control algorithm (Adeli and Kim 2004) is used to find optimal values of the control parameters for an 8-story frame using three different simulated earthquake ground accelerations. The wavelet-hybrid feedback LMS algorithm has also been used for vibration control of cable-stayed bridges under various seismic excitations (Kim and Adeli 2005b).

Recent literature indicates that adaptive/intelligent control algorithms such as those developed by Adeli and Kim (2004) are advantageous over classical feedback control algorithms for active or hybrid control of smart structures for several reasons. First, these algorithms can tolerate the imprecision in the sensed data. Second, they require less prior knowledge of the structural system to be controlled. Third, they can be used to handle nonlinearity. Finally, they usually converge quickly and are therefore practical for online active control of large-scale structures.

The focus of this chapter is active nonlinear control of 3D building structures based on the wavelet neural network model developed by the authors and applied to system identification problems presented in Chapters 13 and 14. Both material and geometrical nonlinearities are considered in modeling the structural response under strong earthquake loadings. Furthermore, the structural modeling takes into account two coupling actions: the coupling action between the actuators and the structure and the coupling between the lateral and torsional motions of 3D irregular structures. A dynamic fuzzy wavelet neural network (WNN) model is developed as a fuzzy wavelet *neuroemulator* to predict structural responses from the immediate past structural responses and actuator dynamics. In the next chapter, an intelligent control algorithm is presented for finding the optimal control forces for active nonlinear control of building structures using the dynamic fuzzy wavelet neuroemulator presented in this chapter.

16.2 Coupled Nonlinear Dynamics of 3D Structures with an Active Control System

16.2.1 Nonlinear Dynamics of Irregular 3D Structures

A 3D high-rising building structure can in general have both plan and elevation irregularities. In general, the center of mass, C_M, does not coincide with the center of resistance, C_R, in each floor (Fig. 16.1a), and the centers of mass and resistance of floors do not lie on the same vertical axes (their locations can vary from floor to floor) due to different story stiffnesses along the two principal directions. This difference may exist even for a structure with a symmetric plan because different members with different stiffnesses may be used in a symmetric floor arrangement. In such situations there is a coupling between lateral and torsional motions. The coupling may result in the maximum lateral displacement in a direction other than the direction of minimum stiffness. In this case, neglecting the coupling between lateral and torsional motions usually underestimates the maximum responses of the structure under earthquake loadings (Kim and Adeli 2005c).

In this chapter, floor diaphragms are assumed to be rigid and axial deformations of the columns are neglected. As such, the irregular building structure is modeled using three displacement degrees of freedom (DOF) for each floor: translations in x- and y-directions and a rotation about the vertical axis z passing through the center of resistance. The total number of degrees of freedom is therefore $m = 3L$ where L is the number of stories. The structural displacement vector at time t is expressed as

$$\mathbf{u}(t) = [u_1(t) \quad v_1(t) \quad \theta_1(t) \quad u_2(t) \quad v_2(t) \quad \theta_2(t) \quad \cdots \quad u_L(t) \quad v_L(t) \quad \theta_L(t)]' \quad (16.1)$$

where $u_i(t)$, $v_i(t)$, and $\theta_i(t)$ are the translations in x- and y-directions and the rotation about the vertical axis z of the i-th floor, respectively. The prime represents the transpose of the matrix. The equation of motions of 3D building structures with an active control system subjected to seismic excitations is written as

$$\mathbf{M\ddot{u}}(t) + \mathbf{C\dot{u}}(t) + \mathbf{R}(x,t) = \mathbf{I}_c\mathbf{F}(t) - \mathbf{M}_0\mathbf{I}_g\ddot{x}_g(t) \quad (16.2)$$

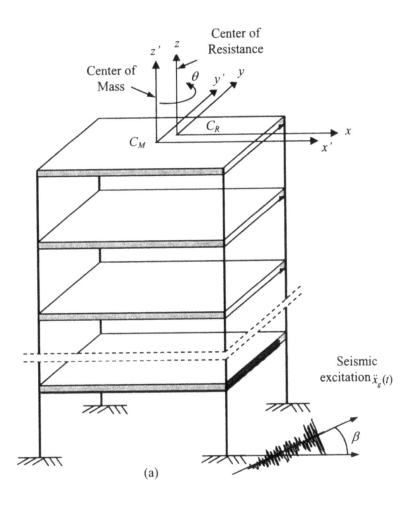

(a)

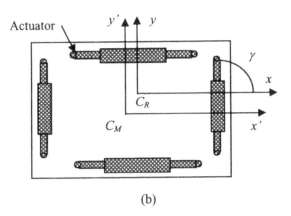

(b)

Figure 16.1 Structural model of a 3D building: (a) 3D building, and (b) typical story with four actuators

where $\mathbf{M}$ and $\mathbf{C}$ represent $m \times m$ mass and damping coefficient matrices of the structure, respectively; $\mathbf{R}(x,t)$ is the $m \times 1$ restoring force vector; $\mathbf{F}(t)$ is the $r \times 1$ control force vector whose elements $F(t)$ are in the form of time series; $\mathbf{I}_c$ is an $m \times r$ location matrix representing the location of the actuators; $\ddot{x}_g(t)$ is the input horizontal earthquake acceleration which can have any arbitrary orientation. In this chapter, we assume that two pairs of actuators are used in every floor of the building along two perpendicular axes x and y (Fig. 16.1b). Each pair consists of two identical actuators, representing the control force $F(t)$ in the corresponding direction. For the sake of simplifying the computations, we assume that the distance between the center of mass (C_M) and the center of resistance (C_R) in each floor is small. Thus, the moment caused by actuators in two directions is ignored. In practical applications, the moment caused by actuators in two directions can be minimized by properly assigning the control force to the two actuators in each direction, as explained below.

We assume that each actuator used in every floor of the building generates a control force F_i with a moment arm a_i ($i = 1, \ldots, 4$). Thus, the total moment caused by the four actuators, M_a, is calculated as follows (Fig. 16.2):

$$M_a = F_1 a_1 - F_2 a_2 + F_3 a_3 - F_4 a_4 \qquad (16.3)$$

$$F_x = F_1 + F_2 \qquad (16.4)$$

$$F_y = F_3 + F_4 \qquad (16.5)$$

where F_x and F_y are the required control forces in x and y directions, respectively, which are obtained by the floating-point genetic algorithm presented in the next chapter.

Obviously, M_a is a function of four control forces and the corresponding moment arms. For a given structure with an active control system, we may not be able to change the moment arm a_i. However, we can divide F_x between F_1 and F_2 for the two actuators in the x-direction or F_y between F_3 and F_4 for the two actuators in the y-direction, such that the condition $M_a = 0$ in Eq. (16.3) is satisfied.

The focus of the current chapter is on novel computational techniques, not actuator placement. Thus the proposed methodologies are not tied to any particular configuration. As such, $r = 2L$. The angle of orientation of the actuators is $\gamma = 0°$ for one pair and $\gamma = 90°$ for the other pair. The total number of actuators used in the building is therefore $4L$.

In Eq. (16.2), $\mathbf{M}_0$ is an $m \times m$ diagonal mass matrix whose diagonal terms are the same as the diagonal terms of the full matrix $\mathbf{M}$, and $\mathbf{I}_g$ is an $m \times 1$ orientation matrix denoting the orientation of the external earthquake excitation:

$$\mathbf{I}_g = [\cos\beta \quad \sin\beta \quad 0 \quad \cos\beta \quad \sin\beta \quad 0 \quad \cdots \quad \cos\beta \quad \sin\beta \quad 0]' \qquad (16.6)$$

where β is the direction angle of the earthquake motion measured from the x-axis (Fig. 16.1). The location matrix $\mathbf{I}_c$ in Eq. (16.2) is expressed as follows:

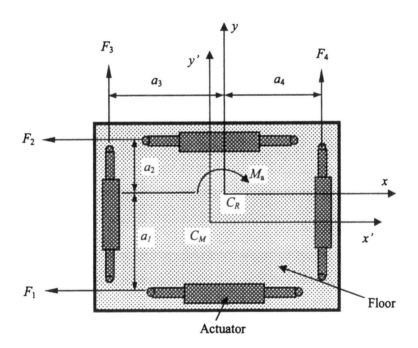

C_R: Center of resistance
C_M: Center of mass
F_i: i-th actuator force in the floor ($i = 1, ..., 4$)
a_i: Moment arm of F_i
M_a: Total moment caused by four actuators

Figure 16.2 Typical story with four actuators

$$\mathbf{I}_c = diag\left(\begin{bmatrix} \cos\gamma_1 & \sin\gamma_1 \\ \sin\gamma_2 & \cos\gamma_2 \\ 0 & 0 \end{bmatrix} \begin{bmatrix} \cos\gamma_3 & \sin\gamma_3 \\ \sin\gamma_4 & \cos\gamma_4 \\ 0 & 0 \end{bmatrix} \cdots \begin{bmatrix} \cos\gamma_{r-1} & \sin\gamma_{r-1} \\ \sin\gamma_r & \cos\gamma_r \\ 0 & 0 \end{bmatrix}\right)_{m\times r} \tag{16.7}$$

Figure 16.3 shows a building structure with an active control system. The earthquake excitation is represented by the horizontal ground acceleration in the form of a time series, $\ddot{x}_g(t)$. The active control system consists of a computer, hydraulic actuators, and sensors. The computer is used to receive, process, and send signals between the actuators and the structure via sensors and cable lines. When the structure is subjected to an earthquake loading, the structural response at the top floor, for example $u(t)$ in the x-direction, is sent to the controller along with the earthquake excitations. Optimal control forces from actuators are needed to minimize the structural displacement at the top of the structure in every time step. Each optimal control force is converted to a corresponding electric control signal $E_s(t)$ based on the properties of the actuator, to be discussed later. Each actuator is then excited by the control signal to generate a required control force.

16.2.2 Nonlinear Hysteretic Effect

In order to represent the material nonlinear and hysteretic behavior of a high-rising building structure, the restoring force of a structural element at time t, $R(x,t)$, is defined by the following equation proposed by Wen (1976):

$$R(x,t) = akx + (1-a)kd_y\gamma \tag{16.8}$$

where x is the displacement of the element, and $R_L(x) = akx$ and $R_{NL}(v) = (1-\alpha)kd_y\gamma$ represent the linear and nonlinear parts of the restoring force, respectively, in which v represents the nonlinear displacement. The parameters k and αk ($\alpha<1$) are the slope coefficients for the linear and nonlinear parts of the load-displacement curve (Fig. 16.4), respectively; and γ is a hysteretic variable in the range $-1 \le \gamma \le 1$, which is determined by solving the following nonlinear differential equation (Wen 1976):

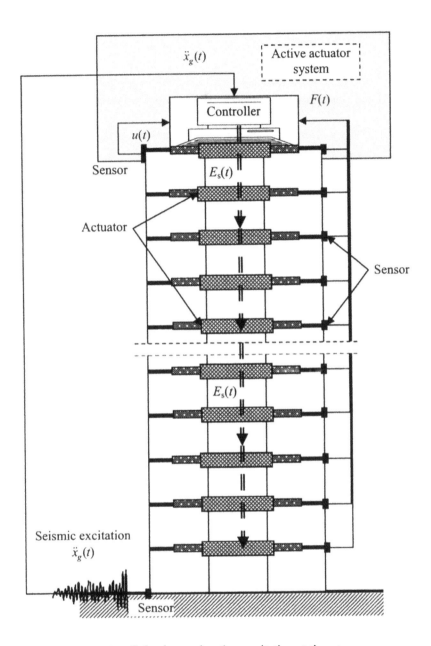

$\ddot{x}_g(t)$ = Seismic acceleration excitation at time t

$x(t)$ = Structural displacement response

$E_s(t)$ = Electric signal to the actuator

$f_c(t)$ = Control force from the actuator

Figure 16.3 Building with an active control system in one direction

$$\dot{v}(t) = \dot{x}\left[1 - \frac{1 + \text{sgn}(\dot{x}\gamma)}{2}|\gamma|^{n}\right] \Big/ d_y \qquad (16.9)$$

where n is an integer number representing the yield exponent. It is used to control the sharpness of the transition from the elastic part of the curve to the plastic region (noted by n in Fig. 16.4). For a bi-linear hysteretic representation $n = 1$. The parameter d_y is the displacement of the element at the yield point when $n = 1$ (for the case of bi-linear hysteretic representation). The notation sgn is the sign function. It is equal to 1 when its argument is positive and -1 otherwise. Equation (16.9) is solved iteratively in each time step of the structural dynamic analysis.

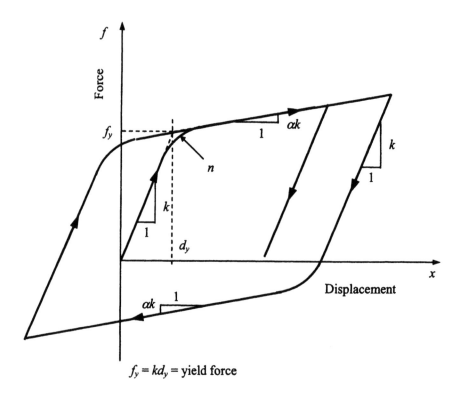

Figure 16.4 Load-displacement curve for an element considering material nonlinearity and hysteretic behavior

Equations (16.8) and (16.9) represent the nonlinear stiffness hysteretic model of a structural element where its restoring force is a function of a displacement variable x and a hysteretic variable v. The restoring force vector for the entire structure is expressed in matrix form as:

$$\mathbf{R}(x,t) = \mathbf{K}_G \mathbf{X}(t) + \mathbf{K}_M \mathbf{V}(t) \qquad (16.10)$$

where $\mathbf{K}_G$ and $\mathbf{K}_M$ are structural stiffness matrices. The stiffness matrix $\mathbf{K}_G$ includes geometric nonlinearity terms. The stiffness matrix $\mathbf{K}_M$ includes both geometric (to be discussed in the next section) and material nonlinearity terms, and $\mathbf{X}(t)$ and $\mathbf{V}(t)$ are the total displacement and nonlinear displacement variable vectors at time t (representing the second term in the right-hand side of Eq. 16.8), respectively.

16.2.3 Geometrical Nonlinearity Effect

Geometrical nonlinearity also significantly affects the structural behavior of a high-rising building structure under a severe earthquake loading. Figure 16.5 shows the P-δ geometrical nonlinearity of a column subjected to a vertical axial load, P, and a horizontal load, H. The additional bending moment due to the horizontal displacement δ can be assumed to be produced by an additional horizontal load at the top of the column equal to $P(\delta/l)$, where l is the height of the column element. The equilibrium equation for the column in the horizontal direction becomes (Fig. 16.5b)

$$k\delta = H + P\delta/l \qquad (16.11)$$

where k is the lateral stiffness of the column. Rearranging Eq. (16.11) yields:

$$k^{\mathit{eff}} = H/\delta = k - P/l \qquad (16.12)$$

where k^{eff} is the effective lateral stiffness of the column element. Thus, the P-δ geometrical nonlinearity for high-rising buildings with rigid floors can be taken into account approximately by reducing the lateral stiffness of the column according to Eq. (16.12). For a given column element, Eq. (16.11) can be expressed in a matrix form as follows:

$$\begin{Bmatrix} f_{xi} \\ f_{xj} \end{Bmatrix} = (k - \frac{P}{l}) \begin{Bmatrix} 1 & -1 \\ -1 & 1 \end{Bmatrix} \begin{Bmatrix} u_i \\ u_j \end{Bmatrix} \qquad (16.13)$$

where u_i and u_j are the displacements of the two ends of the column and f_{xi} and f_{xj} are the nodal forces in the x direction incorporating the geometrical nonlinearity of the column. Equation (16.13) can also be used to consider the geometrical nonlinearity of the column in the y direction using the corresponding displacement variable v in the y direction. The stiffness matrix of each column represented by Eq. (16.13) is assembled into the structural stiffness matrices, $\mathbf{K}_G$ and $\mathbf{K}_M$, in Eq. (16.10).

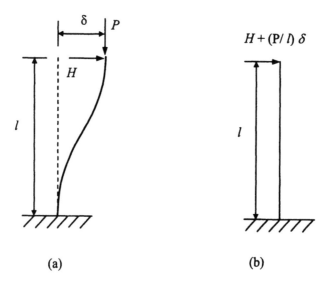

Figure 16.5 P-δ effect in a column: (a) Column deflection, and (b) equivalent horizontal load

16.2.4 Electrohydraulic Actuator

The importance of the actuator-structure coupling in active control of structures has been recognized in several literature (Dyke *et al.* 1995; Ghaboussi and Joghataie 1995; Nikzad *et al.* 1996; Kim and Lee 2001; Bani-Hani and Ghaboussi 1998b). The linear hydraulic actuator is employed in this chapter for active control of a building structure because of its ability to generate relatively large forces with a relatively small response time in the order of a few milliseconds and maintain the force with very little power (Nikzad *et al.* 1996). The level of force produced by a hydraulic actuator can be in the order of 1,000 kilo-Newtons (220 kilo-pounds) (**http://www.nees.buffalo.edu/pdfs/244Actuators.pdf**). The magnitude of the control force, $F(t)$, is found by solving the following differential equation numerically in each time step of the dynamic analysis (Desilva 1989):

$$q(t) = A_p \dot{x}_p + C_l F(t)/A_p + V_c \dot{F}(t)/(2\kappa A_p) \qquad (16.14)$$

where A_p is the effective cross-sectional area of the piston, $\dot{x}_p$ is the piston velocity, C_l is the leakage coefficient, V_c is the volume of the cylinder, and κ is the compressibility coefficient of the fluid. The displacement of the actuator/piston is assumed to be the same as the displacement of the floor it is attached to.

16.2.5 State Space Model

Considering material nonlinearity and hysteretic behavior represented by Eq. (16.8), geometric nonlinear behavior of the structure represented by Eq. (16.13), and the dynamics of the actuator represented by Eq. (16.14), the nonlinear equation of the motion for a structure with an active control system subjected to earthquake loading is expressed in the state space form as follows:

$$\dot{Z}(t) = AZ(t) + BQ_c(t) + BF_g(t) \qquad (16.15)$$

where $Z(t) = [u(t), V(t), F'(t), \dot{u}(t)]'$ is a column vector of $4m$ state space variables, in which $F'(t) = I_c F(t)$. The $4m \times 4m$ matrix A is expressed by

$$A = \begin{bmatrix} 0 & 0 & 0 & I \\ 0 & 0 & 0 & D \\ 0 & 0 & -G_1 & -G_2 \\ -M^{-1}K_G & -M^{-1}K_M & M^{-1}I_c & -M^{-1}C \end{bmatrix} \qquad (16.16)$$

where D, K_G, and K_M are $m \times m$ matrices, whose terms are defined as follows:

$$D_{i,j} = (k_{i,j}^{eff}/f_y)\{[1 - [1 + \text{sgn}(\dot{x}_{i,j}\gamma_{i,j})]/2 \,|\gamma_{i,j}|^n]\} \qquad (16.17)$$

$$(K_G)_{i,j} = \alpha k_{i,j}^{eff} \tag{16.18}$$

$$(K_M)_{i,j} = (1-\alpha)k_{i,j}^{eff} d_{yi,j} \tag{16.19}$$

where k^{eff} represents the effective lateral stiffness of the column element which takes into account the geometrical nonlinearity.

In Eq. (16.16), $\mathbf{0}$ and $\mathbf{I}$ are the $m \times m$ zero and identity matrices, respectively, and $\mathbf{G_1}$ and $\mathbf{G_2}$ are two $m \times m$ diagonal matrices representing the characteristics of hydraulic actuators as follows:

$$\mathbf{G_1} = diag \left(\begin{bmatrix} \dfrac{2\kappa_x C_{lx}}{V_{Cx}} & 0 & 0 \\ 0 & \dfrac{2\kappa_y C_{ly}}{V_{Cy}} & 0 \\ 0 & 0 & 0 \end{bmatrix}, \cdots, \begin{bmatrix} \dfrac{2\kappa_x C_{lx}}{V_{Cx}} & 0 & 0 \\ 0 & \dfrac{2\kappa_y C_{ly}}{V_{Cy}} & 0 \\ 0 & 0 & 0 \end{bmatrix} \right)_{L \times l} \tag{16.20}$$

$$\mathbf{G_2} = diag \left(\begin{bmatrix} \dfrac{2\kappa_x A_{Px}^2}{V_{Cx}} & 0 & 0 \\ 0 & \dfrac{2\kappa_y A_{Py}^2}{V_{Cy}} & 0 \\ 0 & 0 & 0 \end{bmatrix}, \cdots, \begin{bmatrix} \dfrac{2\kappa_x A_{Px}^2}{V_{Cx}} & 0 & 0 \\ 0 & \dfrac{2\kappa_y A_{Py}^2}{V_{Cy}} & 0 \\ 0 & 0 & 0 \end{bmatrix} \right)_{L \times l} \tag{16.21}$$

The matrices $\mathbf{Q_c}(t)$ and $\mathbf{F_g}(t)$ are $4m \times 1$ actuator flow rate and external excitation vectors, respectively, expressed as follows:

$$\mathbf{Q_c}(t) = \begin{bmatrix} \mathbf{0_1} \\ \mathbf{0_1} \\ \mathbf{I_c q}(t) \\ \mathbf{0_1} \end{bmatrix}, \quad \mathbf{F_g}(t) = \begin{bmatrix} \mathbf{0_1} \\ \mathbf{0_1} \\ \mathbf{0_1} \\ \mathbf{M_0 I_g \ddot{x}_g}(t) \end{bmatrix} \tag{16.22}$$

where $\mathbf{q}(t)$ is a $2L \times 1$ vector representing the flow rate of hydraulic actuators (Desilva 1989) and

$$\mathbf{B} = [\mathbf{0_1} \quad \mathbf{0_1} \quad \mathbf{0_1} \quad \mathbf{M^{-1} I_1}] \tag{16.23}$$

is a $4m\times1$ matrix, in which $\mathbf{0}_1 = [0 \quad 0 \quad \cdots \quad 0]'$ and $\mathbf{I}_1 = [1 \quad 1 \quad \cdots \quad 1]'$ are $m\times1$ zero and unit column vectors, respectively.

The nonlinear state space equation represented by Eq. (16.15) is solved numerically using the fourth-order Runge-Kutta method with a proper integration time step (e.g., $\Delta t = 0.01$s). In order to obtain the mass, stiffness, and damping coefficient matrices of the structure, building structures are modeled with finite elements in two steps. In the first step, columns and beams are modeled as three-dimensional frame elements with two end nodes, each node having six DOF (three displacements and three rotations). In the second step, dynamic condensation is applied to reduce the structural model to one with rigid floor diaphragms and no axial column deformations. The resulting structural model has 3 DOF for each floor (translations in x- and y-directions and a rotation about the vertical axis z passing through the center of resistance, shown in Fig. 16.1).

16.3 Dynamic Fuzzy Neuroemulator

16.3.1 Constructing the Neuroemulator

In Chapter 13, a multi-paradigm dynamic time-delay fuzzy wavelet neural network (WNN) model was presented for nonparametric identification of structures using the nonlinear autoregression moving average with exogenous inputs (NARMAX) approach. Furthermore, an adaptive Levenberg-Marquardt-Least-Squares (LM-LS) algorithm with a backtracking inexact linear search scheme was presented in Chapter 14 for training of the dynamic fuzzy WNN model.

In order to control the response of a given structure effectively the structural response in future time steps have to be estimated. This is necessary in order to determine the magnitude of the required control forces. In this chapter, the dynamic fuzzy WNN model presented in Chapter 13 is used as the neuroemulator to predict the nonlinear structural response in future time steps from the immediate past structural response and actuator dynamics. Since the earthquake can occur in any horizontal direction two fuzzy WNN-based emulators, x- and y-neuroemulators, are created to predict the corresponding structural displacement responses in the x- and y-directions. Details of how to construct the dynamic fuzzy WNN model using a given time series were presented in Section 13.3. The procedures to construct the two neuroemulators are similar. The response data in the x-direction are used to construct the x-neuroemulator and those in the y-direction are used for the y-neuroemulator. The resulting numbers of the hidden or wavelet nodes in the hidden layer of the two neuroemulators are in general different.

16.3.2 Training the Neuroemulator

The neuroemulator presented in this chapter is a non-physically-based model, which does not explicitly involve any physical parameters of the structure, in the context of both training and testing. The neuroemulator is trained to map nonlinear relationships between input and output data to represent the physical model of the corresponding structure, and then the trained neuroemulator is used to predict the structural response under various input excitations.

The dynamic fuzzy WNN neuroemulator is trained using the structural displacement response data generated from the nonlinear structural analysis taking into account both geometrical and material nonlinearities as described previously. The training data are generated from the nonlinear analyses of the structure subjected to different earthquake loadings. Figure 16.6 shows the training diagram of the dynamic fuzzy WNN model using the adaptive LM-LS learning algorithm for active nonlinear control of structures. The training is done in three steps, identified by numbers 1) to 3) in Fig. 16.6 as follows:

1) The input-output data sets, $\mathbf{X}_k$ and y_k ($k = 1, 2, ..., N_a$), are formed from structural responses, actuator forces, and earthquake excitations, where the k-th input state space vector (time t during the earthquake excitation), $\mathbf{X}_k$, is constructed as follows:

$$\mathbf{X}_k = [\mathbf{x}_k, \mathbf{f}_k, \mathbf{a}_k]' = [x_k, x_{k+1}, ..., x_{k+\tau_x}, F_k, F_{k+1}, ..., F_{k+\tau_c}, \ddot{x}_{g\,k}, \ddot{x}_{g\,k+1}, ..., \ddot{x}_{g\,k+\tau_g}]' \quad (16.24)$$

in which x_k, F_k, and $\ddot{x}_{g\,k}$ are the k-th structural displacement response, actuator/control force, and earthquake excitation, respectively. The parameters τ_x, τ_c, and τ_g are time delays for the three variables representing dimensions of the inputs (number of the previous discrete time steps used for any input time series). Their optimal dimensions are obtained by the false nearest neighbor (FNN) method, as discussed in Section 4.3.3. The number of data sets N_a is obtained by

$$N_a = N - \max(\tau_x, \tau_c, \tau_g) \quad (16.25)$$

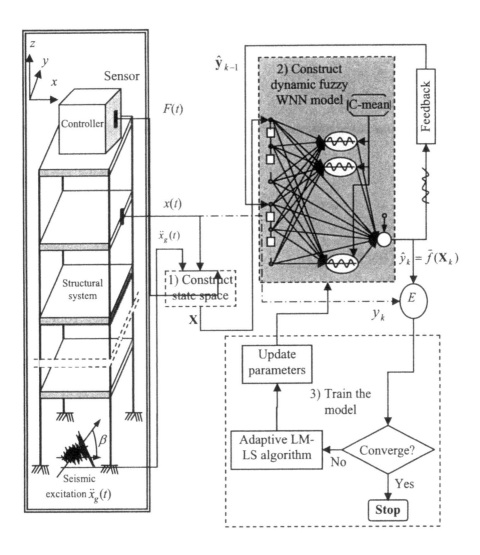

$F(t)$ = actuator/control force at time t

$\ddot{x}_g(t)$ = ground acceleration

$y(t)$ = measured structural response at time t

y_k = k-th measured structural response (k represents time step t)

$\mathbf{X} = [\mathbf{X}_1 \quad \mathbf{X}_2 \quad \cdots \quad \mathbf{X}_k]'$ = input state space vector for all structural response data

$\bar{f}(\mathbf{X}_k)$ = fuzzy WNN mapping function represented by Eq. (13.5)

$\hat{y}_k = \bar{f}(\mathbf{X}_k)$ = k-th computed output

$\hat{\mathbf{y}}_{k-1}$ = feedback structural response

E = error function

Figure 16.6 Training diagram of dynamic fuzzy WNN model using adaptive LM-LS learning algorithm for active nonlinear control of structures

where N is the total number of data points used to train the system.

2) The dynamic fuzzy WNN model with a recurrent feedback topology is created as described in Section 13.3.

3) The dynamic fuzzy WNN model is trained using the adaptive LM-LS algorithm and the reconstructed input-output data sets as described in Section 14.3.

The trained neuroemulator is tested using the data generated from both nonlinear and linear structural analyses. The ability of the trained model to predict both linear and nonlinear responses of the structure is investigated.

16.4 Numerical Example

In order to demonstrate that the dynamic fuzzy wavelet neuroemulator model can accurately predict the dynamic responses of structures with irregularities, two 3D multistory steel building structures, a twelve-story structure with vertical setbacks (Example 1) and an eight-story structure with plan irregularity (Example 2) are studied. Kim and Adeli (2005c) used the same examples in their study of hybrid control of irregular high-rising building structures assuming linear behavior. They report that a 2D dynamic analysis neglecting the effect of coupling between lateral and torsional motions underestimates the maximum response of the structure up to 4% for Example 1 and 7% for Example 2.

The nonlinear dynamic responses of the two example structures subjected to earthquake excitations are obtained by solving their dynamic equation of motion represented by Eq. (16.2). Both geometric and material nonlinearities are considered in the structural analysis. In Eqs. (16.8) and (16.9), steel yield stress of $f_y = 36$ ksi (248.3 N/mm^2), a yield ratio of $\alpha = 0.1$, and a yield exponent of $n = 2$ are used. A damping ratio of 2% is used for both structures.

The original Chichi earthquake and 200% Chichi earthquake record of Richter magnitude 7.3 which occurred at Foothills, Western Taiwan, on September 21, 1999, are applied to the structures. The structure is assumed to behave linearly when the difference of the maximum displacement obtained from the nonlinear and linear analyses is within one percent of the linear maximum displacement (Kim and Adeli 2005c). Based on this definition, under the Chichi earthquake loading, the 12-story structure (Example 1) behaves nonlinearly while the 8-story structure (Example 2) behaves linearly. When the

200% Chichi earthquake loading is applied, both structures behave nonlinearly.

The displacement responses of the uncontrolled structure obtained from 200% Chichi earthquake representing the structural nonlinear behavior are used to train the neuroemulator, while the displacement responses from the original Chichi earthquake excitation are used to test the neuroemulator. Displacement responses are obtained at each floor in the x and y horizontal directions at increments of 0.01 seconds over a period of 40 seconds, equal to the duration of the Chichi earthquake excitation. Thus, the number of sample data, N in Eq. (16.25), is 4,000.

The nonlinear control model is trained using the results obtained when the structure is subjected to a) earthquake excitation (uncontrolled structure) and b) both earthquake excitation and actuator forces simultaneously. This will enhance the adaptability of the fuzzy wavelet neuroemulator because for some cases no control forces may be necessary, for example, when earthquake excitation is small.

16.4.1 Example1: Twelve-Story Irregular Building

<u>Example Structure and its Response</u>

This example is an irregular 12-story three-dimensional steel building structure with vertical setbacks as shown in Fig. 16.7. This structure was created by Adeli and Saleh (1998 & 1999) for the study of structural control and used in Chapter 14 for nonlinear system identification of structures (see Section 14.5.1). It was also used by Kim and Adeli (2005c) for hybrid linear control of the irregular structure. For a description of this example, see Section 14.5.1. In earlier papers (Saleh and Adeli 1998a; Jiang and Adeli 2005a), the structure was modeled as a three-dimensional space frame with 462 DOF. In this chapter, floor diaphragms are assumed to be rigid and axial column deformations are neglected. The rigid-floor structural model has 36 degrees of freedom.

The structure is subjected to the combination of uniformly distributed floor dead and live loads of 60 psf (2.88 Kpa) and 50 psf (2.38 Kpa), respectively, and the Chichi earthquake records. The Chichi earthquake records were used as representative of one of the most destructive earthquakes of the past few decades for illustration. The modal analyses of this structure conducted by Kim and Adeli (2005c) demonstrate that the coupling effect of lateral and torsional vibrations is most significant when the earthquake excitation is applied in the direction with $\beta = -13°$ (Fig. 16.7). Therefore, this is the angle

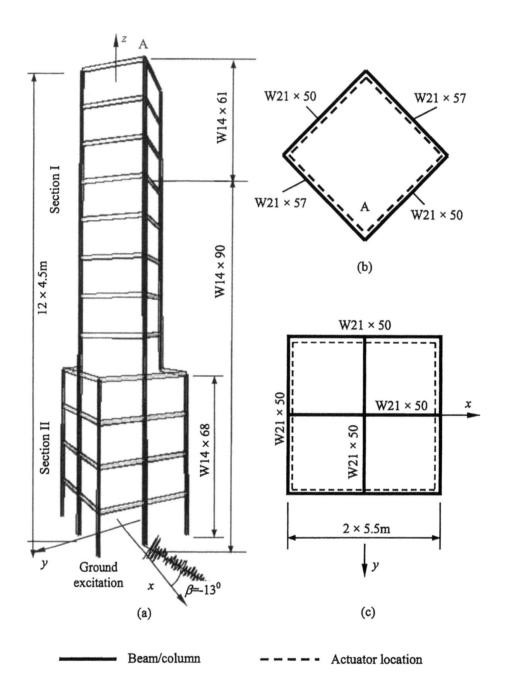

Figure 16.7 Twelve-story steel building with vertical irregularity (Example 1): (a) Perspective view; (b) plan of Section I, and (c) plan of Section II

used in this chapter to perform nonlinear dynamic analyses of the structure for training and testing the fuzzy wavelet neuroemulator. Figure 16.8 shows the x and y displacements of joint A at the top floor of the structure (identified in Fig. 16.7) when the structure is subjected to the original Chichi and 200% of Chichi earthquakes in the direction with $\beta = -13°$.

Actuators

Four double-acting actuators (which are able to create both push and pull forces) with the same properties are installed in every floor along the exterior envelope of the structure as indicated with dashed lines in Figs. 16.7b and 16.7c. The properties of an actuator used in this chapter are based on data provided by one manufacturer (**http://www.mts.com/vehicles/testline/pdfs/100-016-993.pdf**) and summarized in Table 16.1. For training the nonlinear control model, two sets of control forces are randomly generated in the x- and y-directions with a maximum actuator force of 68 kN (Table 16.1) and a uniform distribution of control force in the interval of (-68 kN, 68 kN). Figure 16.9 shows the randomly-generated control forces in the x-direction over the same earthquake duration of 40 s.

Constructing Fuzzy Wavelet Neuroemulators

Three sets of time series data are needed to construct each one of the two fuzzy WNN neuroemulators in the x- and y-directions: the structural displacement response at joint A (Fig. 16.7a), earthquake excitation, and control force. Each set of time series consists of 4,000 data points. Optimum time delay steps of $\tau_x=2$, $\tau_g=5$, and $\tau_u=8$ are obtained for the displacement response at joint A, the earthquake excitation, and the control force using the FNN method (see Section 4.3.3), respectively. The same numbers of time steps are obtained for the data sets in both x- and y-directions. The number of input vectors created is thus equal to $N_a = 4,000 - \max\{2, 5, 8\} = 3,992$ (Eq. 16.25).

A five-level wavelet decomposition was performed on the 3,992 sets of input vectors with the size of $\tau_x+\tau_u+\tau_g = 15$ (Eq. 16.24) using the Mexican hat wavelet function (Fig. 12.1). The empty wavelets whose supports do not contain any data are eliminated, resulting in 109 and 105 non-empty wavelets for x- and y-direction vectors, respectively. Based on the AFPE criterion (see Section 12.3.5) and using the modified Gram-Schmidt

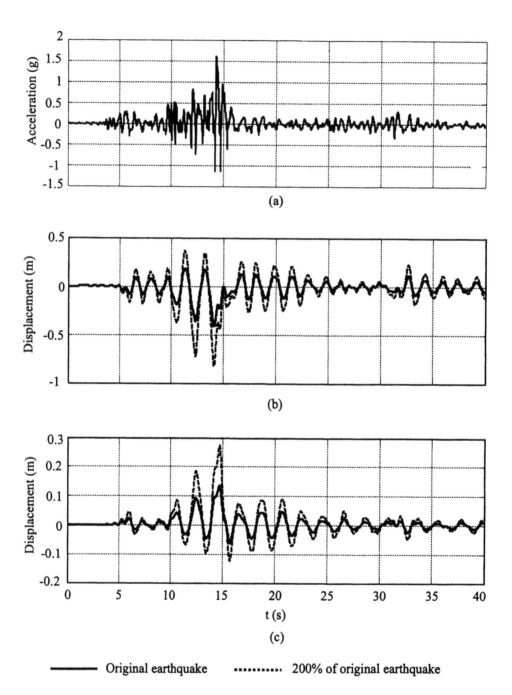

Figure 16.8 Displacements of joint A at the top floor of the structure (identified in Figure 16.7) when the structure is subjected to the original Chichi and 200% Chichi earthquakes in the direction with $\beta = -13°$: (a) 200% Chichi earthquake, Taiwan, Sep. 21, 1999; (b) displacements of joint A in the x-direction, and (c) displacements of joint A in the y-direction

Table 16.1 Properties of the actuator

Variables	Description	Value
A_p	Effective piston area	3.368×10^{-3} m^2
V_c	Chamber volume	1.01×10^{-3} m^3
C_l	Leakage coefficient	0.1×10^{-10} m^5/(N.s)
β	Compressibility coefficient	2.1×10^{10} N/m^2
q_{max}	Maximum flow rate	2.0×10^{-3} m^3/s
F_{max}	Maximum actuator force	68 kN
τ	Servovalve time constant	0.15 s
g	Servovalve constant	2.1×10^{-4} m^3/s/volt

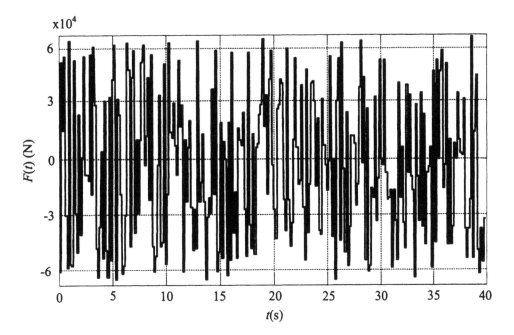

Figure 16.9 Randomly-generated control forces in the x-direction over the same earthquake duration of 40 s for the actuator with properties shown in Table 16.1

algorithm (see Section 12.3.3), it is concluded that 3 out of the 109 and 4 out of the 105 non-empty wavelets are sufficient to construct x- and y-fuzzy WNN neuroemulators, respectively. As such, there are three WNN nodes in the hidden layer of the x-neuroemulator for predicting the x-directional displacement response and four WNN nodes in the hidden layer of the y-neuroemulator for predicting the y-directional displacement response. For a discussion on the concept of empty wavelets and how the necessary number of non-empty wavelets is chosen, refer to Section 12.3.2.

Training Fuzzy Wavelet Neuroemulators

Two sets of data are used to train each neuroemulator, one set at a time, denoted as Set1 and Set2. In Set1, original Chichi earthquake excitations during the previous $\tau_g = 5$ time intervals and the resulting structural displacement responses of joints A (Fig. 16.7a) during the previous $\tau_x = 2$ time intervals are used as inputs. The structural responses in the x-direction (Fig. 16.8b) are used for training the x-neuroemulator and those in the y-direction (Fig. 16.8c) for the y-neuroemulator. In Set2, 200% of Chichi earthquake excitations (scaled) during the previous $\tau_g = 5$ time intervals, the randomly-generated control forces during the previous $\tau_u = 8$ time intervals, and the structural displacement responses of joints A during the previous $\tau_x = 2$ time intervals subjected to combined earthquake and control loadings are used as inputs. The control forces in the x-direction (Fig. 16.9) and their corresponding responses are used for training the x-neuroemulator and those in the y-direction for training the y-neuroemulator. In both sets of training, the current displacement response of joint A is the output of the dynamic fuzzy wavelet neuroemulators. Set1 data are used to illustrate the effectiveness of the proposed neuroemulator in predicting linear response of the structure, while Set2 data are used to illustrate its effectiveness in predicting nonlinear response of the structure.

The training of the model using the adaptive LM-LS learning algorithm (See Section 14.3) converges very fast after only six iterations for training the x-neuroemulator and eight iterations for the y-neuroemulator using each of the two sets of data. The model is implemented in a combination of C++ programming language and MATLAB 6.1 (Mathworks 2001) on a Windows XP Professional platform and a 1.5GHz Intel Pentium 4 processor. The CPU time for training the x-neuroemulator using each set of data is only 53 seconds and for training the y-neuroemulator, only 55 seconds.

Testing Fuzzy Wavelet Neuroemulators

The performance of the trained neuroemulators is evaluated by comparing their predicted displacement response with the finite element simulation results obtained directly by Eq. (16.15). The evaluation is conducted in both time and frequency domains for nonlinear response of the structure subjected to 100% of Chichi earthquake. Each of the trained x- and y-neuroemulators is tested using two sets of data defined previously for 100% of Chichi earthquake. Figure 16.10 shows the predicted and simulated structural displacement responses of joint A of the 12-story structure in the x- and y-directions under the combined Chichi earthquake excitation and actuator forces. The relative root mean square (RRMS) errors between the simulated and predicted results for the four test cases (2 in the x-direction and 2 in the y-direction) are summarized in Table 16.2. The RRMS errors of the displacement responses for all four sets of test data are less than 0.3, which is quite small.

The pseudospectra method, a power density spectrum method, presented in Chapter 15 for damage detection of high-rising buildings, is employed in this chapter to evaluate the accuracy of the trained neuroemulators in the frequency domain. The multiple signal classification (MUSIC) method is employed to compute the pseudospectrum from the structural response time series. Figure 16.11 shows a comparison of pseudospectra of the predicted and simulated displacement responses of joint A of the 12-story structure in the x- and y-directions under the combined earthquake and actuator loadings. The pseudospectra of the predicted and simulated structural responses match quite well for both x- and y-directions, thus demonstrating that the fuzzy WNN neuroemulators provide accurate prediction of structural displacement responses.

Table 16.2 Relative root mean square (RRMS) error of four test cases in Example 1

Loading type	Earthquake excitation	Combined earthquake excitation and actuator forces
x-direction	0.29	0.15
y-direction	0.23	0.24

16.4.2   Example 2: Eight-Story Irregular Building

Example Structure and its Response

This example structure is an 8-story three-dimensional steel building structure with a plan irregularity and a height of 36 m shown in Fig. 16.12. This structure was created by Kim and Adeli (2005b) for the study of hybrid control of 3D irregular buildings. It consists of 208 members and 99 nodes. The members of the structures are various W shapes as described in Kim and Adeli (2005b). Floor diaphragms are assumed to be rigid and axial column deformations are neglected. The resulting structural model has 24 degrees of freedom, assuming rigid floors.

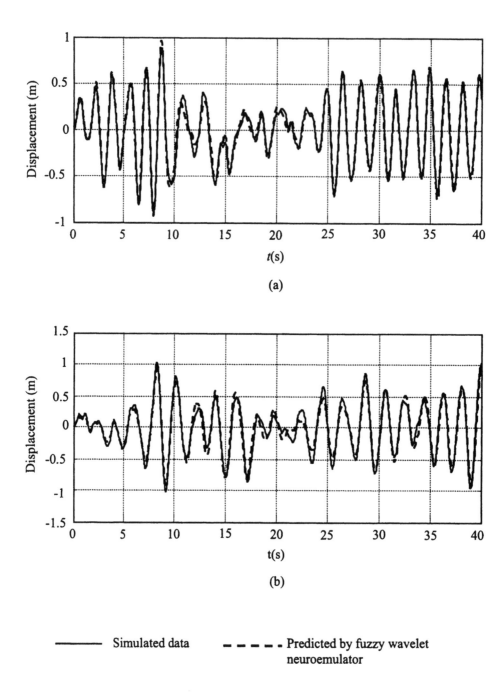

Figure 16.10 Predicted and simulated displacement responses of joint A of 12-story structure (Example 1) subjected to combined Chichi earthquake excitation and actuator forces: (a) Displacements of joint A in the x-direction, and (b) displacements of joint A in the y-direction

The structure is subjected to the combination of uniformly distributed floor dead and live loads of 100 psf (4.78 Kpa) and 70 psf (3.35 Kpa), respectively, and the Chichi earthquake records. The modal analyses of this structure conducted by Kim and Adeli (2005b) demonstrate that the coupling effect of lateral and torsional vibrations is most significant when the earthquake excitation is applied in the direction with $\beta = 83.4°$ (Fig. 16.12). Therefore, this is the angle used in this chapter to perform nonlinear dynamic analyses of the structure for training and testing the fuzzy wavelet neuroemulator.

Actuators

Four double-acting actuators with the same properties are installed in every floor as indicated with dashed lines in Fig. 16.12b. Their properties are the same as those employed in Example 1 and summarized in Table 16.1. The control forces are randomly generated and used similar to that explained for Example 1.

Constructing Fuzzy Wavelet Neuroemulators

Following the procedure for constructing the neuroemulators explained for Example 1, optimum time delay steps of $\tau_x = 2$, $\tau_g = 5$, and $\tau_u = 14$ are obtained for the displacement response at joint A shown in Fig. 16.12a, the earthquake excitation, and the control force using the FNN method (see Section 4.3.3), respectively. The number of input vectors created is equal to $N_a = 4,000 - \max\{2, 5, 14\} = 3,986$ (Eq. 16.25). The 3,986 sets of input vectors are used to construct the neuroemulators. The resulting x-neuroemulator has 21 input nodes in the input layer ($\tau_x + \tau_u + \tau_g = 21$ in Eq. 16.24) and 2 hidden/wavelet nodes in the hidden layer; while the resulting y-neuroemulator has 21 input nodes in the input layer and 3 hidden/wavelet nodes in the hidden layer.

Training Fuzzy Wavelet Neuroemulators

Two sets of data, denoted by Set1 and Set2, similar to those explained for Example 1, are used to train each neuroemulator, one set at a time. The training of the model using the adaptive LM-LS learning algorithm converges after only four iterations for training the x-neuroemulator and six iterations for the y-neuroemulator using each of the two sets of data.

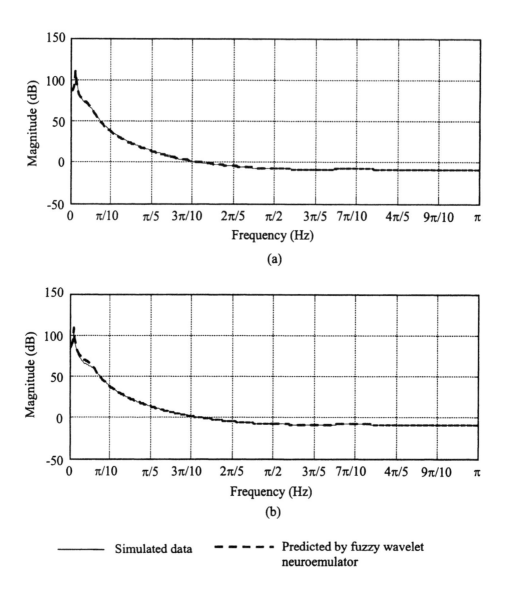

Figure 16.11 Comparison of pseudospectra of predicted and simulated displacement responses of joint A of 12-story structure (Example 1) under combined Chichi earthquake and actuator loadings: (a) Comparison of displacements in the *x*-direction, and (b) Comparison of displacements in the *y*-direction

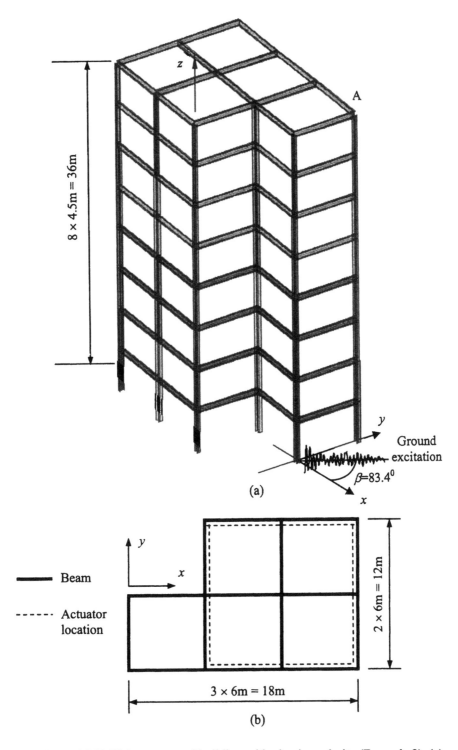

Figure 16.12 Eight-story steel building with plan irregularity (Example 2): (a) Perspective view, and (b) plan view

Testing Fuzzy Wavelet Neuroemulators

Testing is performed similar to that for Example 1 under 100% of Chichi earthquake. This structure, however, behaves linearly under this earthquake loading as mentioned earlier. Figure 16.13 shows the predicted and simulated displacement responses of joint A of the structure in the x- and y-directions under the combined Chichi earthquake excitation and actuator forces. The RRMS errors between the simulated and predicted results for the four test cases (2 in the x- and 2 in the y-direction) are summarized in Table 16.3. The RRMS errors of the displacement responses for all four sets of test data are less than 0.25, which is quite small. The performance of the trained neuroemulators is also evaluated using the pseudospectra method in the frequency domain for the Chichi earthquake. The pseudospectra of the predicted and simulated structural responses match quite well for both x- and y-directions.

16.5   Concluding Remarks

An effective model was presented for active nonlinear control of large 3D structures. Both geometric and material nonlinearities are included in the control formulation. Further, coupling between lateral and torsional motions of the structure as well as the actuator dynamics are taken into account in the control model.

It was demonstrated that the dynamic fuzzy wavelet neuroemulator provides accurate prediction of structural displacement responses in both linear and nonlinear ranges. In the next chapter, a floating point genetic algorithm is presented for finding the optimal control forces needed for active nonlinear control of building structures.

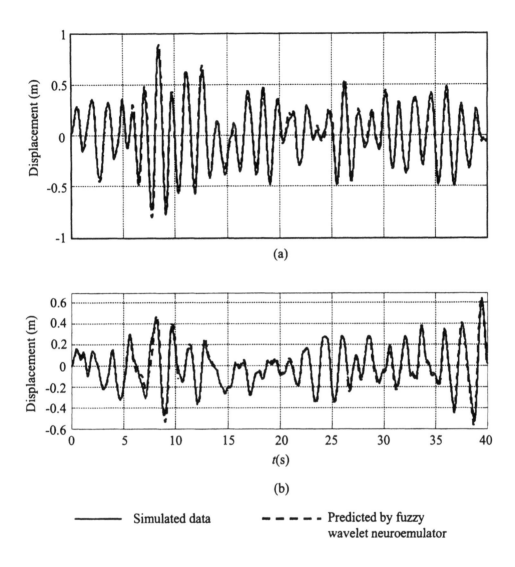

Figure 16.13 Predicted and simulated displacement responses of joint A of 8-story building (Example 2) subjected to combined Chichi earthquake excitation and actuator forces: (a) Displacements of joint A in the *x*-direction, and (b) displacements of joint A in the *y*-direction

Table 16.3 Relative root mean square (RRMS) error of four test cases in Example 2

Loading type	Earthquake excitation	Combined earthquake excitation and actuator forces
x-direction	0.22	0.14
y-direction	0.20	0.19

Chapter 17

NEURO-GENETIC ALGORITHM FOR NONLINEAR CONTROL OF SMART STRUCTURES

17.1 Introduction

In the previous chapter, a nonlinear control model was presented for active control of three-dimensional high-rising building structures. Both geometrical and material nonlinearities were included in the structural control formulation. Two dynamic coupling actions were taken into account simultaneously in the control model: a) coupling between lateral and torsional motions of the structure and b) coupling between the actuator and the structure. A dynamic fuzzy wavelet neural network (WNN) model was developed as a neuroemulator to predict structural responses in future time steps from the immediate past structural responses. The neuroemulator was trained using the adaptive Levenberg-Marquardt-Least-Squares (LM-LS) algorithm presented in Section 14.3. The neuroemulator was validated using two three-dimensional (3D) steel building structures with vertical and plan irregularities. It was demonstrated that the neuroemulator provides accurate prediction of structural displacement responses, which is required in neural network models for active control of structures.

In the existing neural network-based control algorithms (e.g., Chen *et al.* 1995; Ghaboussi and Joghataie 1995; Bani-Hani and Ghaboussi 1998a&b; Hung *et al.* 2000), two neural network models are used for active control of structures. A neural network-based emulator is created to predict the structural responses and a neural network-based controller is created to determine the control forces required to satisfy the control criteria. Both neural network models need to be trained using structural response data. These published works are applied mostly to small two-dimensional frame structures. The approach of using a neurocontroller along with a neuroemulator is effective for small structures with a few members.

For large three-dimensional structures such as high-rising buildings, however, the network size of the neurocontroller becomes increasingly large and the training of the resulting network becomes prohibitively expensive. It should be pointed out that training of a neuroemulator is based on a supervised learning algorithm such as the backpropagation (BP) algorithm (Rumelhart *et al.* 1986) and the adaptive LM-LS algorithm developed by the authors (Jiang and Adeli 2005a) and presented in Section 14.3. The training of a neurocontroller, on the other hand, is based on the trained neuroemulator and some kind of unsupervised learning algorithm (Ghaboussi and Joghataie 1995; Bani-Hani and Ghaboussi 1998b; Hung *et al.* 2000) because the magnitudes of control forces are problem- (structure- and earthquake-) dependent and therefore unknown. Training of a neurocontroller requires a considerable amount of processing time, especially for large structures such as high-rising building structures.

High-rising building structures may experience yielding and nonlinear behavior (geometrical or material nonlinearity or both) under severe earthquakes. Even the controlled structure may behave nonlinearly under insufficient control forces, especially in the beginning of a control loop. In such cases, a computational model based on the assumption that the controlled structure behaves linearly would be inadequate for representing the actual behavior of the structure. The classical linear control theories such as the linear quadratic regulator feedback control algorithm (Soong 1990; Adeli and Saleh 1999) and the linear quadratic Gaussian control algorithm (Stein and Athans 1987) are not effective for active nonlinear control of high-rising building structures subjected to extreme loadings.

In this chapter, an intelligent control algorithm is presented for nonlinear control of structures using a floating point genetic algorithm (GA). It is shown that the algorithm is particularly effective for active control of large 3D high-rising building structures. The dynamic fuzzy wavelet neuroemulator presented in Chapter 16 is used to predict future nonlinear structural displacement responses from the immediate past structural responses and actuator dynamics. A genetic algorithm is developed to search for the optimal control forces in every control loop during the earthquake excitation. It should be pointed out that GA has been used in structural engineering since 1993 mostly in the area of structural optimization, such as minimum weight design (e.g., Jenkins 1992; Adeli and Cheng 1993, 1994a&b; Adeli and Kumar 1995a&b), minimum cost design (e.g., Kim and Adeli 2001; Sarma and Adeli 2000a&b, 2001, 2002, 2003) and machine learning (Adeli and Hung 1995). In the past few years GAs have been used by a number of researchers to find the optimum control forces in structural control problems (e.g., Abullah *et al.* 2001; Kim and Ghaboussi 1999 & 2001; Li *et al.* 1998, 2000 & 2002; Tan *et al.* 2005; Yoshida and Dyke 2005). In this chapter, fuzzy logic, wavelets and genetic algorithm are integrated for nonlinear active control of structures. The two irregular 3D steel building structures described in the previous chapter are used to validate the control methodology under three different earthquake loadings.

17.2 Control Strategy for High-rising Buildings

In each time step of the earthquake excitation a control loop is created to find the required optimal control forces iteratively. In the *j*-th time step the control loop is ended when the following control criteria are satisfied:

$$\overline{d}_j \leq d_{\max} \tag{17.1}$$

and

$$\overline{d}_j \leq \overline{d}_R \tag{17.2}$$

where $d_{\max}$ is the maximum allowable displacement of the high-rising building, usually at the top floor, and $\overline{d}_j$ is the average displacement of the floor of maximum lateral displacement at current and p future time steps computed by

$$\overline{d}_j = \sum_{i=0}^{p} |d^i(\mathbf{F}_j)| \Big/ (p+1) \tag{17.3}$$

where $d^i(\mathbf{F}_j)$ is the displacement at the *i*-th prediction (future) time step of the control loop in the *j*-th time step of the earthquake loading obtained by

$$d^i(\mathbf{F}_j) = \sqrt{[\hat{x}^i(\mathbf{F}_j)]^2 + [\hat{y}^i(\mathbf{F}_j)]^2} \, , \, i = 1, 2, ..., p \tag{17.4}$$

where $\hat{x}^i(\mathbf{F}_j)$ and $\hat{y}^i(\mathbf{F}_j)$ are the predicted structural displacements along two convenient *x*- and *y*-axes, respectively, obtained using the *x*- and *y*-neuroemulators described in Chapter 16. The variable $\mathbf{F}_j = [F(t)_1 \quad F(t)_2 \quad \cdots \quad F(t)_{2L}]'$ is the $2L \times 1$ column matrix of control forces at time *t* (the *j*-th time step), in which *L* is the number of stories of the building structure (two sets of identical actuators are used in every floor, one set along the x axis and the other along the y axis, as explained in the previous chapter).

The parameter $\bar{d}_R = C_0 \bar{d}_0$ is the average displacement of the floor of maximum lateral displacement of the uncontrolled structure, in which $\bar{d}_0$ is an average displacement defined similar to Eq. (17.3) but for the uncontrolled structure, and $C_0 \leq 1.0$ is a reduction factor used to guarantee that the average maximum displacement of the controlled structure is not larger than that of the uncontrolled structure. Actuators apply forces in the opposite direction which create the potential for overshooting and creating unacceptably large displacements in the opposite direction. Equation (17.2) is included in the control strategy to avoid this possibility. The control designer can choose any value less than or equal to one for parameter C_0 (a value of 0.9 is chosen in the examples presented in this chapter).

The number of prediction time steps, p in Eq. (17.3), plays an important role in the effectiveness of the control algorithm. If p is too small the control algorithm may not be able to provide accurate results for values of required control forces. If p is too large the computer processing time may be excessive. Further, since the oncoming earthquake is unknown, it is impossible to accurately predict structural responses in a large number of future time steps (for example, 30 steps or 0.3 second). As such, control algorithms requiring a large p in general are not effective. On the other hand, p must be large enough for two things to take place during the p time steps: the computer must compute the control forces using the control algorithm and actuators must respond within their response time which may be several centi-seconds.

A trial-and-error approach is usually used to find the most suitable value for the number of prediction time steps in the neural network-based control algorithms (Bani-Hani and Ghaboussi 1998b). However, the trial-and-error approach a) does not provide a rational basis for the selection, b) is cumbersome, c) is computationally time-consuming, and d) does not guarantee accurate active control results.

In this work, earthquake motion is considered to be chaotic. Concepts of attractor and time lag from chaos theory are used for finding the value of p (see Chapter 4). In the chaos theory, the set of points used to simulate the evolution trajectory in the state space is referred to as an *attractor*. The purpose of the attractor is to unfold the time series back to a multivariate state space representing the original physical system. An attractor is the geometric invariance in the state space representation of a time series. The invariants of the dynamic system producing the time series are preserved if the time series is transformed into a sufficiently large reconstructed state space (see Section 4.1). In this chapter, for active control of large three-dimensional structures such as high-rising buildings, the time lag required to reconstruct the attractor from a given time series with sufficient accuracy is computed for any given earthquake ground acceleration using the FNN method described in Section 4.3.3. Then, the number of prediction time steps is the computed time lag divided by the selected time step (0.01 second in the examples presented in this chapter).

The objective of active control of a given structure is to find optimal actuator forces in every time step during the earthquake loading, minimizing displacements at points of maximum displacement, such as the top of a high-rising building structure, and satisfying the control criteria defined by Eqs. (17.1) and (17.2). As such, the optimal control problem is formulated as a constrained nonlinear optimization problem as follows:

Minimize

$$f(\mathbf{F}_j) = \bar{d}_j - \min\{\bar{d}_R, d_{\max}\} = \frac{\sum_{i=0}^{p}\sqrt{[\hat{x}^i(\mathbf{F}_j)]^2 + [\hat{y}^i(\mathbf{F}_j)]^2}}{p+1} - \min\{\bar{d}_R, d_{\max}\} \qquad (17.5)$$

Subjected to the following constraints:

$$-F_{k,\max} \le F_{k,j} \le F_{k,\max}, \; k = 1, 2, ..., 2L \qquad (17.6)$$

$$\bar{d}_j \le d_{\max} \qquad (17.1) \text{ (Repeated)}$$

$$\bar{d}_j \le \bar{d}_R \qquad (17.2) \text{ (Repeated)}$$

where F_{kj} is the k-th control force in the j-th time step and $F_{k,\max}$ is the control force capacity of the k-th actuator.

17.3 Neuro-Genetic Algorithm for Active Control of Smart Structures

17.3.1 Genetic Algorithm for Structural Control

A floating point genetic algorithm is developed to solve the formulated structural control optimization problem and find the optimal control forces. The algorithm does not need computation of any first-order gradient or second-order Hessian matrix required in most nonlinear optimization algorithms such as the Newton-Gaussian method. The genetic algorithm requires the determination of six fundamental issues: 1) chromosome representation, 2) initial population, 3) evaluation/fitness function, 4) selection function, 5) genetic operator, and 6) termination criteria.

1) Representation Scheme

A proper representation scheme is required for the solution of the structural control problem formulated in the previous section. Each chromosome is made up of a sequence of *genes*, which could consist of binary digits (0 and 1), floating point numbers, integers or symbols (i.e., A, B, C ...). The floating-point representation scheme is used in this chapter for solution of the nonlinear control problem because it provides a more natural representation of the problem than other schemes. This representation scheme is based on real-valued floating point numbers with values within upper and lower bounds of control force variables which are the actuator capacities, i.e., $-F_{k,\max}$ and $F_{k,\max}$ ($k = 1, 2, ..., 2L$).

2) Initial Population

Genetic algorithms require an initial population for the solution of an optimization problem. The initiation population P_0 is generated through a random search process. A larger value for the population size, N_p, will result in a larger computational cost for the GA approach, while a smaller value will lead to inadequate computational accuracy. The range of 50 to 150 for N_p was found to balance out the tradeoff between the computational requirement and accuracy (Pham and Karaboga 2000). An initial population with N_p chromosomes is generated for every actuator/control force variable using the uniform distribution function between $-F_{k,\max}$ and $F_{k,\max}$ ($k = 1, 2, ..., 2L$).

3) Evaluation/Fitness Function

Equation (17.5) is the fitness or objective function for the genetic control algorithm. This equation is also a displacement-based performance function for active control of a high-rising building structure.

4) Selection Function

The control force variables collectively make a chromosome. The selection of chromosomes to mate and produce successive generations plays a significant role in moving the search towards a promising space and finding the global optimum solution quickly. A probabilistic ranking method is used in the floating point genetic control algorithm to select the chromosomes for the next generation based on the following three steps:

a) A probability of selection, p_j, is computed for each chromosome j based on the normalized geometric ranking of its fitness as follows (Goldberg 1989; Joines and Houck 1994):

$$p_j = \frac{q(1-q)^{r-1}}{1-(1-q)^{N_p}}, j = 1, 2, ..., N_p \tag{17.7}$$

where q is the preselected probability of selecting the fittest chromosome and N_p is the population size. A value of $q = 0.05$ is used for the probability of selecting the fittest chromosome in the examples presented in this chapter. The parameter r is the rank of the jth chromosome obtained by sorting the chromosomes in the increasing order of their fitness function value.

b) The cumulative probability of each chromosome is calculated by

$$C_j = \sum_{i=1}^{j} p_i, j = 1, 2, ..., N_p \tag{17.8}$$

c) The j-th chromosome is selected and copied into the new population P_g (g represents the generation number) if it satisfies the following criterion:

$$C_{j-1} < c_0 \leq C_j, j = 1, 2, ..., N_p \tag{17.9}$$

where c_0 is a value between 0 and 1 generated randomly.

5) Genetic Operator

A genetic operator is used for mating to create improved chromosomes or solutions to be used in the new population. There are two categories of operators: *crossover* (a typical mating operator) and *mutation* (a typical perturbation operator). In this chapter, a nonuniform mutation operator is used as the genetic operator to find an improved solution for the new generation g as follows:

$$F_j' = \begin{cases} F_j + (F_{max} - F_j)h(g) & \text{if } r_1 \geq 0.5 \\ F_j & \text{otherwise} \end{cases} \quad j = 1, 2, ..., N_p \quad (17.10)$$

where F_j is the value of the j-th variable in the chromosome in the current population and F_j' is the improved value of the same variable in the new generation P_g, and $h(g) = [r_2(1 - g/g_{max})]^b$ is a mutation probability function in which g_{max} is the maximum number of generations, b is a positive integer, called the shape parameter, determining the degree of non-uniformity (a value of $b = 3$ is chosen in the examples presented in this chapter), and r_1 and r_2 are randomly generated numbers with values in the range $(0,1)$ (Michalewicz 1996). The operation represented by Eq. (17.10) allows for local tuning as it searches the local space uniformly through the use of random numbers r_1 and r_2.

6) Termination Criteria

Termination criteria are needed to stop selecting and reproducing parents for mating from generation to generation. Pre-assigned maximum number of generations, g_{max}, and minimum change in the fitness function between two generations, ε, are used in this chapter to terminate the control loop in every time step during the earthquake loading as follows:

$$\Delta f(\mathbf{F}^g) = | f(\mathbf{F}^g) - f(\mathbf{F}^{g-1})| \leq \varepsilon \quad (17.11)$$

$$g > g_{max} \quad (17.12)$$

Note that either Eq. (17.11) or Eq. (17.12) will terminate the control loop. The selection of g_{max} and ε depends on the desirable accuracy for the given problem. A smaller value of ε will result in higher accuracy for the optimal control forces $\mathbf{F}^g$, at the cost of additional iterations and computation time for the GA to converge. The flow diagram for the genetic algorithm for active control of building structures is shown in Fig. 17.1. The iterative procedure is applied to find the optimal control forces in every time step j during the earthquake excitation.

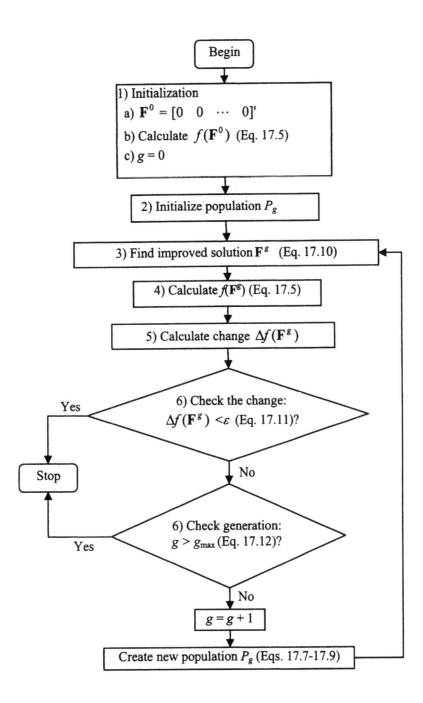

Figure 17.1 Flow diagram for the genetic algorithm for active control of building structures

17.3.2 Integration of GA with Dynamic Fuzzy Wavelet Neuroemulator

The neuro-genetic active control methodology consists of the dynamic fuzzy wavelet neuroemulator and the floating point genetic control algorithm. Figure 17.2 shows how the GA is integrated with the dynamic fuzzy wavelet neuroemulator to find the optimal control forces. The inputs include the earthquake excitations, control forces, and structural displacement responses in previous time steps. The outputs of the model are the optimal actuator control forces and the displacements of the controlled structure. Three steps are identified in Fig. 17.2 by numbers 1 to 3 and described in this section.

Step 1: Construct state space vectors

Both structural response data and control forces are required for constructing and training two dynamic fuzzy wavelet neuroemulators in x- and y-directions. Structural response data are created from nonlinear dynamic analyses, taking into account both geometrical and material nonlinearities. Control forces in the range of the capacities of hydraulic actuators are initially generated randomly using a uniform distributed function.

At each time step j two input state space vectors, $\mathbf{X}_j^x$ and $\mathbf{X}_j^y$, are constructed for x- (Step 1a) and y- (Step 1b) directions, respectively. Every input state vector consists of three parts: structural displacement responses, actuator forces, and earthquake excitations. Construction of each input state vector is described in the previous chapter.

Step 2: Predict structural displacement responses

Structural displacement responses at the j-th time step in both x- and y-directions, $\hat{y}_j^x = \bar{f}(\mathbf{X}_j^x)$ and $\hat{y}_j^y = \bar{f}(\mathbf{X}_j^y)$, are predicted using the trained x- and y-neuroemulators, respectively, where $\bar{f}(\mathbf{X}_j^x)$ and $\bar{f}(\mathbf{X}_j^y)$ are the general discrete input-output mapping functions for the x- and y-neuroemulators, respectively, as described in the previous chapter.

Step 3: Find optimal control forces

The floating-point genetic algorithm described previously (Fig. 17.1) is applied to find the optimum control forces for any time step during the earthquake excitation, which result in a minimum objective value represented by Eq. (17.5). Next, the optimal control

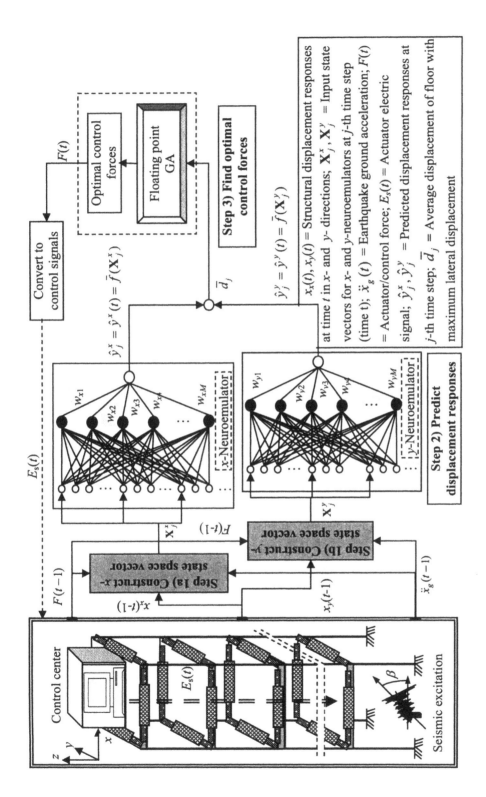

Figure 17.2 Integration of GA with dynamic fuzzy wavelet neuroemulator to find the optimal control forces

forces are converted to the corresponding electric signals based on the properties of the actuator, as described in the previous chapter. The electric signals are sent to the actuators to generate the required control forces for the controlled structure.

17.4 Numerical Example

Two three-dimensional example structures are used to validate the nonlinear control methodology, a twelve-story structure with vertical setbacks (Example 1) and an eight-story structure with a plan irregularity (Example 2), as described in Sections 16.4.1 and 16.4.2. The horizontal components of three different earthquake loadings are used in the numerical validation: 1) Chichi earthquake of Richter magnitude 7.3 recorded at Foothills, Western Taiwan, on September 21, 1999 (Case 1), 2) the same recorded amplified by a factor of 2 and referred to as 200% Chichi (Case 2), and 3) Kern County N90E earthquake of Richter magnitude 7.7, which occurred at Kern County, California, on July 21, 1952 (Case 3). The Chichi earthquake record was presented in the previous chapter. Figure 17.3 shows the Kern County earthquake excitation at increments of 0.02 seconds over a period of 82 seconds.

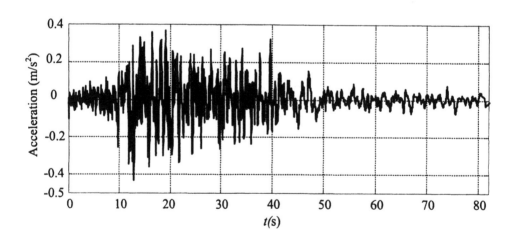

Figure 17.3 Kern County earthquake excitation of July 21, 1952

Displacement responses are obtained at the top floor of each example structure in both x and y directions at increments of 0.01 seconds for the Chichi earthquake and at increments of 0.02 seconds for the Kern County earthquake. Thus, the number of sample data is 4,000 for both Chichi and Kern County earthquakes. The controlled responses of the two example structures are found for three aforementioned earthquake loading cases and compared with the uncontrolled responses.

The same population size of $N_p = 100$ is used for both examples. Values of $\varepsilon = 0.0001$ and $g_{max} = 30$ are used for the termination criteria. These values are selected by a trial-and-error approach and found to provide sufficient accuracy in the examples. For training the nonlinear control model, two sets of control forces are randomly generated in the x- and y-directions with a maximum actuator force of $F_{max} = 68$ kN and a uniform distribution in the interval of (-68 kN, 68 kN). Prediction time step of $p = 5$ is obtained for the Chichi earthquake excitation and $p = 4$ for the Kern County earthquake excitation for both example structures using the FNN method described in Section 4.3.3.

17.4.1   Example 1: Twelve-Story Irregular Building

In this example, the earthquake excitations are applied to the structure with an angle of $\beta = -13°$ with respect to the x-direction (Fig. 16.7). This incidence angle is used because it produces the maximum coupling between the lateral and torsional variations of the structure as explained in the previous chapter.

Case 1: 100% of Chichi Earthquake

In this case as well as Case 2, $H/500$ is used as the maximum allowable displacement at the top floor of the controlled structure where $H = 54$ m is the height of the building. Thus, the maximum allowable displacement at the top floor of the 12-story structure is $d_{max} = 0.108$ m. In this case, the uncontrolled structure behaves nonlinearly but the controlled structure behaves linearly. Figure 17.4 shows the time histories of the top floor displacement of the uncontrolled and controlled structures subjected to 100% Chichi earthquake. Figures 17.4a and 17.4b show the displacements in the x- and y-directions, respectively. Figure 17.4c shows the total displacement. The maximum values of the displacement at the top floor are also summarized in Table 17.1. The reduction of maximum displacement in the controlled structure is 74% in the x-direction and 63% in

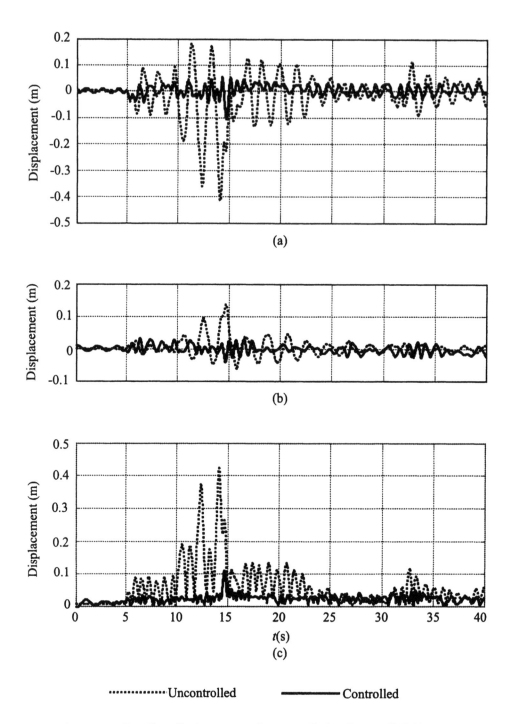

Figure 17.4 Top floor displacements of uncontrolled and controlled 12-story structures (Example 1) subjected to 100% Chichi earthquake (Case 1): (a) *x*-direction; (b) *y*-direction, and (c) total displacement

the y-direction.

The neuro-genetic algorithm is implemented in MATLAB 6.1 (Mathworks 2001) on a Windows XP Professional platform. The CPU time for finding the optimal control forces in each time step (0.01 seconds) depends on the speed of the CPU processor. For example, on a 1.5GHz Intel Pentium 4 processor with 512M RAM memory, it takes 64 ms to perform the control operations, while on a workstation with two 2.4G Hz Intel Pentium 4 processors and one 1G RAM memory, it takes less than 3 ms. It will take only a fraction of a millisecond to perform the control operations using the existing high-performance computer technology.

Table 17.1 Maximum displacements of top floor of 12-story structure (Example 1) subjected to three earthquake ground accelerations

Earthquake	Displacement Response	Uncontrolled (m)	Controlled (m)	Reduction (%)
Case 1 100% Chichi	x	0.416	0.107	74
	y	0.137	0.051	63
	Total	0.424	0.108	74
Case 2 200% Chichi	x	0.829	0.106	87
	y	0.271	0.074	73
	Total	0.846	0.108	87
Case 3 Kern County	x	0.073	0.029	60
	y	0.014	0.001	96
	Total	0.075	0.029	61
	(Kim and Adeli 2005a)	(0.074)	(0.039)	(47)

Case 2: 200% of Chichi Earthquake

In this case, the uncontrolled structure behaves nonlinearly with significant geometric nonlinearity as noted by the maximum displacement of the top floor. The controlled structure, however, behaves linearly as in Case 1. Figure 17.5 shows the time histories of the top floor displacement of the uncontrolled and controlled structures subjected to 200% Chichi earthquake. The maximum values of the displacement at the top floor are also summarized in Table 17.1. The reduction of maximum displacement in the controlled structure is 87% in the x-direction and 73% in the y-direction. As an example, the variation of the value of the optimal control force for the actuator in the x-direction at

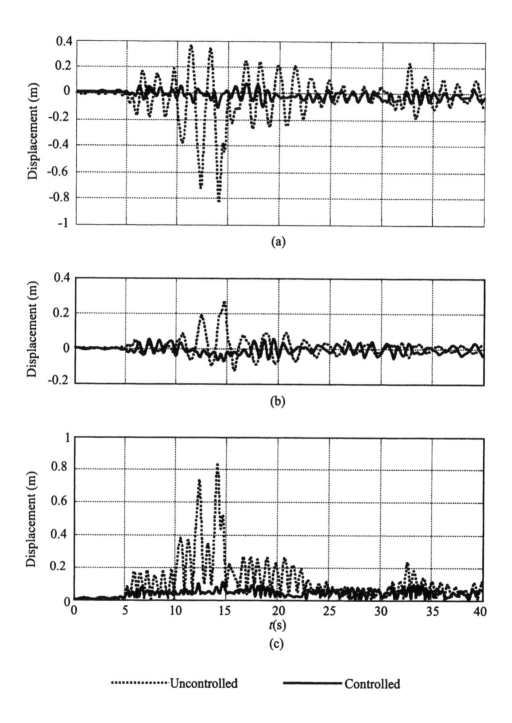

Figure 17.5 Top floor displacements of uncontrolled and controlled 12-story structures (Example 1) subjected to 200% Chichi earthquake (Case 2): (a) *x*-direction; (b) *y*-direction, and (c) total displacement

the top floor of the building is shown in Fig. 17.6.

Figure 17.7 shows sample convergence curves for the floating point GA at three sample times $t = 6.1$ s $(j = 610)$, 14.2 s $(j = 1,420)$, and 18.6 s $(j = 1,860)$. It is shown that the minimum value of the average displacements of the top floor at the three time steps are $\overline{d}_{610} = 0.023$ m, $\overline{d}_{1420} = 0.051$ m, and $\overline{d}_{1860} = 0.084$ m, respectively. The GA is terminated at generations $g = 27, 24,$ and 30 for the three cases based on the two terminating criteria. Figure 17.7 shows the proposed GA for structural control problems has good convergence properties.

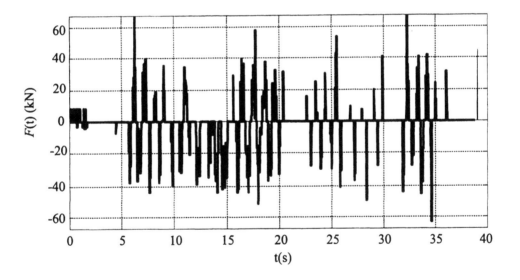

Figure 17.6 Variation of the value of the optimal control force for the actuator in the *x*-direction at the top floor of the 12-story building (Example 1)

Figure 17.8 shows maximum accelerations and displacements along the height of the structure for both controlled and uncontrolled structures subjected to 100% and 200% Chichi earthquake ground accelerations. Substantial reduction of the lateral displacements in the controlled structure is noticeable.

Case 3: Kern County Earthquake

The ground accelerations of the Kern County earthquake are rather small. In this case the uncontrolled structure behaves linearly and the maximum displacement at the top floor of the uncontrolled structure is only 0.074 m. This earthquake record is chosen for the sake of comparison with the work of Kim and Adeli (2005a), who used the same record in their hybrid damper-semi-active tuned liquid column damper (TLCD) control system. A value of d_{max} = 0.03 m is chosen for the maximum allowable displacement at the top floor of the controlled structure.

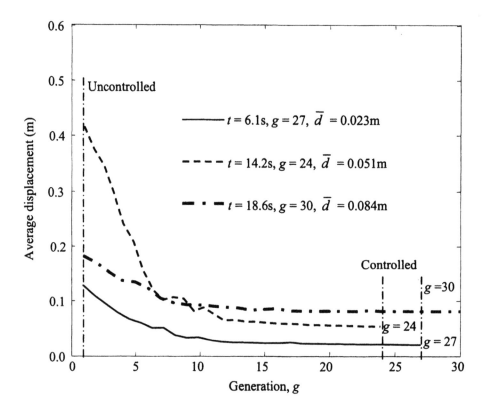

Figure 17.7 Sample convergence curves for the floating point GA at three sample times t = 6.1 s (j = 610), 14.2 s (j = 1,420), and 18.6 s (j = 1,860) for Case 2 of Example 1

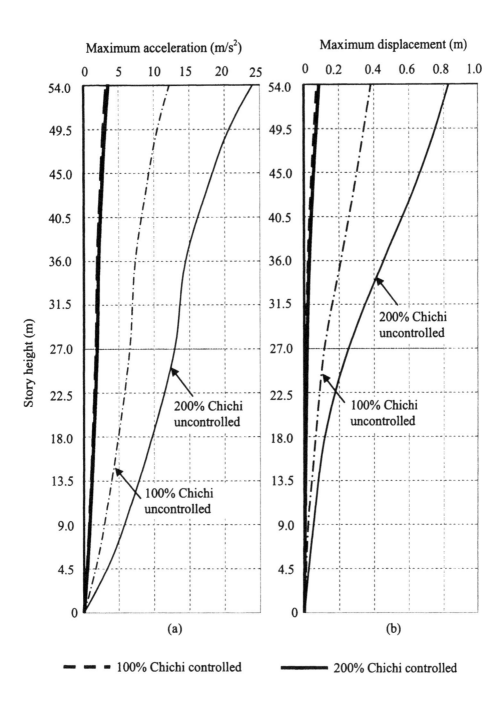

Figure 17.8 Maximum accelerations and displacements along the height of the 12-story structure for both controlled and uncontrolled structures subjected to 100% and 200% Chichi earthquakes: (a) Maximum acceleration (m/s²), and (b) maximum displacement (m)

Figure 17.9 shows the time histories of the top floor displacement of the uncontrolled and controlled structures subjected to the Kern County earthquake. The maximum values of the displacement at the top floor are also summarized in Table 17.1. The reduction of maximum displacement in the controlled structure is 60% in the x-direction and 96% in the y-direction. The values in the parentheses of the last line in Table 17.1 are from Kim and Adeli (2005c) based on their hybrid damper-TLCD control system.

17.4.2 Example 2: Eight-Story Irregular Building

In this example, the earthquake excitations are applied to the structure with an angle of β = 83.4° with respect to the x-direction (Fig. 16.12). This incidence angle is used because it produces the maximum coupling between the lateral and torsional variations of this building structure as explained in the previous chapter.

In this case as well as Case 2, $H/500$ is used as the maximum allowable displacement at the top floor of the controlled structure where $H = 36$ m is the height of the building. Thus, the maximum allowable displacement at the top floor of the 8-story structure is $d_{max} = 0.072$ m for Cases 1 and 2. In Case 3, a smaller value of $d_{max} = 0.03$ m is chosen for the maximum allowable displacement at the top floor of the controlled structure similar to that of Example 1.

In Cases 1 and 2, the uncontrolled structure behaves nonlinearly but the controlled structure behaves linearly. In Case 3, both uncontrolled and controlled structures behave linearly. Figure 17.10 shows the time histories of the top floor displacement of the uncontrolled and controlled structures subjected to 200% Chichi earthquake (Case 2). Similar results are obtained for the other two cases but are not presented in this chapter for the sake of brevity. The maximum values of the total displacement at the top floor for all three cases are summarized in Table 17.2. The reductions of maximum total displacements are 82%, 93%, and 73% for the three cases. The values in the parentheses of the last line in Table 17.2 are from Kim and Adeli (2005c) based on their hybrid damper-TLCD control system.

17.5 Concluding Remarks

The neuro-genetic control methodology is based on the integration of a floating point

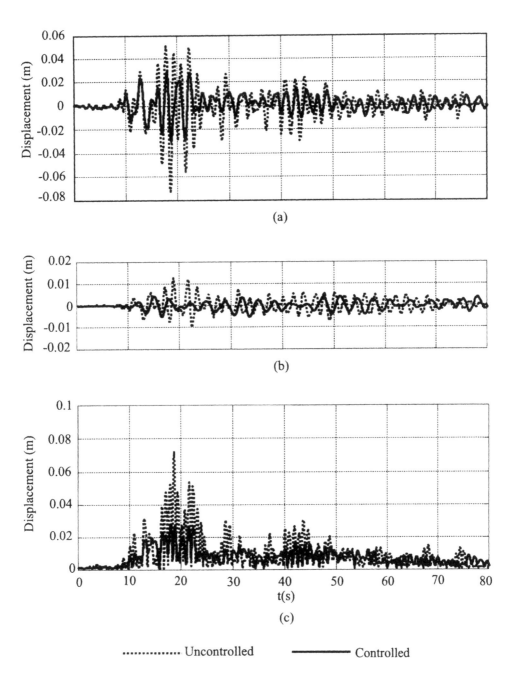

Figure 17.9 Top floor displacements of uncontrolled and controlled 12-story structures (Example 1) subjected to Kern County earthquake (Case 3): (a) *x*-direction; (b) *y*-direction, and (c) total displacement

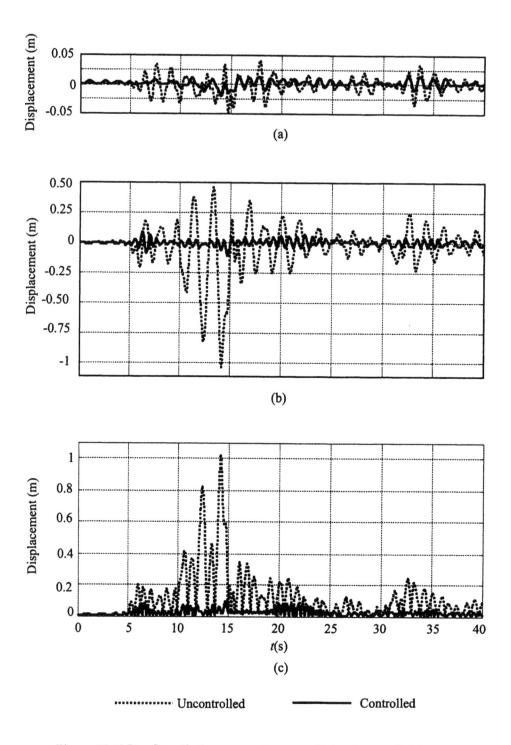

Figure 17.10 Top floor displacements of uncontrolled and controlled 8-story structures (Example 2) subjected to 200% Chichi earthquake (Case 2): (a) *x*-direction; (b) *y*-direction, and (c) total displacement

Table 17.2 Maximum displacements of top floor of 8-story structure (Example 2) subjected to three earthquake ground accelerations

Earthquake	Uncontrolled (m)	Controlled (m)	Reduction (%)
Case 1	0.411	0.072	82
Case 2	1.027	0.0719	93
Case 3	0.112	0.030	73
(Kim and Adeli 2005c)	(0.111)	(0.037)	(67)

genetic optimization algorithm with the dynamic fuzzy wavelet neuroemulator presented in the previous chapter. The dynamic fuzzy wavelet neuroemulator is created to predict structural displacement responses from the immediate past structural response and actuator dynamics. The genetic control algorithm is developed to find the optimal control forces for any time step during the earthquake excitation. The algorithm does not need the pre-training required in a neural network-based controller, which improves the efficiency of the general control methodology significantly.

Two irregular 3D steel building structures were used to validate the neuro-genetic control algorithm under three seismic excitations. Validation results demonstrate that the control methodology is effective in significantly reducing the response of large three dimensional building structures subjected to seismic excitations including structures with plan and elevation irregularities.

In all six cases for the two example structures the controlled structure behaves linearly. It should be pointed out, however, that in every time step optimal control forces are obtained iteratively. During this control loop it is necessary to include nonlinear behavior of the structure. In each time step, the neuro-genetic algorithm starts with the displacements of the uncontrolled structure in the initial population and yields the control forces required for the controlled structure in the final population.

Bibliography

Abarbanol, H. D. I. (1993), "The analysis of observed chaotic data in physical systems." *Review of Modern Physics*, 65(4), 1331–1392.

Abarbanel, H. D. I. (1996), *Analysis of Observed Chaotic Data*. Springer-Verlag, New York.

Abullah, M., Richardson, A., and Hanif, J. (2001), "Placement of sensors/actuators on civil structures using genetic algorithms." *Earthquake Engineering and Structural Dynamics*, 30(8), 1167–1184.

Ackley, D.H., Hinton, G.E., and Sejnowski, T.J. (1985), "A learning algorithm for Boltzmann machines." *Cognitive Science*, 9, 147–169.

Adeli, H. (2000), "High-performance computing for large-scale analysis, optimization, and control." *Journal of Aerospace Engineering*, ASCE, 13(1), 1–10.

Adeli, H. (2001), "Neural networks in civil engineering: 1989-2000." *Computer-Aided Civil and Infrastructure Engineering*, 16(2), 126–142.

Adeli, H., and Cheng, N.T. (1993), "Integrated genetic algorithm for optimization of space structures." *Journal of Aerospace Engineering*, ASCE, 6(4), 315–328.

Adeli, H., and Cheng, N.T. (1994a), "Augmented Lagrangian genetic algorithm for structural optimization." *Journal of Aerospace Engineering*, ASCE, 7(1), 104–118.

Adeli, H., and Cheng, N.T. (1994b), "Concurrent genetic algorithms for optimization of large structures." *Journal of Aerospace Engineering*, ASCE, 7(3), 276–296.

Adeli, H., and Ghosh-Dastidar, S. (2004), "Mesoscopic-wavelet freeway work zone flow and congestion feature extraction model." *Journal of Transportation Engineering*, ASCE, 130(1), 94–103.

Adeli, H., Ghosh-Dastidar, S., and Dadmehr, N. (2005a), "Alzheimer's disease and models of computation: imaging, classification, and neural models." *Journal of Alzheimer's Disease*, 7(3), 187–199.

Adeli, H., Ghosh-Dastidar, S., and Dadmehr, N. (2005b), "Alzheimer's disease: Models of computation and analysis of EEGs." *Clinical EEG and Neuroscience*, 36(3), 131–140.

Adeli, H., Ghosh-Dastidar, S., and Dadmehr, N. (2007), "A wavelet-chaos methodology for analysis of EEGs and EEG subbands to detect seizure and epilepsy." *IEEE Transactions on Biomedical Engineering*, 54(2), 205–211.

Adeli, H., and Hung, S.L. (1993), "A fuzzy neural network learning model for image recognition." *Integrated Computer-Aided Engineering*, 1(1), 43–55.

Adeli, H., and Hung, S.L. (1994), "An adaptive conjugate gradient algorithm for efficient training of neural networks." *Applied Mathematics and Computation*, 62, 81–102.

Adeli, H., and Hung, S.L. (1995), *Machine Learning - Neural Networks, Genetic Algorithms, and Fuzzy Sets*. Wiley, New York.

Adeli, H., and Jiang, X. (2003), "Neuro-fuzzy logic model for freeway work zone capacity estimation." *Journal of Transportation Engineering*, ASCE, 129(5), 484–493.

Adeli, H., and Jiang, X. (2006), "Dynamic fuzzy wavelet neural network model for structural system identification." *Journal of Structural Engineering*, ASCE, 132(1), 102–111.

Adeli, H., and Kao, W.-M. (1996), "Object-oriented blackboard models for integrated design of steel structures." *Computers and Structures*, 61(3), 545–561.

Adeli, H., and Karim, A. (1997), "Scheduling/cost optimization and neural dynamics model for construction." *Journal of Construction Management and Engineering*, ASCE, 123(4), 450–458.

Adeli, H., and Karim, A. (2000), "Fuzzy-wavelet RBFNN model for freeway incident detection." *Journal of Transportation Engineering*, ASCE, 126(6), 464–471.

Adeli, H., and Karim, A. (2001), *Construction Scheduling, Cost Optimization, and Management – A New Model Based on Neurocomputing and Object Technologies*. Spon Press, London.

Adeli, H., and Karim, A. (2005), *Wavelets in Intelligent Transportation Systems*. Wiley, Hoboken, New Jersey.

Adeli, H., and Kim, H. (2000), "Web-based interactive courseware for steel design using Java." *Computer-Aided Civil and Infrastructure Engineering*, 15(2), 158–166.

Adeli, H., and Kim, H. (2004), "Wavelet-hybrid feedback-least mean square algorithm for robust control of structures." *Journal of Structural Engineering*, ASCE, 130(2), 128–137.

Adeli, H., and Kumar, S. (1995a), "Distributed genetic algorithms for structural optimization." *Journal of Aerospace Engineering*, ASCE, 8(3), 156–163.

Adeli, H., and Kumar, S. (1995b), "Concurrent structural optimization on a massively parallel supercomputer." *Journal of Structural Engineering*, ASCE, 121(11), 1588–1597.

Adeli, H., and Kumar, S. (1999), *Distributed Computer-Aided Engineering for Analysis, Design, and Visualization*. Taylor & Francis, CRC Press, Boca Raton, Florida.

Adeli, H., and Park, H.S. (1998), *Neurocomputing for Design Automation*. Taylor & Francis, CRC Press, Boca Raton, Florida.

Adeli, H., and Sarma, K. (2006), *Cost Optimization of Structures – Fuzzy Logic, Genetic Algorithms, and Parallel Computing*. John Wiley and Sons, West Sussex, United Kingdom.

Adeli, H., and Saleh, A. (1997), "Optimal control of adaptive/smart bridge structures." *Journal of Structural Engineering*, ASCE, 123(2), 218–226.

Adeli, H., and Saleh, A. (1999), *Control, optimization, and smart structures—High-performance bridges and buildings of the future*. Wiley, New York.

Adeli, H., and Samant, A. (2000), "An adaptive conjugate gradient neural network - wavelet model for traffic incident detection." *Computer-Aided Civil and Infrastructure Engineering*, 13(4), 251–260.

Adeli, H., and Soegiarso, R. (1999), *High-Performance Computing in Structural Engineering*. Taylor & Francis, CRC Press, Boca Raton, Florida.

Adeli, H., and Wu, M. (1998), "Regularization neural network for construction cost estimation." *Journal of Construction Engineering and Management, ASCE*, 124(1), 18–24

Adeli, H., and Yu, G. (1993), "An object-oriented data management model for numerical analysis in computer-aided engineering." *Microcomputers in Civil Engineering*, 8, 199–209.

Adeli, H., Zhou, Z., and Dadmehr, N. (2003), "Analysis of EEG records in an epileptic patient using wavelet transform." *Journal of Neuroscience Methods*, 123(1), 69–87.

Agrawal, A.K., and Yang, J.N. (1996), "Optimal polynomial control of seismically excited linear structures." *Journal of Engineering Mechanics*, ASCE, 122(8), 753–761.

Aguirre, L.A., Donoso-Garcia, P.F., and Santos-Filho, R. (2000), "Use of *a priori* information in the identification of global nonlinear models – A case study using a buck converter." *IEEE Transactions on Circuits and Systems-I: Fundamental Theory and Applications*, 47(7), 1081–1085.

Aktan, A.E., Farhey, D.N., Helmicki, A.J., Brown, D.L., Hunt, V.J., Lee, K.L., and Levi, A. (1997), "Structural identification for condition assessment: Experimental arts." *Journal of Structural Engineering*, ASCE, 123(12), 1674–1684.

Al-Dawod, M., Smali, B., Kwok, K., and Naghdy, F. (2004), "Fuzzy controller for seismically excited nonlinear buildings." *Journal of Engineering Mechanics*, ASCE, 130(4), 407–415.

Al-Kaisy, A., and Hall, F. (2001), "Effect of darkness on the capacity of long-term freeway reconstruction zones." *Proceedings of 4th International Symposium on Highway Capacity,* Transportation Research Circular E-C018, Maui, Hawaii, pp. 164–174.

Al-Kaisy, A., Zhou, M., and Hall, F. (2000), "New insights into freeway capacity at work zones: empirical case study." *Transportation Research Record* 1710, Transportation Research Board, National Research Council, Washington, D.C., pp. 154–160.

Aleksander, I., and Morton, H. (1991), *"An Introduction to Neural Computing."* Chapman and Hall, London.

Aleksander, I., and Morton, H. (1993), *Neurons and Symbols: The Stuff That Mind Is Made of.* Chapman and Hall, London.

Al-Sultan, K.S., and Fediki, C. A. (1997), "A Tabu search-based algorithm for the fuzzy clustering problem." *Pattern Recognition*, 30(12), 2023–2030.

Amin, S.M., Rodin, E.Y., and Garcia-Ortiz, A. (1998), "Traffic prediction and management via RBF neural nets and semantic control." *Computer-Aided Civil and Infrastructure Engineering*, 13(5), 315–327.

Anderson, J.A. (1995), *An Introduction to Neural Networks*. MIT Press, Cambridge, Massachusetts.

Arbib, M.A. (1995), *The Handbook of Brain Theory and Neural Networks*. MIT Press, Cambridge, Massachusetts.

Bani-Hani, K., and Ghaboussi, J. (1998a), "Neural networks for structural control of a benchmark problem, active tendon system." *Earthquake Engineering and Structural Dynamics*, 27(11), 1225–1245.

Bani-Hani, K., and Ghaboussi, J. (1998b), "Nonlinear structural control using neural networks." *Journal of Engineering Mechanics*, ASCE, 124(3), 319–327.

Barron, A., Birgé, L., and Massart, P. (1999), "Risk bounds for model selection via penalization." *Probability Theory and Related Fields*, 113, 301–413.

Baumer, D., Gryczan, G., Knoll, R., Lilienthal, C., Riehle, D., and Zullighoven, H. (1997), "Framework development for large systems." *Communications of the ACM*, 40(10), 52–59.

Beck, J.L., Au, S.K., and Vanik, M.W. (2006), "Monitoring structural health using a probabilistic measure." *Computer-Aided Civil and Infrastructure Engineering*, 16(1), 1–11.

Benekohal, R.F., Kaja-Mohideen, A.-Z., and Chitturi, M.V. (2003), "Evaluation of construction work zone operational issues: capacity, queue, and delay." *Report No. ITRC FR 00/01-4*, Illinois Transportation Research Center, Illinois Department of Transportation, Edwardsville, Illinois.

Benzi, R., Marrocu, M., Mazzino, A., and Trovatore, E. (1999), "Characterization of the long-time and short-time predictability of low-order models of the atmosphere." *Journal of the Atmospheric Sciences*, 56(20), 3495–3507.

Bezdek, J.C. (1981), *Pattern Recognition with Fuzzy Objective Function Algorithms*. Plenum, New York.

Billings, L., Bollt, E.M., and Schwartz, I.B. (2002), "Phase-space transport of stochastic chaos in population dynamics of virus spread." *Physical Review Letters*, 88(23), 2341011–2341014.

Billings, S.A., and Tsang, K.M. (1989), "Spectral analysis for nonlinear system, Part I: Parametric nonlinear spectral analysis." *Mechanical System and Signal Processing*, 3(4), 319–339.

Billings, S.A., Tsang, K.M., and Tomlinson, G.R. (1990), "Spectral analysis for nonlinear system, Part III: Case study examples." *Mechanical System and Signal Processing*, 4(1), 3–21.

Bishop, C. (1995), *Neural Networks for Pattern Recognition*. University Press, Oxford.

Bjørnstad, O.N., Finkenstädt, B.F., and Grenfell, B.T. (2002), "Dynamics of measles epidemics: Estimating scaling of transmission rates using a time series SIR model." *Ecological Monographs*, 72(2), 169–184.

Blanas, P., Wenger, M.P., Shuford, R.J., and Das-Gupta, D.K. (1997), "Active composite materials and damage monitoring." In *Structural Health Monitoring: Current Status and Perspectives, Proceedings of the International Workshop on Structural Health Monitoring*, Stanford University, 1997, September 18–20, pp. 199–207, Technomic Publishing Co., Inc., Lancaster, Pennsylvania.

Bloomberg, L., and Dale, J. (2000), "Comparison of VISSIM and CORSIM traffic simulation models on a congested network." *Transportation Research Record*, TRB 1727, National Research Council, Washington D.C., pp. 52–60.

Bolton, R., Stubbs, N., Park, S., Sikorsky, C., and Choi, S. (2006), "Documentation of changes in modal properties of a concrete box girder bridge due to environmental and internal conditions." *Computer-Aided Civil and Infrastructure Engineering*, 16(1), 42–57.

Bose, N.K., and Liang, P. (1996), *Neural Network Fundamentals with Graphs, Algorithms and Applications*. McGraw-Hill, New York.

Box, G.E.P., Jenkins, G.M., and Reinsel, G.C. (1994), *Time Series Analysis: Forecasting and Control*. Prentice Hall, Englewood Cliffs, New Jersey.

Brockwell, P.J., and Davis, R.A. (2002), *Introduction to Time Series and Forecasting*. Springer-Verlag, New York.

Brown, A.S., and Yang, H.T.Y. (2001), "Neural networks for multi-objective adaptive structural control." *Journal of Structural Engineering*, ASCE, 127(2), 203–210.

Burrus, C.S., Gopinath, R.A., and Guo, H. (1998), *Introduction to wavelets and wavelet transforms: a primer*. Prentice Hall, New Jersey.

Buzug, Th. M., and Pfister, G. (1992), "Optimal delay time and embedding dimension for delay time coordinates by the analysis of global statical and local dynamical behaviour of strange attractors." *Physical Review A*, 45, 7073–7084.

Cannon, R.L., Dave, J.V., and Bezdek, J.C. (1986), "Efficient Implementation of the fuzzy c-means clustering algorithms." *IEEE Transactions on Pattern Analysis and Machine Intelligence*, 8(2), 248–255.

Carpenter, W.C., and Barthelemy, J.F. (1994), "Common misconceptions about neural networks as approximators." *Journal of Computing in Civil Engineering*, ASCE, 8(3), 345–358.

Casdagli, M., Eubank, S., Farmer, J. D., and Gibson, J. (1991), "State space reconstruction in the presence of noise." *Physica D*, 51, 52–98.

Cassidy, M.J., and Bertini, R.L. (1999), "Some traffic features at freeway bottlenecks." *Transportation Research B*, 33, 25–42.

Catbas, F.N., and Aktan, A.E. (2002), "Condition and damage assessment: Issues and some promising indices." *Journal of Structural Engineering*, ASCE, 128(8), 1026–1038.

Chang, C.C., and Yang, H.T.Y. (1995), "Control of building using active tuned mass dampers." *Journal of Engineering Mechanics*, ASCE, 121(3), 355–366.

Chang, C.C., Sun, Z., and Li, N. (2003), "Identification of structural dynamic properties using wavelet transform." In Wu, Z. and Abe, M. Eds., *Structural Health Monitoring and Intelligent Infrastructure*, 2, Balkema Publishing, Lisse, the Nethelan, pp. 1243–1248.

Chassiakos, A.G., and Masri, S.F. (1996), "Modeling unknown structural system through the use of neural networks." *Earthquake Engineering and Structural Dynamics*, 25(2), 117–128.

Chassiakos, A.P., and Stephanedes, Y.J. (1993), "Smoothing algorithms for incident detection." *Transportation Research Record*, TRB 1394, National Research Council, Washington D.C., pp. 8–16.

Chen, H.M., Tsai, K.H., Qi, G.Z., Yang, J.C.S., and Amini, F. (1995), "Neural network for structure control." *Journal of Computing in Civil Engineering*, ASCE, 9(2), 168–176.

Chien, S., and Schonfeld, P. (2001), "Optimal work zone segment lengths for four-lane highways." *Journal of Transportation Engineering*, ASCE , 127(2), 124–131.

Ching, J., Muto, M., and Beck, J.L. (2006), "Structural model updating and health monitoring with incomplete modal data using Gibbs sampler." *Computer-Aided Civil and Infrastructure Engineering*, 21(4), 242–257.

Chiu, S. (1994), "Fuzzy model identification based on cluster estimation." *Journal of Intelligent & Fuzzy Systems*, 2(3), 267–278.

Chong, K.P., Carino, N.J., and Washer, G. (2003), "Health monitoring of civil infrastructures." *Smart Materials and Structures*, 12(3), 483–493.

Chong, K.P., Dillon, O.W., Scalzi, J.B., and Spitzig, W.A. (1994), "Engineering research in composite and smart structures." *Composites Engineering*, 4(8), 829–852.

Chui, C. K. (1992), *An Introduction to Wavelets*. Academic Press, San Diego, California.

Chung, D.D.L. (2001), "Structural health monitoring by electrical resistance measurement." *Smart Materials and Structures*, 10(4), 624–636.

Coifman, R. R., and Wickerhauser, M. V. (1992), "Entropy-based algorithms for best basis selection." *IEEE Transaction on Information Theory*, 38(2), 713–718.

Connor, J.J. (2003), *Introduction to Structural Motion Control*. Prentice Hall, New Jersey.

Cook, A.R., and Cleveland, D.E. (1974), "Detection of freeway capacity reducing incidents by traffic stream measurements." *Transportation Research Record*, TRB 495, National Research Council, Washington D.C., pp. 1–11.

396

Courage, K.G., and Levin, M. (1968), *A freeway corridor surveillance information and control system*. Research Report No. 488-8, Texas Transportation Institute, College Station, Texas.

Daganzo, C.F. (1997), "A continuum theory of traffic dynamics for freeways with special lanes." *Transportation Research B*, 31(2), 83–102.

Daganzo, C.F., Lin, W., and Castillo, J.M. (1997), "A simple physical principle for the simulation of freeways with special lanes and priority vehicles." *Transportation Research B*, 31(2), 103–125.

Daubechies, I. (1988), "Orthonormal bases of compactly supported wavelets." *Communication on Pure and Applied Mathematics*, 41, 909–996.

Daubechies, I. (1992), *Ten Lectures on Wavelets*. Society for Industrial and Applied Mathematics, Phildelphia, Pennsylvania.

Davis, G.A., and Nihan, N.L. (1991), "Nonparametric regression and short-term freeway traffic forecasting." *Journal of Transportation Engineering*, ASCE, 177(2), 178–188.

Dendrinos, D.S. (1994), "Traffic-flow dynamics: a search for chaos." *Chaos, Solitons and Fractals*, 4(4), 605–617.

Dennis, J.E., and Schnable, R.B. (1983), *Numerical Methods for Unconstrained Optimization and Nonlinear Equations*. Prentice Hall, New Jersey.

Desilva, C.W. (1989), *Control sensors and actuators*. Prentice Hall, New Jersey.

Dharia, A., and Adeli, H. (2003), "Neural network model for rapid forecasting of freeway link travel time." *Engineering Applications of Artificial Intelligence*, 16(7-8), 607–613.

Dillon, W.R., and Goldstein, M. (1984), *Multivariate Analysis: Methods and Applications*. Wiley, New York.

Disbro, J.E., and Frame, M. (1989), "Traffic flow theory and chaotic behavior." *Transportation Research Record*, TRB 1225, National Research Council, Washington, D.C., pp. 109–115.

Dixon, K.K., and Hummer, J.E. (1995), *Capacity and Delay in Major Freeway Construction*. Center for Transportation Engineering Studies, North Carolina State University, Raleigh, North Carolina.

Dixon, K.K., Hummer, J.E., and Lorscheider, A.R. (1997), "Capacity for North Carolina freeway work zones." *Transportation Research Record* 1529, Transportation Research Record, National Research Council, Washington, D. C., pp. 27–34.

Donoho, D.L. (1995), "De-noising by soft-thresholding." *IEEE Transactions on Information Theory*, 413(3), 613–627.

Draper, N.R., and Smith, H. (1998), *Applied Regression Analysis*. Wiley, New York.

Dudek, C.L., and Richards, S.H. (1981), *Traffic Capacity through Work Zones on Urban Freeways*. Report FHWA/TX-81/28+228-6. Texas Department of Transportation, Austin, Texas.

Duffin, J., and Schaeffer, A.C. (1952), "A class of nonharmonic Fourier series." *Transactions of American Mathematic Society*, 72, 341–336.

Dyke, S.J., Spencer, B.F., and Sain, M.K. (1995), "Role of control-structure interaction in protective system design." *Journal of Engineering Mechanics*, ASCE, 121(2), 322–338.

Edwards, C., and Spurgeon, S. (1998), *Sliding Mode Control: Theory and Applications*. Taylor & Francis, CRC Press, Boca Raton, Florida.

Elkordy, M.F., Chang, K.C., and Lee, G.C. (1993), "Neural networks trained by analytically simulated damage states." *Journal of Computers in Civil Engineering*, ASCE, 7(2), 130–145.

FHWA. (1991), *Freeway Incident Management Handbook*. U.S. Department of Transportation, McLean, Virginia.

Florio, L., and Mussone, L. (1996), "Neural network models for classification and forecasting of freeway traffic flow stability." *Control Engineering Practice*, 4(2), 153–164.

Fraser, A.M., and Swinney, H.L. (1986), "Independent coordinates for strange attractors from mutual information." *Physical Review A*, 33(2), 1134–1140.

Fukunaga, K. (1990), *Introduction to Statistical Pattern Recognition*. 2nd Ed., Academic Press, New York.

Fur, L.S, Yang, H.T.Y., and Ankireddi, S. (1996), "Vibration control of tall buildings under seismic and wind loads." *Journal of Structural Engineering*, ASCE, 122(8), 948–957.

Furukawa, A., Kiyono, J., and Otsuka, H. (2006), "Structural damage detection method using uncertain frequency response functions." *Computer-Aided Civil and Infrastructure Engineering*, 21(4), 292–305.

Furuta, H., He, J., and Watanabe, E. (1996), "A fuzzy expert system for damage assessment using genetic algorithms and neural networks." *Microcomputers in Civil Engineering*, 11(1), 37–45.

Galka, A. (2000), *Topics in Nonlinear Time Series Analysis with Implication for EEG analysis*. World Scientific, Hackensack, New Jersey.

398

Gallager, R.G. (1968), *Information Theory and Reliable Communication*. Wiley, New York.

Ghaboussi, J., and Joghataie, A. (1995), "Active control of structures using neural networks." *Journal of Engineering Mechanics*, ASCE, 121(4), 555–567.

Ghanem, R., and Romeo, F. (2000), "A wavelet-based approach for the identification of linear time-varying dynamical systems." *Journal of Sound and Vibration*, 234(4), 555–576.

Ghanem, R., and Shinozuka, M. (1995), "Structural-system identification, I and II." *Journal of Engineering Mechanics*, ASCE, 121(2), 255–273.

Ghosh-Dastidar, S., and Adeli, H. (2003), "Wavelet-clustering-neural network model for freeway incident detection." *Computer-Aided Civil and Infrastructure Engineering*, 18(5), 325–338.

Ghosh-Dastidar, S., and Adeli, H. (2006), "Neural network-wavelet micro-simulation model for delay and queue length estimation at freeway work zones." *Journal of Transportation Engineering*, ASCE, 132(4), 331–341.

Ghosh-Dastidar, S. and Adeli, H. (2007), "Improved spiking neural networks for EEG classification and epilepsy and seizure detection." *Integrated Computer-Aided Engineering*, 14(3), 187–212.

Ghosh-Dastidar, S., Adeli, H., and Dadmehr, N. (2007), "Mixed-band wavelet-chaos-neural network methodology for epilepsy and epileptic seizure detection." *IEEE Transactions on Biomedical Engineering*, 54(9), 1545–1551.

Ghosh-Dastidar, S., Adeli, H., and Dadmehr, N. (2008), "Principal component analysis-enhanced cosine radial basis function neural network for robust epilepsy and seizure detection." *IEEE Transactions on Biomedical Engineering*, 55(2), 512–518.

Gleick, J. (1987), *Chaos: making a new science*. Vintage, New York.

Goldberg, D.E. (1989), *Genetic Algorithms in Search, Optimization, and Machine Learning*. Addison-Wesley, Reading, Massachusetts.

Golden, R.M. (1996), *Mathematical Methods for Neural Network Analysis and Design*. MIT Press, Cambridge, Massachusetts.

Goswami, J.C., and Chan, A.K. (1999), *Fundamentals of Wavelets*. Wiley, New York.

Guo, H.Y., Zhang, L., Zhang, L.L., and Zhou, J.X. (2004), "Optimal placement of sensors for structural health monitoring using improved genetic algorithms." *Smart Materials and Structures*, 13(3), 528–534.

Gurley, K., and Kareem, A. (1999), "Applications of wavelet transforms in earthquake, wind and ocean engineering." *Engineering Structures*, 21(2), 149–167.

Hagan, M.T., Demuth, H.B., and Beale, M. (1996), *Neural Network Design*. PWS Publishing Company, Boston, Massachusetts.

Hans, S., Ibraim, E., Pernot, S., Boutin, C., and Lamarque, C.H. (2000), "Damping identification in multi-degree-of-freedom systems via a wavelet-logarithmic decrement – Part 2: Study of a civil engineering building." *Journal of Sound and Vibration*, 235(3), 375–403.

Hasebe, K., Nakayama, A., and Sugiyama, Y. (1999), "Exact traveling cluster solutions of differential equations with delay for a traffic flow model." In Helbing, D., Herrmann, H.J., Schreckenberg, M., and Wolf, D.E., Eds., *Traffic and Granular Flow*. Springer, New York, pp. 413–418.

Hassoun, .H. (1995), *Fundamentals of Artificial Neural Networks*. MIT Press, Cambridge, Massachusetts.

Haykin, S. (1998), *Neural Networks: A Comprehensive Foundation*. 2nd Ed., Prentice Hall, Upper Saddle River, New Jersey.

HCM. (1985), *Highway Capacity Manual*. Special Report 209, Transportation Research Record, National Research Council, Washington, D.C.

HCM. (2000), *Highway Capacity Manual*. Special Report 209, Transportation Research Record, National Research Council, Washington, D.C.

Hegazy, T., Fazio, P., and Moselhi, O. (1994), "Developing practical neural network applications using backpropagation." *Microcomputers in Civil Engineering*, 9(2), 145–159.

Hénon, M. (1976), "A two-dimensional mapping with a strange attractor." *Communications in Mathematical Physics*, 50, 69–77.

Holschneider, M. (1995), *Wavelets: An Analysis Tool*. Oxford University Press, Oxford.

Hong, J.C., Kim, Y.Y., Lee, H.C., and Lee, Y.W. (2002), "Damage detection using the Lipschitz exponent estimated by the wavelet transform: application to vibration modes of a beam." *International Journal of Solids and Structures*, 39(7), 1803–1816.

Hooshdar, S., and Adeli, H. (2004), "Toward intelligent variable message signs in freeway work zones: Neural network model." *Journal of Transportation Engineering*, ASCE, 130(1), 83–93.

Hou, H., Hera, A., and Shinde, A. (2006), "Wavelet-based structural health monitoring of earthquake excited structures." *Computer-Aided Civil and Infrastructure Engineering*, 21(4), 268–279.

Housner, G.W., Bergman, L.A., Caughey, T.K., Chassiakos, A.G., Claus, R.O., Masri, S.F., Skelton, R.E., Soong, T.T., Spencer, B.F., and Yao, J.T.P. (1997), "Structural control: past, present, and future." *Journal of Engineering Mechanics*, ASCE, 123(9), 897–971.

Hsiao, C., Lin, C., and Cassidy, M. (1994), "Applications of fuzzy logic and neural networks to automatically detect freeway traffic incidents." *Journal of Transportation Engineering*, ASCE, 120(5), 753–772.

Hua, J. and Faghri, A. (1994), "Application of artificial neural network system to intelligent vehicle highway system." *Transportation Research Record*, TRB 1453, National Research Council, Washington D.C., pp. 83–90.

Huang, C.-C., and Loh, C.H. (2006), "Nonlinear identification of dynamic systems using neural networks." *Computer-Aided Civil and Infrastructure Engineering*, 16(1), 28–41.

Hung, S.L., and Adeli, H. (1994), "A parallel genetic/neural network algorithm for MIMD shared memory machines." *IEEE Transactions on Neural Networks*, 5(6), 900–909.

Hung, S.L., Huang, C.S., Wen, C.M., and Hsu, Y.C. (2003), "Nonparametric identification of a building structure from experimental data using wavelet neural network." *Computer-Aided Civil and Infrastructure Engineering*, 18(5), 358–370.

Hung, S.L., and Kao, C.Y. (2002), "Structural damage detection using the optimal weights of the approximating artificial neural networks." *Earthquake Engineering & Structural Dynamics*, 31(2), 217–234.

Hung, S.L., Kao, C.Y., and Lee, J.C. (2000), "Active pulse structural control using artificial neural networks." ASCE *Journal of Engineering Mechanics*, 126(8), 839–849.

Hurson, A.R., Pakzad, S., and Jin, B. (1994), "Automated knowledge acquisition in a neural network-based decision support system for incomplete database systems." *Microcomputers in Civil Engineering*, 9(2), 129–143.

Hsieh, D.A. (1991), "Chaos and nonlinear dynamics: Applications to financial models." *Journal of Finance*, 46, 1839–1878.

Jayawardena, A.W., and Fernando, D.A.K. (1998), "Use of radial basis function type artificial neural networks for runoff simulation." *Computer-Aided Civil and Infrastructure Engineering*, 13(2), 91–99.

Jenkins, W.M. (1992), "Plane frame optimum design environment based on genetic algorithms." *Journal of Structural Engineering*, ASCE, 118(11), 3103–3112.

Jiang, Y. (1999a), "A model for estimating excess user costs at highway work zones." *Transportation Research Record*, TRB 1657, National Research Council, Washington D.C., pp. 31–41

Jiang, Y. (1999b), "Traffic capacity, speed and queue-discharge rate of Indiana's four-lane freeway work zones." *Transportation Research Record* No. 1657, Transportation Research Record, National Research Council, Washington, D. C., pp. 10–17.

Jiang, X., and Adeli, H. (2003a), "Freeway work zone traffic delay and cost optimization model." *Journal of Transportation Engineering*, ASCE, 129(3), 230–241.

Jiang, X., and Adeli, H. (2003b), "Fuzzy clustering approach for accurate embedding dimension identification in chaotic time series." *Integrated Computer-Aided Engineering*, 10(3), 287–302.

Jiang, X, and Adeli, H. (2004a), "Wavelet packet-autocorrelation function method for traffic flow pattern analysis." *Computer-Aided Civil and Infrastructure Engineering*, 19(5), 324–337.

Jiang, X., and Adeli, H. (2004b), "Object-oriented model for freeway work zone capacity and queue delay estimation." *Computer-Aided Civil and Infrastructure Engineering*, 19(2), 144–156.

Jiang, X., and Adeli, H. (2004c), "Clustering-neural network models for freeway work zone capacity estimation." *International Journal of Neural Systems*, 14(3), 147–163.

Jiang, X., and Adeli, H. (2005a), "Dynamic fuzzy wavelet neural network for nonlinear identification of highrise buildings." *Computer-Aided Civil and Infrastructure Engineering*, 20(5), 316–330.

Jiang, X., and Adeli H. (2005b), "Dynamic wavelet neural network model for traffic flow forecasting." *Journal of Transportation Engineering*, ASCE, 131(10), 771–779.

Jiang, X., and Adeli, H. (2007), "Psuedospectra, MUSIC, and dynamic wavelet neural network for damage detection of high-rise buildings." *International Journal for Numerical Methods in Engineering*, 71(5), 606–629.

Jiang, X., and Adeli, H. (2008a), "Dynamic fuzzy wavelet neuroemulator for active nonlinear control of structures." *International Journal for Numerical Methods in Engineering*, 74(7), 1045–1066.

Jiang, X., and Adeli, H. (2008b), "Neuro-genetic algorithm for active nonlinear control of structures." *International Journal for Numerical Methods in Engineering*, DOI: 10.1002/nme.2274.

Jiang, X., Chen, H., and Liew, J.Y.R. (2002), "Spread-of-plasticity analysis of three-dimensional steel frames." *Journal of Constructional Steel Research*, 58, 193–212.

Jiang, X., and Mahadevan, S. (2008), "Bayesian wavelet methodology for structural damage detection." *Structural Control and Health Monitoring*, DOI: 10.102/stc.230.

Jiang, X., Mahadevan, S., and Adeli, H. (2007), "Bayesian wavelet packet denoising for structural system identification." *Structural Control and Health Monitoring*, 14(2), 333–356.

Joines, J.A., and Houck, C.R. (1994), "On the use of non-stationary penalty functions to solve nonlinear constrained optimization problems with GA's." *Proceedings of the 1st IEEE International Conference on Evolutionary Computation*. 1, Orlando, Florida, pp. 579–584.

Juang, J.N. (1994), *Applied System Identification*. Prentice Hall, New Jersey.

Kachigan, S.K. (1986), *Statistical Analysis: An Interdisciplinary Introduction to Univariate and Multivariate Methods*. Radius Press, New York.

Kao, C.Y., and Hung, S.L. (2003), "Detection of structural damage via free vibration responses generated by approximating artificial neural networks." *Computers & Structures*, 81(28-29), 2631–2644.

Kanamori, H. (1996), "Initiation process of earthquakes and its implications for seismic hazard reduction strategy." *Proceeding of National Academy of Science in Irvine*, 93, California, pp. 3726–3731.

Karim, A., and Adeli, H. (1999a), "Object-oriented information model for construction project management." *Journal of Construction Engineering and Management*, ASCE, 125(5), 361–367.

Karim, A., and Adeli, H. (1999b), "CONSCOM: an OO construction scheduling and change management system." *Journal of Construction Engineering and Management*, ASCE, 125(5), 368–376.

Karim, A., and Adeli, H. (1999c), "A new generation software for construction scheduling and management." *Engineering Construction and Architectural Management*, 6(4), 380–390.

Karim, A., and Adeli, H. (2002a), "Comparison of the fuzzy – wavelet RBFNN freeway incident detection model with the California algorithm." *Journal of Transportation Engineering*, ASCE, 128(1), 21–30.

Karim, A., and Adeli, H. (2002b), "Incident detection algorithm using wavelet energy representation of traffic patterns." *Journal of Transportation Engineering*, ASCE, 128(3), 232–242.

Karim A, and Adeli H. (2002c), "CBR model for freeway work zone traffic management." *Journal of Transportation Engineering*, ASCE, 129(2), 134–145.

Karim, A., and Adeli, H. (2002d), "Comparison of the fuzzy – wavelet RBFNN freeway incident detection model with the California algorithm." *Journal of Transportation Engineering*, ASCE, 128(1), 21–30.

Karim, A., and Adeli, H. (2003a), "A radial-basis function neural network model for work zone capacity and delay estimation." *Journal of Transportation Engineering*, ASCE, 129(5), 494–503.

Karim, A., and Adeli, H. (2003b), "Fast automatic incident detection on urban and rural freeways using wavelet energy algorithm." *Journal of Transportation Engineering*, ASCE, 129(1), 57–68.

Kellert, S.H. (1993), *In the Wake of Chaos*. University of Chicago Press, Chicago.

Kennel, M.B., Brown, R., and Abarbanel, H.D.I. (1992), "Determining embedding dimension for phase-space reconstruction using a geometrical construction." *Physical Review A*, 45, 3403–3411.

Kerner, B.S. (1999), "Theory of congested traffic flow: observation and theory." *Transportation Research Record*, TRB 1678, National Research Council, Washington D.C., pp. 160–167.

Kijewski, T., and Kareem, A. (2003), "Wavelet transforms for system identification in civil engineering." *Computer-Aided Civil and Infrastructure Engineering*, 18(5), 339–355.

Kim, H., and Adeli, H. (2001), "Discrete cost optimization of composite floors using a floating point genetic algorithm." *Engineering Optimization*, 33(4), 485–501.

Kim, H., and Adeli, H. (2004), "Hybrid feedback-least mean square algorithm for structural control." *Journal of Structural Engineering*, ASCE, 130(2), 120–127.

Kim, H., and Adeli, H. (2005a), "Hybrid control of smart structures using a novel wavelet-based algorithm." *Computer-Aided Civil and Infrastructure Engineering*, 20(1), 7–22.

Kim, H., and Adeli, H. (2005b), "Wavelet hybrid feedback-LMS algorithm for robust control of cable-stayed bridges." *Journal of Bridge Engineering*, ASCE, 10(2), 116–123.

Kim, H., and Adeli, H. (2005c), "Hybrid control of irregular steel highrise building structures under seismic excitations." *International Journal for Numerical Methods in Engineering*, 63(12), 1757–1774.

Kim, Y.J., and Ghaboussi, J. (1999), "A new method of reduced order feedback control using genetic algorithms." *Earthquake Engineering and Structural Dynamics*, 28(2), 193–212.

Kim, Y.J., and Ghaboussi, J. (2001), "Direct use of design criteria in genetic algorithm-based controller optimization." *Earthquake Engineering and Structural Dynamics*, 30(9), 1261–1278.

Kim, J.T., Jung, H.J., and Lee, I.W. (2000), "Optimal structural control using neural networks." *Journal of Engineering Mechanics*, ASCE, 126(2), 201–205.

Kim, D.H., and Lee, I.W. (2001), "Neurocontrol of seismically excited steel structures through sensitivity evaluation scheme." *Earthquake Engineering and Structural Dynamics*, 30(9), 1361–1377.

Kim, T., Lovell, D.J., and Paracha, J. (2001), "A new methodology to estimate capacity for freeway work zones." *Transportation Research Board Annual Meeting*, Washington D.C.

Kim., D.H., Seo, S.H., and Lee, I.W. (2004), "Optimal neurocontroller for nonlinear benchmark structure." *Journal of Engineering Mechanics*, ASCE, 130(4), 424–429.

Kirkpatrick, S., Gelatt, C.D., and Vecchi, M.P. (1983), "Optimization by simulated annealing." *Science*, 220, 671–680.

Ko, J.M., and Ni, Y.Q. (2005), "Technology developments in structural health monitoring of large-scale bridges." *Engineering Structures*, 27(12), 1715–1725.

Krammes, R.A., and Lopez, G.O. (1994), "Updated capacity values for short-term freeway work zone lane closure." *Transportation Research Record* No. 1442, Transportation Research Board, National Research Council, Washington, D.C., pp. 49–56.

Lam, H.-F., Yuen, K.-V., and Beck, J.L. (2006), "Structural health monitoring via measured Ritz vectors utilizing artificial neural networks." *Computer-Aided Civil and Infrastructure Engineering*, 21(4), 232–241.

Lamarque, C. H., Pernot, S., and Cuer, A. (2000), "Damping identification in multi-degree-of-freedom systems via a wavelet-logarithmic decrement – Part 1: Theory." *Journal of Sound and Vibration*, 235(3), 361–374.

Lee, W., and Fambro, D.B. (1999), "Application of subset autoregressive integrated moving average model for short-term freeway traffic volume forecasting." *Transportation Research Record*, TRB 1678, National Research Council, Washington, D.C., pp. 160–167.

Li, Q.S., Liu, D.K., and Fang, J.Q. (1998), "Optimum design of actively controlled structures using genetic algorithms." *Advances in Structural Engineering*, 2(2), 109–118.

Li, Q.S., Liu, D.K., Leung, A.Y.T., Zhang, N., Tam, C.M., and Yang, L.F. (2000), "Modelling of structural response and optimization of structural control system using neural network and genetic algorithm." *The Structural Design of Tall Buildings*, 9(4), 279–293.

Li, Q.S., Liu, D.K., Leung, A.Y.T., Zhang, N., and Luo, Q.Z. (2002), "A multilevel genetic algorithm for the optimum design of structural control systems." *International Journal for Numerical Methods in Engineering*, 55(7), 817–834.

Li, H., Ou, J., Zhao, X., Zhou, W., Li, H., Zhou, Z., and Yang, Y. (2006), "Structural health monitoring system for the Shandong Binzhou Yellow River highway bridge." *Computer-Aided Civil and Infrastructure Engineering*, 21(4), 306–317.

Liew, J.Y.R., Chen, H., and Shanmugam, N.E. (2001), "Inelastic analysis of steel frames with composite beams." *Journal of Structural Engineering*, ASCE, 127(2), 194–202.

Liew, K.M., and Wang, Q. (1998), "Application of wavelet theory for crack identification in structures." *Journal of Engineering Mechanics*, ASCE, 124(2), 152–157.

Ljung, L. (1999), *System Identification - Theory for the User*. 2ⁿᵈ ed., Prentice Hall, Upper Saddle River, New Jersey.

Ljung, L., and Glad, T. (1994), *Modeling of Dynamic Systems*. Prentice Hall, Upper Saddle River, New Jersey.

Loh, C.H., Lin, C.Y., and Huang, C.C. (2000), "Time domain identification of frames under earthquake loadings." *Journal of Engineering Mechanics*, ASCE, 126(7), 693–703.

Lorenz, E.N. (1963), "Deterministic nonperiodic flow." *Journal of the Atmospheric Sciences*, 20, 130–141.

Lu, L.T., Chiang, W.L., and Tang, J.P. (1998), "LQG/LTR control methodology in active structural control." *Journal of Engineering Mechanics*, ASCE, 124(4), 446–454.

Mallat, S.G. (1989), "A theory for multiresolution signal decomposition: The wavelet representation." *IEEE Transactions on Pattern Analysis and Machine Intelligence*, 11(7), 674–693.

Mallat, S.G. (1999), *A Wavelet Tour of Signal Processing*. Academic Press, San Diego, California.

Mandelbrot, B.B. (1982), *The Fractal Geometry of Nature*. W.H. Freeman and Company, San Francisco, California.

Marwala, T. (2000), "Damage identification using committee of neural networks." *Journal of Engineering Mechanics*, ASCE, 126(1), 43–50.

Masri, S.F., Bekey, G.A., and Caughey, T.K. (1982), "On-line control of nonlinear flexible structures." *Journal of Applied Mechanics*, ASME, 49(12), 871–884.

Masri, S.F., Chassiakos, A.G., and Caughey, T.K. (1993), "Identification of nonlinear dynamic systems using neural networks." *Journal of Applied Mechanics*, ASME, 60(1), 123–133.

Masri, S.F., Nakamura, M., Chassiakos, A.G., and Caughey, T.K. (1996), "Neural network approach to the detection of changes in structural parameters." *Journal of Engineering Mechanics*, ASCE, 122(4), 350–360.

Masri, S.F., Smyth, A.W., Chassiakos, A.G., Caughey, T.K., and Hunter, N.F. (2000), "Application of neural networks for detection of changes in nonlinear systems." *Journal of Engineering Mechanics*, ASCE , 126(7), 666–676.

Masri, S.F., Smyth, A.W., Chassiakos, A.G., Nakamura, M., and Caughey, T.K. (1999), "Training neural networks by adaptive random search techniques." *Journal of Engineering Mechanics*, ASCE, 125(2), 123–132.

Mathworks. (2001), *The Language of Technical Computing*. Mathworks Inc., http://www.mathworks.com.

Mathworks. (2000a), *Wavelet Toolbox for Use with MATLAB: User's Guide* Version 2, Mathworks Inc., http://www.mathworks.com.

Mathworks. (2000b), *Statistics Toolbox for Use with MATLAB: User's Guide* Version 3, Mathworks Inc., http://www.mathworks.com.

Mathworks. (2000c), *Neural Network Toolbox for use with MATLAB: User's Guide* Version 4. Mathworks Inc., http://www.mathworks.com.

May, A.D. (1990), *Traffic Flow Fundamental*. Prentice Hall, Upper Saddle River, New Jersey.

May, R.M. (1976), "Simple mathematical models with very complicated dynamics." *Nature*, 261(5560), 459–467.

McCoy, P.T., and Mennenga, D.J. (1998), "Optimum length of single-lane closures in work zones on rural four-lane freeways." *Transportation Research Record*, TRB 1650, National Research Council, Washington D.C., pp. 55–61.

McLachlan, G.J. (1992), *Discriminant Analysis and Statistical Pattern Recognition*. Wiley, New York.

Mehrotra, K., Mohan, C.K., and Ranka, S. (1997), *Elements of Artificial Neural Networks*. MIT Press, Cambridge, Massachusetts.

Melhem, H., and Kim, H. (2003), "Damage detection in concrete by Fourier and wavelet analysis." *Journal of Engineering Mechanics*, ASCE, 129(5), 571–577.

Memmott, J.L., and Dudek, C.L. (1984), "Queue and cost evaluation of work zones (QUEWZ)." *Transportation Research Record* 979, Transportation Research Board, National Research Council, Washington, D.C., pp. 12–19.

Michalewicz, Z. (1996), *Genetic Algorithms + Data Structures = Evolution Programs*. Springer-Verlag, New York.

Miller, R.K., Masri, S.F., Dehghanyar, T., and Caughey, T.K. (1988), "Active vibration control of large civil structures. " *Journal of Engineering Mechanics,* ASCE, 114(9), 1542–1570.

MITRETEK (2000), *QuickZone Delay Estimation Program-user Guide*. MITRETEK Systems Inc.

Moody, J., and Darken, C.J. (1989), "Fast learning in networks of locally-tuned processing units." *Neural Computation*, 1, 281–294.

Nakamura, M., Masri, S.F., Chassiakos, A.G., and Caughey, T.K. (1998), "A method for non-parametric damage detection through the use of neural networks." *Earthquake Engineering & Structural Dynamics*, 27(9), 997–1010.

Neubert, L., Santen, L., Schadschneider, A., and Schreckenberg, M. (1999), "Statistical analysis of freeway traffic." In Helbing, D., Herrmann, H.J., Schreckenberg, M. and Wolf, D.E., Eds., *Traffic and Granular Flow*. Springer-Verlag, New York, pp. 307–314.

Newell, G.F. (1998), "A moving bottleneck." *Transportation Research B*, 32(8), 531–537.

Newell, G.F. (1999), "Delays caused by a queue at a freeway exit ramp." *Transportation Research B*, 33, 337–350.

Ni, Y.Q., Zhou, X.T., and Ko, J.M. (2006), "Experimental investigation of seismic damage identification using PCA-compressed frequency response functions and neural networks." *Journal of Sound and Vibration*, 290(1–2), 242–263.

Nikzad, K., Ghaboussi, J., and Paul, S.L. (1996), "Actuator dynamics and delay compensation using neurocontrollers." *Journal of Engineering Mechanics*, ASCE, 122(10), 966–975.

Ohtori, Y., Christenson, R.E., Spencer, B.F., and Dyke, S.J. (2004), "Benchmark control problems for seismically excited nonlinear buildings." *Journal of Engineering Mechanics*, ASCE, 130(4), 366–385.

Okutani, I., and Stephanedes, Y.J. (1984), "Dynamic prediction of traffic volume through Kalman filtering theory." *Transportation Research B*, 18(1), 1–11.

Panakkat, A., and Adeli, H. (2007), "Neural network models for earthquake magnitude prediction using multiple seismicity indicators." *International Journal of Neural Systems*, 17(1), 13–33.

Papadimitriou, C. (2004), "Optimal sensor placement methodology for parametric identification of structural systems." *Journal of Sound and Vibration*, 278(4/5), 923–947.

Papageorgiou, M., Blosseville, J., and Hadj-Salem, H. (1990), "Modeling and real-time control of traffic flow on the southern part of boulevard peripherique in Paris Part I: Modeling." *Transportation Research A*, 24A(5), 345–359.

Park, B. (2002), "Hybrid neuro-fuzzy application in short-term freeway traffic volume forecasting." *Transportation Research Record*, TRB 1802, National Research Council, Washington D.C., pp. 190–196.

Park, H.S., Lee, H.M., Adeli, H., and Lee, I. (2007), "A new approach for health monitoring of structures: Terrestrial laser scanning." *Computer-Aided Civil and Infrastructure Engineering*, 22(1), 19–30.

Park, B., Messer, C.J., and Urbanik, T.II. (1998), "Short-term freeway traffic volume forecasting using radial basis function neural network." *Transportation Research Record*, TRB 1651, National Research Council, Washington, D.C., pp. 39–47.

Park, S., Stubbs, N., Bolton, R., Choi, S., and Sikorsky, C. (2006), "Field verification of the damage index method in a concrete box girder bridge via visual inspection." *Computer-Aided Civil and Infrastructure Engineering*, 16(1), 58–70.

Patil, D.J., Hunt, B.R., Kalnay, E., Yorke, J.A., and Ott, E. (2001), "Local low dimensionality of atmospheric dynamics." *Physical Review Letters*, 86(26), 5878–5881.

Payne, H.J. and Tignor, S.C. (1978), "Freeway incident detection algorithms based on decision tree with states." *Transportation Research Record*, TRB 682, National Research Council, Washington D.C., pp. 378–382.

Pei, J.S., Smyth, A.W., and Kosmatopoulos, E.B. (2004), "Analysis and modification of Volterra/Wiener neural networks for identification of nonlinear hysteretic dynamic systems." *Journal of Sound and Vibration*, 275(3-5), 693–718.

Percieval, D.B., and Walden, A. T. (2000), *Wavelet Methods for Time Series Analysis*, Cambridge University Press, London.

Persaud, B.N., Hall, F.L., and Hall, L.M. (1990), "Congestion identification aspects of the McMaster incident detection algorithm." *Transportation Research Record*, TRB 1287, National Research Council, Washington D.C., pp.167–175.

Pham, D.T., and Karaboga, D. (2000), *Intelligent Optimisation Techniques: Genetic Algorithms, Tabu Search, Simulated Annealing and Neural Networks*. Springer-Verlag, London.

Pickover, C.A. (1990), *Computers, Pattern, Chaos, and Beauty: Graphics from an Unseen World*. St. Martin's Press, New York.

Poggio, T., and Girosi, F. (1990), "Networks for approximation and learning." *Proceedings of the IEEE*, 78, 1481–1497.

Quek, S.T., Wang, Q., Zhang, L., and Ong, K.H.P. (2001), "Practical issues in the detection of damage in beams using wavelets." *Smart Materials and Structures*, 10(5), 1009–1017.

Ross, P. (1982), "Exponential filtering of traffic data." *Transportation Research Record*, TRB 869, National Research Council, Washington, D.C., pp. 43–49.

Rössler, O. E. (1976), "An equation for continuous chaos." *Physics Letters A*, 57, 397–398.

Rudomin, P., Arbib, M.A., Cervantes-Perez, F., and Romo, R. (1993), *Neuroscience: From Neural Networks to Artificial Intelligence*, Springer-Verlag, Berlin.

Rumelhart, D.E., Hinton, G.E., and Williams, R.J. (1986), "Learning internal representation by error propagation." In Rumelhart DE et al., Eds. *Parallel Distributed Processing*. MIT Press, Cambridge, Massachusetts, pp. 318–362.

Saeki, M., and Hori, M. (2006), "Development of an accurate positioning system using low-cost I1 GPS receivers." *Computer-Aided Civil and Infrastructure Engineering*, 21(4), 258–267.

Salawu, O.S. (1997), "Detection of structural damage through changes in frequency: a review." *Engineering Structure*, 19(9), 718–723.

Saleh, A., and Adeli, H. (1998a), "Optimal control of adaptive/smart multistory building structures." *Computer-Aided Civil and Infrastructure Engineering*, 13(6), 389–403.

Saleh, A., and Adeli, H. (1998b), "Optimal control of adaptive building structures under blast loading." *Mechatronics*, 8(8), 821–844.

Samant, A., and Adeli, H. (2000), "Feature extraction for traffic incident detection using wavelet transform and linear discriminant analysis." *Computer-Aided Civil and Infrastructure Engineering*, 15(4), 241–250.

Samant, A., and Adeli, H. (2001), "Enhancing neural network incident detection algorithms using wavelets." *Computer-Aided Civil and Infrastructure Engineering*, 16(4), 239–245.

Sanayei, M., Wadia-Fascetti, S., Arya, B., and Santini, E.M. (2006), "Significance of modeling error in structural parameter estimation." *Computer-Aided Civil and Infrastructure Engineering*, 16(1), 12–27.

Sarbadhikari, S. N., and Chakrabarty, K. (2001), "Chaos in the brain: A short review alluding to epilepsy, depression, exercise and lateralization." *Medical Engineering and Physics*, 23(7), 445–455.

Sarma, K.C., and Adeli, H. (2000a), "Fuzzy genetic algorithm for optimization of steel structures." *Journal of Structural Engineering*, ASCE, 126(5), 596–604.

Sarma, K.C., and Adeli, H. (2000b), "Fuzzy discrete multicriteria cost optimization of steel structures." *Journal of Structural Engineering*, ASCE, 126(11), 1339–1347.

Sarma, K.C., and Adeli, H. (2001), "Bi-Level parallel genetic algorithms for optimization of large steel structures." *Computer-Aided Civil and Infrastructure Engineering*, 16(5), 295–304.

Sarma, K.C., and Adeli, H. (2002), "Life-cycle cost optimization of steel structures." *International Journal for Numerical Methods in Engineering*, 55(12), 1451–1462.

Sarma, K.C., and Adeli, H. (2003), "Data parallel fuzzy genetic algorithms for cost optimization of steel structures." *International Journal of Space Structures*, 18(3), 195–205.

Schaefer, W.E. (1969), *California Freeway Surveillance System*. California Division of Highways, Department of Public Works, Sacramento, CA.

Schuster, H.G. (1995), *Deterministic Chaos: An Introduction*. Wiley, New York.

Senouci, A.B. and Adeli, H. (2001), "Resource scheduling using neural dynamics model of Adeli and Park." *Journal of Construction Engineering and Management*, ASCE, 127(1), 28–34.

Sierpiński, W. (1915), "Sur une courbe dont tout point est un point de ramification." *C. R. Academic Science Paris*, 160, 302–305.

Sjöberg, J., Zhang, Q., Ljung, L., Benveniste, A., Delyon, B., Glorennec, P., Hjalmarsson, H., and Juditsky, A. (1995), "Nonlinear black-box modeling in system identification: a unified overview." *Automatica*, 31(12), 1691–1724.

Son, Y.T. (1999), "Queuing delay models for two-lane highway work zones." *Transportation Research B*, 33, 459–471.

Soong, T.T. (1990), *Active Structural Control: Theory and Practice*. Longman's, New York.

Smith, B.L., and Demetsky, M.J. (1994), "Short-term traffic flow prediction: neural network approach." *Transportation Research Record*, TRB 1453, National Research Council, Washington, D.C., pp. 98–104.

Smith, B.L., and Demetsky, M.J. (1997), "Traffic flow forecasting: Comparison of modeling approaches." *Journal of Transportation Engineering*, ASCE, 123(4), 261–266.

Smith, B.L., Williams, B.M., and Oswald, R.K. (2002), "Comparison of parametric and nonparametric models for traffic flow forecasting." *Transportation Research Part C*, 10, 302–321.

Soegiarso, R., and Adeli, H. (1998), "Parallel-vector algorithms for optimization of large steel structures." *Computer-Aided Civil and Infrastructure Engineering*, 13(3), 207–217.

Staszewski, W.J. (1997), "Identification of damping in MDOF systems using time-scale decomposition." *Journal of Sound and Vibration*, 203(2), 283–305.

Staszewski, W.J. (1998), "Identification of nonlinear systems using multi-scale ridges and skeletons of the wavelet transform." *Journal of Sound and Vibration*, 214(4), 639–658.

Stein, G., and Athans, M. (1987), "The LQG/LTR procedure for multivariable feedback control design." *IEEE Transaction of Automatic Control*, AC32(2), 105–114.

Stephanedes, Y.J., Chassiakos, A.P., and Michalopoulos, P.G. (1992), "Comparative performance evaluation of incident detection algorithms." *Transportation Research Record*, TRB 1360, National Research Council, Washington, D.C., pp. 50–57.

Stephanedes, Y.J., and Liu, X. (1995), "Artificial neural networks for freeway incident detection." *Transportation Research Record*, TRB 1494, National Research Council, Washington, D.C., pp. 91–97.

Stoica, P., and Moses, R.L. (1997), *Introduction to Spectral Analysis*. Prentice Hall, Upper Saddle River, New Jersey.

Sugeno, M., and Kang, G.T. (1988), "Structure identification of fuzzy model." *Fuzzy Sets and System*, 28, 15–34.

Suzuki, H., Nakatsuji, T., Tanaboriboon, Y., and Takahashi, K. (2000), "Dynamic estimation of origin-destination travel time and flow on a long freeway corridor: Neural Kalman filter." *Transportation Research Record*, TRB 1739, National Research Council, Washington, D.C., pp. 67–75.

Szewezyk, P., and Hajela, P. (1994), "Damage detection in structures based on feature-sensitivity neural networks." *Journal of Computers in Civil Engineering*, ASCE, 8(2), 163–179.

Takens, F. (1981), "Detecting strange attractors in turbulence." In *Dynamical Systems and Turbulence*, Rand, D.A. and Young L.-S., Eds. Lecture notes in mathematics, 898, Springer, Berlin, pp. 366–381.

Tan, P., Dyke, S., Richardson, A., and Abdullah, M. (2005), "Integrated device placement and control design in civil structures using genetic algorithms." *Journal of Structural Engineering*, ASCE, 131(10), 1489–1496.

Theiler, J. (1990), "Estimating fractal dimension." *Journal of the Optical Society of America A*, 7(6), 1055–1073.

Thomson, M., Schooling, S.P., and Soufian, M. (1996), "The practical application of a nonlinear identification methodology." *Control Engineering Practice*, 4(3), 295–306.

Thunberg, H. (2001), "Periodicity versus chaos in one-dimensional dynamics." *SIAM Review*, 43(1), 3–30.

Tomasula, D.P., Spencer, B.F., and Sain, M.K. (1996), "Nonlinear control strategies for limiting dynamic response extremes." *Journal of Engineering Mechanics*, ASCE, 122(3), 218–229.

Topping, B.H.V., and Bahreininejad, A. (1997), *Neural Computing for Structural Mechanics*. Saxe-Coburg Publications, Edinburgh.

Trefethen, L.N. (1992), "Pseudospectra of matrices." In *Numerical Analysis*, Griffiths, D. F. and Watson, G. A., Eds., Longman Scientific and Technical, Harlow.

Trefethen, L.N. (1999), "Computation of pseudospectra." *Acta Numerica*, 8, 247–295.

Treiber, M., Hennecke, A., and Helbing, D. (2000), "Congested traffic states in empirical observations and microscopic simulations." *Physical Review E*, 62(2), 1805–1824.

TSIS. (1999), *Traffic Software Integrated System: User Guide*, Version 4.3. FHWA, US Department of Transportation, McLean, VA.

Tsonis, A.A. (1992), *Chaos: From Theory to Applications*. Plenum Press, New York.

Turcotte, D.L. (1992). *Fractals and Chaos in Geology and Geophysics*. Cambridge University Press, Cambridge, Massachusetts.

Wen, Y.K. (1976), "Method for random vibration for hysteretic structures." *Journal of the Engineering Mechanics Division*, ASCE, 102(EM2), 249–263.

Wickerhauser, M.V. (1994), *Adapted Wavelet Analysis from Theory to Software*, A K Peters Ltd, Massachusetts.

Williams, B.M., Durvasula, P.K., and Brown, D.E. (1998), "Urban freeway traffic flow prediction: Application of seasonal autoregressive integrated moving average and exponential smoothing models." *Transportation Research Record*, TRB 1644, National Research Council, Washington, D.C., pp. 132–141.

Wong, F.S., and Yao, J.T.P. (2006), "Health monitoring and structural reliability as a value chain." *Computer-Aided Civil and Infrastructure Engineering*, 16(1), 71–78.

Wu, M., and Adeli, H. (2001), "Wavelet-neural network model for automatic traffic incident detection." *Mathematical & Computational Applications*, 6(2), 85–96.

Wu, Z.S., Xu, B., and Yokoyama, K. (2002), "Decentralized parametric damage detection based on neural networks." *Computer-Aided Civil-Infrastructure Engineering*, 17(3), 175–184.

Xu, B., Wu, Z.S., and Yokoyama, K. (2000), "Decentralized identification of large-scale structure-AMD coupled system using multi-layer neural networks." *Transactions of the Japan Society for Computational Engineering and Science*, 2, 187–197.

Yager, R.R., and Filev, D.P. (1994), "Approximate clustering via the mountain method." *IEEE Transactions on Systems, Man and Cybernetics*, 24(8), 1279–1284.

Yang, J.N., Long, F.X., and Wong, D. (1988), "Optimal control of nonlinear structures." *Journal of Applied Mechanics*, 55(4), 931–938.

Yang, J.N., Danielians, A., and Liu, S.C. (1992), "Hybrid control of nonlinear and hysteretic systems I&II." *Journal of Engineering Mechanics*, ASCE, 118, 1423–1439 & 1441–1457.

Yasdi, R. (1999), "Prediction of road traffic using a neural network approach." *Neural Computing & Application*, 8, 135–142.

Yeh, I.C. (1998), "Structural engineering applications with augmented neuron networks."" *Computer-Aided Civil and Infrastructure Engineering*, 13(2), 83–90.

Yen, G.G. (1994), "Identification and control of large structures using neural networks." *Computers and Structures*, 52(5), 859–870.

Yin, H., Wong, S.C. , Xu, J., and Wong, C.K. (2002), "Urban traffic flow prediction using a fuzzy-neural approach." *Transportation Research Part C*, 10, 85–98.

Yoon, D.J., Weiss, W.J., and Shah, S.P. (2000), "Assessing damage in corroded reinforced concrete using acoustic emission." *Journal of Engineering Mechanics*, ASCE, 126(3), 273–283.

Yoshida, O., and Dyke, S.J. (2005), "Response control of full-scale irregular buildings using magnetorheological dampers." *Journal of Structural Engineering*, ASCE, 131(5), 734–742.

Yu, G., and Adeli, H. (1991), "Computer-aided design using object-oriented programming paradigm and blackboard architecture." *Microcomputers in Civil Engineering*, 6, 177–189.

Yu, G., and Adeli, H. (1993), "Object-oriented finite element analysis using an EER model." *Journal of Structural Engineering*, ASCE, 119(9), 2763–2781.

Yuen, K.V., and Katafygiotis, L.S. (2006), "Substructure identification and health monitoring using noisy response measurements only." *Computer-Aided Civil and Infrastructure Engineering*, 21(4), 280–291.

Yun, S.Y., Namkoong, S., Rho, J.H., Shin, S.W., and Choi, J.U. (1998), "A performance evaluation of neural network models in traffic volume forecasting." *Mathematical Computer Modelling*, 27(9-11), 293–310.

Zadeh, L.A. (1978), "Fuzzy set as a basis for a theory of possibility." *Fuzzy Sets and Systems*, 1(1), 3–28.

Zhang, Q. (1997), "Using wavelet network in nonparametric estimate." *IEEE Transactions on Neural Networks*, 8(2), 227–236.

Zhang, Q., and Benveniste, A. (1992), "Wavelet networks." *IEEE Transactions on Neural Networks*, 3(6), 889–898.

Zhang, H., Ritchie, S.G., and Lo, Z. (1997), "Macroscopic modeling of freeway traffic using an artificial neural network." *Transportation Research Record*, TRB 1588, National Research Council, Washington D.C., pp. 110–119.

Zhou, Z., and Adeli, H. (2003), "Time-frequency signal analysis of earthquake records using Mexican hat wavelets." *Computer-Aided Civil and Infrastructure Engineering*, 18(5), 379–389.

Index

Milton Keynes UK
Ingram Content Group UK Ltd.
UKHW050456071024
449327UK00015B/401

9 780367 386719